Hydrodynamic Cavitation

Hydrodynamic Cavitation

Devices, Design, and Applications

Vivek V. Ranade
Vinay M. Bhandari
Sanjay Nagarajan
Varaha P. Sarvothaman
Alister T. Simpson

WILEY-VCH

Authors

Professor Vivek V. Ranade
University of Limerick
Bernal Chair of Process Engineering
Limerick V94
Ireland

Dr. Vinay M. Bhandari
CSIR – National Chemical Laboratory
Dr. Homi Bhabha Road
Pune 411008
India

Dr. Sanjay Nagarajan
University of South Wales
Sustainable Environment Research Centre
9 Graig Fach
Pontypridd CF37 4BB
UK

Dr. Varaha P. Sarvothaman
Queen's University Belfast
School of Chemistry and Chemical Engineering
39 Stranmillis Road
Belfast BT95AG
UK

Dr. Alister T. Simpson
Wilde Analysis Ltd.
Whitworth House, 28 Charles Street
Stockport, Cheshire SK1 3JR
UK

Library of Congress Card No.: applied for

British Library Cataloguing-in-Publication Data
A catalogue record for this book is available from the British Library.

Bibliographic information published by the Deutsche Nationalbibliothek
The Deutsche Nationalbibliothek lists this publication in the Deutsche Nationalbibliografie; detailed bibliographic data are available on the Internet at <http://dnb.d-nb.de>.

Print ISBN: 978-3-527-34643-1
ePDF ISBN: 978-3-527-82287-4
ePub ISBN: 978-3-527-34645-5
oBook ISBN: 978-3-527-34644-8

Typesetting Straive, Chennai, India
Printing and Binding CPI Group (UK) Ltd, Croydon, CR0 4YY

C095259_051022

Contents

Preface *xi*

Part I Introduction *1*

1 **Hydrodynamic Cavitation** *3*
1.1 Hydrodynamic Cavitation *5*
1.2 Hydrodynamic Cavitation Devices *7*
1.3 Applications of Hydrodynamic Cavitation *11*
1.4 Organization of the Book *11*
References *15*

Part II Hydrodynamic Cavitation Devices *21*

2 **Hydrodynamic Cavitation Devices Based on Axial/Linear Flow** *23*
2.1 Introduction *23*
2.2 Orifice-Based Devices *24*
2.3 Venturi-Based Devices *34*
2.4 Enhancing Performance of Orifice/Venturi-Based Hydrodynamic Cavitation Devices *41*
2.5 Summary and Outlook *44*
Notations *45*
Greek Symbols *45*
References *45*

3 **Hydrodynamic Cavitation Devices Based on Rotational/Swirling Flows** *49*
3.1 Rotor-Stator Hydrodynamic Cavitation Devices *49*
3.2 Vortex-Based Cavitation Devices *52*
3.2.1 Vortex Cavitation *52*
3.2.2 Vortex Models *53*

3.2.3 Vortex Cavitation Devices *56*
3.3 Devices Based on Combinations of Linear and Swirling Flows *59*
3.4 Summary and Outlook *63*
Notations *68*
Greek Symbols *68*
References *68*

Part III Characterizing and Modeling of Cavitation Devices *71*

4 Experimental Characterization of Hydrodynamic Cavitation Devices *73*
4.1 Experimental Set-up for Characterization of Hydrodynamic Cavitation Devices *73*
4.1.1 Holding Tank *74*
4.1.2 Pump *76*
4.1.3 Hydrodynamic Cavitation Device *78*
4.1.4 Piping Arrangements/Fittings *79*
4.1.5 In-line Sensors *79*
4.2 Identification of Inception of Hydrodynamic Cavitation *80*
4.3 Characterizing Overall Process Performance *84*
4.4 Conclusions *89*
Notations *90*
References *90*

5 Modeling of Hydrodynamic Cavitation-Based Processes *95*
5.1 Introduction *95*
5.2 Empirical Models *98*
5.2.1 Pseudo-reaction Kinetics Model *98*
5.2.2 Per-pass Performance Model *101*
5.2.3 Data-Driven Models *104*
5.3 Physics-Based Models *106*
5.3.1 Cavity Dynamics Models *108*
5.3.1.1 Model Equations Governing Single-Bubble Dynamics *109*
5.3.1.2 Estimation of Generation of Hydroxyl Radicals *110*
5.3.1.3 Estimation of Hammer Pressure/Jet Velocity Due to Collapse *111*
5.3.1.4 Illustrative Results from Cavity Dynamics Models *112*
5.3.2 Multi-scale/Multi-layer Models for Simulating Performance of Cavitation Processes *114*
5.4 Modeling of Heterogeneous Systems Treated with HC *122*
5.5 Summary and Outlook *125*
Abbreviations *127*
Symbols *128*
Greek Symbols *130*

Subscripts 130
Superscripts 131
5.A CFD Modeling of Cavitating Flows 131
References 136

Part IV Applications of Hydrodynamic Cavitation 145

6 Disinfection of Water 147
6.1 Introduction 147
6.2 Conventional Methods of Disinfection 149
6.2.1 Major Drawbacks in Continuing the Use of Conventional Methods 152
6.2.2 Emerging Newer Methods of Disinfection 153
6.3 Disinfection of Water by Cavitation 154
6.3.1 Cavitation Process Principle 154
6.3.2 Present Status 158
6.3.3 Hydrodynamic Cavitation and Cavitation Devices/Reactors 159
6.3.4 Kinetics of Disinfection in Hydrodynamic Cavitation 162
6.4 Hybrid Methods of Disinfection Involving Cavitation 165
6.4.1 Process Integration–Conventional 168
6.4.2 Cavitation with Hydrogen Peroxide Addition 168
6.4.3 Cavitation with Ozone Addition 169
6.4.4 Cavitation with Aeration or Oxygen 169
6.5 Hybrid Hydrodynamic Cavitation Technology Using Natural Oils 170
6.5.1 Mechanism of Disinfection in Hydrodynamic Cavitation- Conventional vs. Hybrid Processes 174
6.5.2 Effect of Temperature 176
6.6 Process Economics 177
6.6.1 Cost Comparison of Different Processes 177
6.6.2 Typical Cost Calculation for Vortex Diode as Reactor in Hybrid Process Using Natural Oils 178
6.7 New Developments and Future Potential 179
6.7.1 Applications in Drinking Water Treatment 179
6.7.2 Applications in Sewage Water Treatment 180
6.7.3 Applications in Ballast Water Treatment 181
6.8 Summary 183
References 184

7 Wastewater Treatment 193
7.1 Introduction 193
7.2 Hydrodynamic Cavitation for Wastewater Treatment 196
7.3 Performance of Hydrodynamic Cavitation-based Wastewater Treatment 201
7.3.1 Influence of Device Design 203
7.3.2 Influence of Operating Parameters 206

7.3.2.1 *Inlet Concentration of Pollutant* *206*
7.3.2.2 *Pressure Drop Across Cavitation Device* *208*
7.3.2.3 *Downstream Pressure* *209*
7.3.2.4 *Operating pH* *210*
7.3.2.5 *Operating Temperature* *213*
7.3.2.6 *Influence of Dissolved Gases/Sparged Gases* *214*
7.4 Enhancing the per-pass Performance: Augmentation by Hybrid Processes *217*
7.4.1 Coupling of HC with AOPs Using Alternative Energy Sources *218*
7.4.1.1 *UV-assisted HC* *218*
7.4.1.2 *Plasma-based HC* *219*
7.4.2 Coupling of HC with Chemical-based AOPs *220*
7.4.2.1 *Hydrogen Peroxide (H_2O_2) Treatment* *221*
7.4.2.2 *Ozone (O_3) Treatment* *222*
7.4.2.3 *Peroxonation (Hydrogen Peroxide—H_2O_2 + Ozone—O_3)* *224*
7.4.3 Augmenting Hydrodynamic Cavitation by Catalyst-based AOPs *225*
7.4.3.1 *Fenton's Process* *225*
7.4.3.2 *Photocatalysis* *226*
7.5 Summary and Outlook *229*
Notations *229*
References *230*

8 Pre-treatment of Biomass for Enhancing Biofuel Yields *239*
8.1 Introduction *239*
8.2 Hydrodynamic Cavitation for Enhancing Bioethanol Yield *244*
8.3 Hydrodynamic Cavitation for Enhancing Biogas Production *250*
8.3.1 Wastewater and Sludge *253*
8.3.2 Lignocellulosic Biomass (LCB) *260*
8.4 Net Energy Gains *264*
8.5 Summary and Path Forward *269*
Notations *271*
Symbols *272*
Abbreviations *272*
References *273*

9 Other Applications of Hydrodynamic Cavitation *283*
9.1 Introduction *283*
9.2 Gas–Liquid Applications *283*
9.3 Liquid–Liquid Applications *289*
9.3.1 Oxidative Desulfurization *289*
9.3.2 Emulsification *294*
9.3.3 Microalgal Oil Extraction *296*
9.3.4 Transesterification of Oils to Produce Biodiesel *297*
9.3.5 Food (Juice and Milk) Sterilization *300*
9.4 Solid–Liquid Applications *301*

9.4.1 Beer Brewing *301*
9.4.2 Bioactive Compound Extraction *302*
9.4.3 Particle Size Reduction *304*
9.5 Summary, Outlook, and Conclusions *305*
References *306*

Part V Status and Path Forward *313*

10 Summary and Outlook *315*
10.1 Devices, Experimental Characterization, and Modeling of Hydrodynamic Cavitation *316*
10.2 Applications of Hydrodynamic Cavitation *323*
Notations *330*
References *330*

Index *335*

Preface

The first author (VR) was introduced to hydrodynamic cavitation while researching on fluidic devices needed for the nuclear industry while working at CSIR—National Chemical Laboratory (NCL), Pune, India. The focus then was on avoiding hydrodynamic cavitation. Later, VR got involved in a large project on ballast water treatment being carried out in a sister CSIR Laboratory in Goa, India. That was the beginning of his work on harnessing hydrodynamic cavitation for beneficial applications. The work resulted in a patent on a new family of vortex-based hydrodynamic cavitation devices (US patent 7585416, 2009). VR then teamed up with VB and together they developed applications of hydrodynamic cavitation for water treatment at CSIR-NCL. The work resulted in an edited book entitled *Industrial Wastewater Treatment, Recycling, and Reuse* in 2014. The research in those early years was more focussed on the development of applications using experiments. VR moved to Queen's University Belfast in 2016 and started collaborating with SN, VS, and AS. Work prior to that had revealed significant gaps in understanding the basics of hydrodynamic cavitation and a lack of reliable predictive models for simulating and optimizing hydrodynamic cavitation devices and processes/applications based on those. That was the time we as a team realized that despite so many publications on hydrodynamic cavitation, there is no single source which presents the basics of hydrodynamic cavitation and links those with the device design and applications. The challenges we faced in understanding and quantitatively simulating hydrodynamic cavitation devices as well as in using the developed understanding and models for applications motivated us to develop this book.

Hydrodynamic cavitation which generates locally extreme conditions while operating under bulk ambient conditions has been harnessed for realizing numerous beneficial physicochemical transformations and innovative products. Hydrodynamic cavitation has also been used for intensifying a wide range of processes. Despite the intense research and a handful of start-up companies, the full potential of hydrodynamic cavitation is not yet realized. It is therefore essential to develop a thorough understanding, suitable devices and appropriate models for enabling process innovations and translating laboratory research to practice. Several excellent reviews and a few books on hydrodynamic cavitation are available. However, we felt that there is a need for a compiled resource that bridges the gap between device design, experiments, models, and practice. The proposed book is aimed at filling this void.

We have made an attempt to provide a systematic basis for (i) the design of hydrodynamic cavitation devices, (ii) the design of experimental setup and procedures for investigating different applications, and (iii) developing computational models for interpreting and simulating experiments and facilitating translation to practice. In the next part of the book, three major applications of hydrodynamic cavitation namely, water disinfection, water treatment, and biomass pre-treatment are discussed. A brief summary of few other applications of hydrodynamic cavitation is provided as well. The goal of the this part was to illustrate the development of applications of hydrodynamic cavitation. An attempt is made to convey general guidelines, so that the book is useful even for applications not presented here. Our comments on the state of the art and future prospects are included at the end. We hope that this book will stimulate further work in this very important area from an industrial as well as from a scientific point of view.

The intended users of this book are chemical and process engineers working in chemical and allied industries and industrial R&D laboratories as well as researchers and research students working on realizing process innovations and productivity enhancements. Hydrodynamic cavitation offers an attractive technology platform which may be used to realize significant improvements in productivity while reducing energy consumption. The book will be directly useful for researchers and practitioners working in water and bioenergy sectors. Some of the new applications discussed in the book like emulsions will be useful to harness hydrodynamic cavitation for producing high-value products. We hope that the material included in this book will be useful in several different ways and at various stages of process innovations projects involving hydrodynamic cavitation. Some prior background in fluid mechanics, transport phenomena, and reaction engineering is assumed for readers. It may be used as a basic resource of methodologies for investigating hydrodynamic cavitation devices and processes. The content may also be useful as study material for an in-house course, or as a companion book while solving practical problems in applying hydrodynamic cavitation. We hope that this book will help chemical and process engineers to realize the true potential of hydrodynamic cavitation.

Several students, associates, collaborators, and funding institutes have contributed to the research that has led to this book. We are grateful for their support. VR would like to acknowledge the support provided by the Queen's University of Belfast and the University of Limerick for his research on hydrodynamic cavitation. VR would also like to acknowledge various funding agencies such as the Council of Scientific and Industrial Research (CSIR), India; Department of Biotechnology, India; Innovate UK; BBSRC, UK, Science Foundation of Ireland for supporting his research on hydrodynamic cavitation. VR would like to thank Ansys Inc. for supporting his CFD work at the University of Limerick via academic partnership. VR would like to thank his wife, Dr Nanda Ranade for carefully reading the entire manuscript and providing suggestions to enhance the clarity of the presentation. VB would like to acknowledge and thank the support from CSIR and the Department of Science and Technology,

India for their financial support to various projects. VB also expresses his profound gratitude to Prof. V.A. Juvekar for being a constant source of inspiration in research. VB would like to thank Ms. Maya Mane, Ms. Divya Dixit and Dr. Kshama Balapure for the development of SWASTIIK Technology (Safe Water and Sustainable Technology Initiative from Indian Knowledgebase). We also wish to thank the editorial team at Wiley for their patience and understanding during the long process of developing this book.

The book project took much longer than what was anticipated. Partly it may be blamed on the COVID-19 pandemic. Such a long, arduous and demanding project is impossible to complete without wholehearted support from the family. VR wishes to acknowledge his wife Nanda for her patience, understanding, and enthusiastic support. VB wishes to acknowledge the support from his parents and his son, Omkar. SN would like to acknowledge the support of his wife Eswarimeena for her understanding and support. SN would also like to acknowledge his daughters Saiseerthi and Keerthi for keeping him always on his toes. VS wishes to acknowledge the support and encouragement received from his wife Namana, parents—Shanthi and Sarvothaman, and friends—Shekhar, Sebastien, Gunjan, and Shubham during this book project.

January 2022

Vivek V. Ranade, Limerick, Ireland
Vinay M. Bhandari, Pune, India
Sanjay Nagarajan, Pontypridd, UK
Varaha P. Sarvothaman, Belfast, UK
Alister T. Simpson, Belfast, UK

Part I

Introduction

1

Hydrodynamic Cavitation

Hydrodynamic cavitation is a phenomenon of formation, growth and collapse of vapor filled cavities (micro bubbles) within a liquid due to variation of local pressure. The cavities are generated in a low pressure region. When these cavities travel to a region of higher pressure, they implode (collapse). The cavity collapse under certain conditions results in very high pressures and temperatures near the location of collapse [1–3]. These extreme temperatures (>2000 K) and pressures (>100 MPa) result in the generation of highly reactive radical species (from water and dissolved gases). In addition, the imploding cavities also result in high velocity jets and intense shear. These extreme physio-chemical effects produced by cavitation have been of interest to engineers and scientists for over a century, largely due to its potential to damage equipment. The mechanisms of cavitation damage have been and continue to be hotly debated, and to this day the causes and effects of cavitation still present fundamental scientific challenges to engineers; for example, continuing issues in the field of water turbines described by the "three gorges puzzle" [4].

The extreme conditions generated during cavitation can also be harnessed for realizing beneficial physico-chemical transformations. In the last couple of decades, the application of cavitation has been extensively explored for a variety of physical, chemical and biological processes. Cavitation offers a novel way of intensifying these processes in an energy efficient manner [5, 6]. The in situ generation of strong oxidants like OH· radicals, local hot spots and intense shear, has a potential to become a very promising technology platform for realizing various transformations (see Figure 1.1). It can be harnessed for waste water treatment [7–12]; microbial disinfection [13–17]; desulphurisation of fuels [18–20]; biomass pre-treatment [21–23]; biodiesel synthesis [24, 25]; and in food and beverage production [26, 27]; esterification reactions [28]; and many other process intensification applications [3, 5, 6, 29, 30].

The references cited here represent only a tiny fraction of illustrative examples from published laboratory studies on applications of hydrodynamic cavitation. There are also several patents and start-up/spin-out companies commercializing cavitation based technologies and applications. See for example, reviews by Carpenter et al. [5] and Holkar et al. [6]. Despite such a large number of publications

Hydrodynamic Cavitation: Devices, Design, and Applications, First Edition.
Vivek V. Ranade, Vinay M. Bhandari, Sanjay Nagarajan, Varaha P. Sarvothaman and Alister T. Simpson.

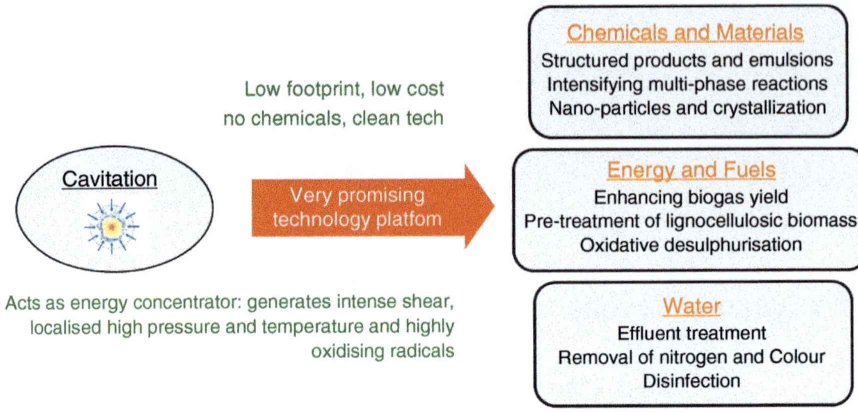

Figure 1.1 Cavitation for beneficial physico-chemical transformations.

on cavitation (thousands of research papers per year – Google Scholar), the promise of hydrodynamic cavitation as a technology platform is still largely unfulfilled. One of the key reasons of holding back the realization of this promise is inadequate understanding of inception as well as resulting physico-chemical effects of cavitation. Systematic methodology for designing hydrodynamic cavitation devices and generalized framework for development and optimisation of hydrodynamic cavitation processes is still not established and available.

Mathematical modeling of hydrodynamic cavitation and the related physico-chemical effects is rather complex even for the simplest devices and systems. Dynamics of a single cavity and subsequent collapse when exposed to adequately strong pressure fluctuations has been studied for long since Lord Rayleigh [31]. The book by Brennen [1] provides detailed models as well as results of cavity dynamics. The quest for developing better models for simulating cavity dynamics is still continuing (see for example, Pandit et al. [32] and references cited therein). Though such single cavity models are available, use of these models for simulating overall performance is still far from satisfactory. The book by Shah et al. [2] has attempted to present a methodology for systematic analysis of cavitation processes and discussed possible pathways for connecting single cavity dynamics to overall performance. Franc and Michel [33] published a book on fundamentals of cavitation. However, the focus was mainly on hydrodynamics and various applications were not discussed. About a decade ago, Ozonek [34] attempted to present a methodology for applying hydrodynamic cavitation primarily to water treatment applications. There are couple of reviews from the group of Professor Pandit [30, 35] with discussion on design aspects of orifice and venturi based hydrodynamic cavitation.

None of the available resources provide systematic basis for designing and evaluating different hydrodynamic cavitation devices or for simulating overall performance and scale-up of processes based on such devices. This book attempts to bridge this gap. This chapter provides a general introduction to hydrodynamic cavitation and devices to realize hydrodynamic cavitation. It also presents overall structure of the book to facilitate better usage of the presented contents.

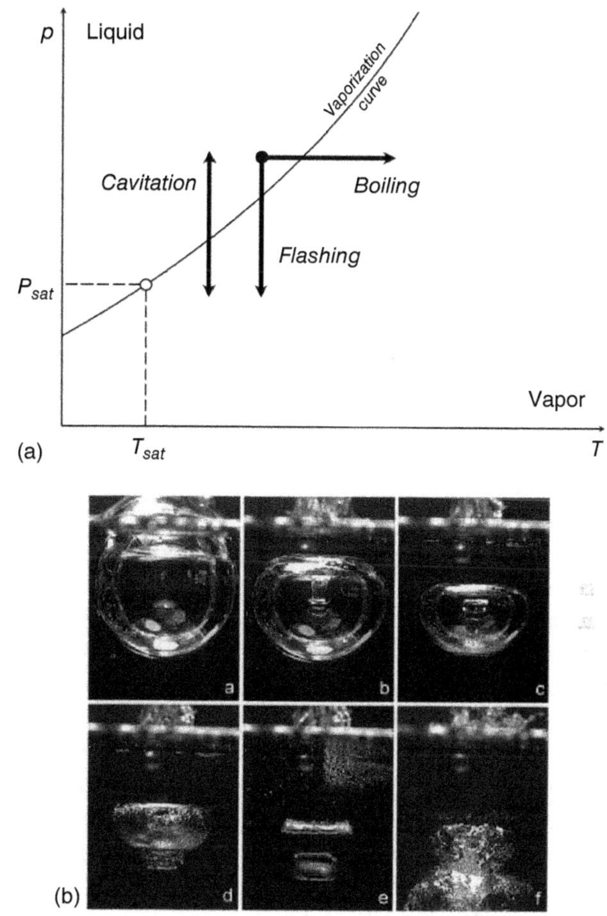

Figure 1.2 (a) Boiling and cavitation illustrated in thermodynamic diagram (Source: Reprinted from [36], with permission from Meijn/Delft University of Technology) and (b) Graphic illustration of growth, implosion and collapse of bubbles in the cavitation process (Source: Reprinted from [37], with permission from Elsevier).

1.1 Hydrodynamic Cavitation

Cavitation in a simple sense is the phenomenon of formation, growth and collapse of gaseous pockets in a liquid medium due to a dynamic pressure change in the bulk of the medium. When a liquid is subjected to a pressure below its vapor pressure, there is a possibility of generating a cavity or a gaseous bubble which is called as cavitation (see Figure 1.2a). In absence of any nuclei, liquid phase is able to withstand negative pressure (with respect to its vapor pressure) without rupturing or forming vapor cavities. In such cases, cavitation may occur via homogeneous nucleation (formation of small nuclei – transient gaps between molecules caused by random motions of the molecules). However, in most of the real life cases, liquid contains tiny suspended particles or some dissolved gases. These particles or desorption of dissolved gases provide nucleation sites for cavitation. In the presence of such heterogeneous nucleation, cavitation may be assumed to occur when local pressure falls below vapor pressure of liquid.

When a cavity generated in a manner described above is subjected to varying pressure field, complex process of expansion and contraction may occur eventually leading to the collapse of the cavity. Cavity dynamics is generally quite fast and has been extensively investigated. The experimentally visualized growth and collapse of a typical cavity is shown in Figure 1.2b. The collapse of cavities give rise to high velocity jets, very high localized pressure and temperature as well as results in the generation of hydroxyl radicals [3, 32]. The intense shear and hydroxyl radicals generated by collapsing cavities and resulting physico-chemical transformations have been of interest to scientists and engineers for over a century to either develop approaches for avoiding them or harnessing them for variety of desired applications [38].

The history of this phenomenon can be traced back to it being proposed as the possible reason for reduced efficiency of a warship – HMS *Darling* back in 1885. The phenomenon was thought to be responsible for the formation of gaseous pockets at very high speeds in naval structures and thereby reducing its efficiency. Barnaby [39, 40] wrote papers to describe the phenomenon of creation of voids and cavities, below a certain low pressure. Lord Rayleigh formalized these discussions and theoretically proposed mechanisms for the cavitation phenomenon in 1917, which has been improved until now to develop the current understanding on the mechanism of cavitation. There is evidence that the phenomenon was anticipated as early as 1704 by Newton and further in the works of Euler and Reynolds in the nineteenth century [41]. A detailed historical account of the origin prior to Lord Rayleigh's work is succinctly provided by Young. The historical aspects such as the coinage of the word and earlier theoretical descriptions of existence of such a phenomenon exist prior to Lord Rayleigh's work [33], but for all practical purposes, theoretical investigation and the foundation of the current understanding was from the early twentieth century.

Depending on the way of generation, four types of cavitation processes are known: optical, particle, acoustic and hydrodynamic cavitation. The cavitation phenomena realized by optic and particle methods are not widely investigated, as they are not suitable for effecting change in bulk solutions and have very limited applicability to chemical processing [42, 43]. Acoustic cavitation, as evident from its name, uses ultrasound for generating cavitation. When high frequency sound waves (20 kHz to 200 MHz) of adequate intensity are passed through a liquid, local pressure may dip below the vapor pressure and cavitation may occur. Ultrasonic cavitation was first used to accelerate chemical reactions in 1927 [44]. Since then, several papers have been published on a wide range of applications of ultrasonic cavitation. For example, chemical reactions [45, 46], homogenization [47, 48], cell disruption [49, 50], dentistry [51] among other niche applications such as imaging of foetus [52] and physiotherapy [53]. A recent book edited by Ashokkumar [54] provides an excellent compilation on ultrasonic cavitation and its applications. Despite several published studies on numerous applications, industrial applications of ultrasonic cavitation are rather limited barring a few exceptions. This may be because of inability to distribute cavitation zones in a reactor volume, metal contamination in horn type of reactors, very low energy efficiency and high capital as well as operating costs [55].

Unlike ultrasonic cavitation, hydrodynamic cavitation by its nature is more amenable for scaling up and for industrial scale implementation. Hydrodynamic cavitation is realized by forcing liquid to flow through specifically designed fluidic devices which are configured in such a way that a single or multiple low pressure zones are realized. Cavities are generated in such low pressure regions and travel with the flowing liquid. Typically, flow in these devices is in the turbulent regime. The inception of cavitation may therefore happen even when time averaged local pressure is higher than vapor pressure. When cavities traveling with liquid are exposed to higher pressures and turbulent pressure fluctuations, these cavities collapse and can be harnessed for achieving desired physico-chemical transformations. Hydrodynamic cavitation has all the desired attributes required for successful implementation in industrial processes. In last two decades, there is tremendous interest in developing and applying hydrodynamic cavitation based processes. Several examples of such applications are cited at the beginning of this chapter. Though the idea of generating low pressure zones and thereby realizing cavities which may be harnessed for desired application looks simple, the lack of systematic basis for designing hydrodynamic cavitation devices and lack of understanding of how to simulate physico-chemical effects of generated cavities in such devices is a bottleneck. This book attempts to provide systematic framework for developing devices, design approaches and applications of hydrodynamic cavitation. The following section provides a general introduction to hydrodynamic cavitation devices.

1.2 Hydrodynamic Cavitation Devices

Hydrodynamic cavitation devices can be broadly classified into two types, namely, with and without moving parts (Figure 1.3). Devices with moving parts usually involve a rotor – an impeller rotating at high speed and a stator with a small clearance between rotor and stator. The surface of rotor or stator may have dents or dimples of appropriate size, shape and depth which facilitate generation of low pressure regions and consequently, cavitation. The inception and extent of cavitation depends on specific configuration of rotor – stator, rotating speed and net flow rate of liquid through the device. See for example Petkovšek et al. [56] and Garlicka et al. [57]. Rotor–stator based cavitation devices are expensive, energy intensive and require maintenance leading to higher operational costs. Despite the limitations, rotor-stator based devices have been proposed and used for wastewater treatment [58] and for biomass pre-treatment [59].

Hydrodynamic cavitation devices without any moving parts are designed to generate low pressure zones within the device by manipulating the flow field. Typically, realizing high velocity zones by virtue of geometric configuration of device generates desired low pressure. Broadly speaking, such passive hydrodynamic cavitation devices either use small constrictions and linear flow through these constrictions or use strongly swirling flows for generating low pressure regions. Typical examples of such devices and corresponding pressure profiles are shown schematically in Figure 1.4. The presence of small constrictions used in linear flow based device may

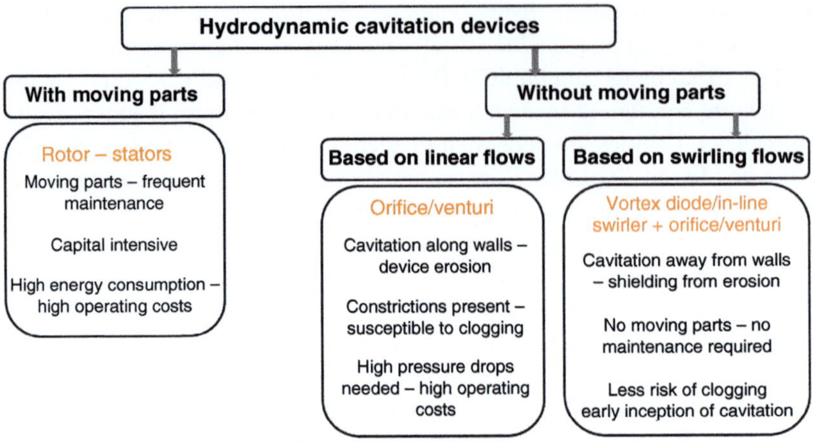

Figure 1.3 Classification of hydrodynamic cavitation devices.

pose a risk of clogging when handling liquids containing suspended solids (which is not uncommon for many wastewater or other biomass pre-treatment applications). In such devices, the cavitation bubble collapse occurs near the walls and therefore devices are susceptible to erosion [38]. Swirling flow based devices developed by Ranade et al. [61] overcome some of these disadvantages. In these vortex based devices, the flowing fluid enters the circular chamber tangentially, leading to a free vortex flow. When the free vortex is converted to a forced vortex at the axial port exit, cavitation is generated. The mode of generation of cavities in vortex based devices involve a swirl flow and therefore does not require very small constrictions and generate cavities in the core of the flowing liquid [38]. Such devices are much less susceptible to clogging than conventional orifice or venturi based devices. More importantly, the cavity collapse occur in the vortex core of the flowing fluid. This naturally shields the device walls from erosion. It is also possible to develop passive hydrodynamic cavitation devices by combining swirling and linear flows. For example, a commercially available cavitation device, Dynaswirl [62] or use of inline swirlers by Simpson and Ranade [38]. However, such devices still are based on small constrictions and therefore may inherit some of the disadvantages of conventional linear flow based devices.

Newer hydrodynamic cavitation devices are being developed and several attempts are being made to commercialize these devices. For example, Biobang/Cavimax (www.biobang.com) is commercializing rotor – stator based devices, HyCa Technologies (www.hycator.com) is commercializing linear flow based devices, VIVIRA Process Technologies (www.vivira.in) is commercializing vortex diode as well as hybrid devices based on in-line swirler designs. It is important to develop systematic basis for comparing such different devices.

Apart from the geometric configuration of the hydrodynamic cavitation devices, selecting an appropriate range for operation is also important. In all such hydrodynamic cavitation devices, the inception of cavitation occurs at a certain flow rate which is a function of device design, characteristic of liquid, presence of suspended solids or dissolved gases and operating temperature. As the flow rate increases,

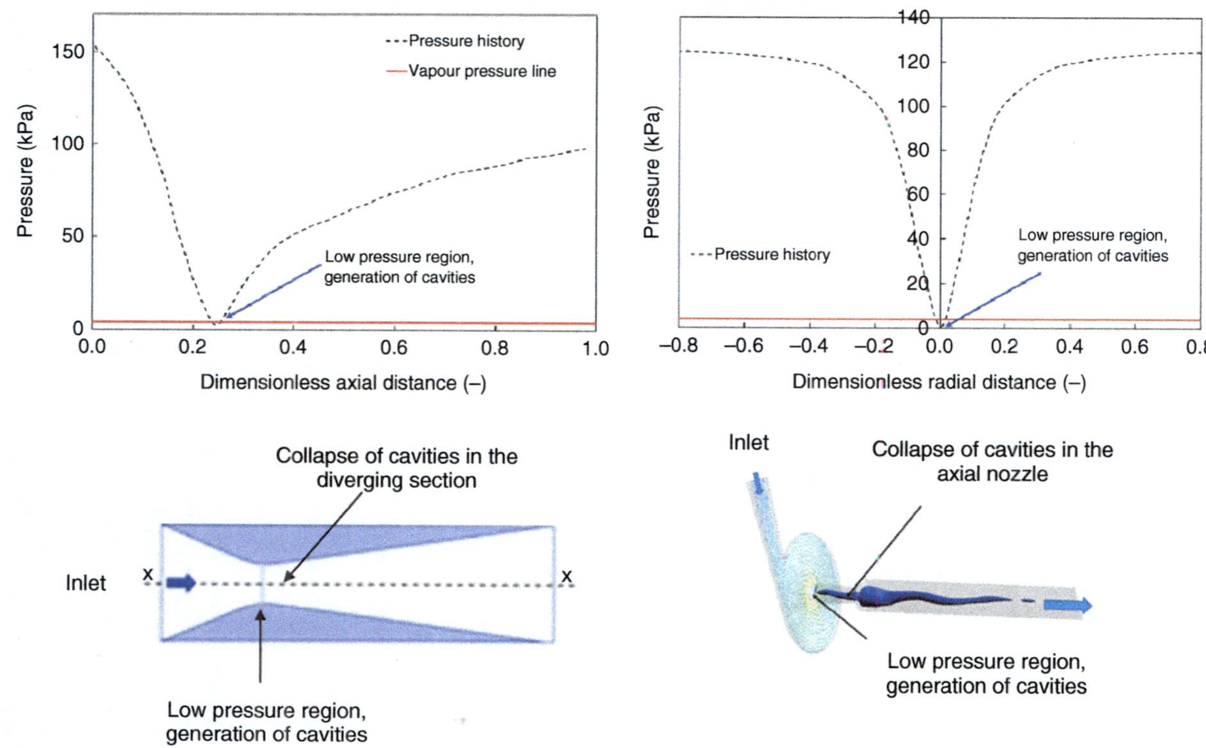

Figure 1.4 Pressure profile in linear flow based cavitation device and swirling flow based cavitation device (Source: Adapted from [38, 60]).

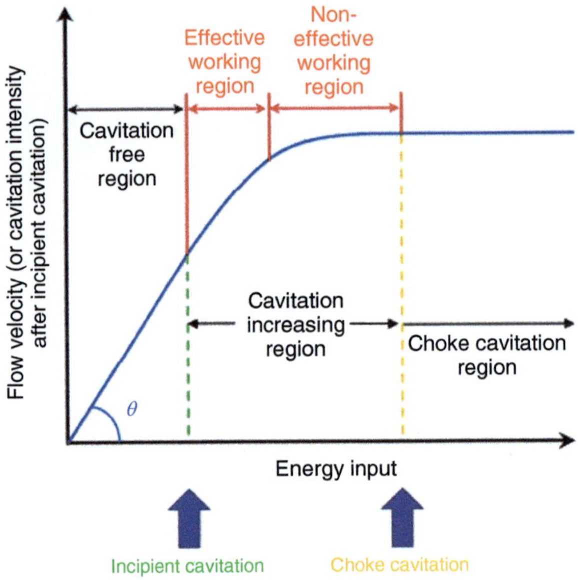

Figure 1.5 Different cavitation regimes as a function of flow rate or energy input (Source: From [37]/with permission of Elsevier).

extent of cavitation increases up to a certain limit. Beyond a certain threshold, choked cavitation is realized. This is schematically shown in Figure 1.5.

For appropriate selection of geometric configuration and operating conditions of hydrodynamic cavitation devices, it is essential to develop an ability to quantitatively simulate cavitating flows in such devices and thereby estimate overall process performance, facilitate optimisation and scale-up. Development of computational models for simulating performance of cavitation devices is not straight forward primarily because of existence of a wide range of spatio-temporal scales (single cavity with characteristic scales of $\sim 10^{-6}$ m, $\sim 10^{-4}$ s to cavitation device scale of $\sim 10^{0}$ m, $\sim 10^{0}$ s). Different types of models have been proposed and used in previous studies. Flow models range from a single cavity (for example, Gireesan and Pandit [63], Pawar et al. [64], and Pandit et al. [32]) to device scale computational fluid dynamics models (for example, Simpson and Ranade [65]). For estimating overall performance, typically reaction engineering models are used (Holkar et al. [6], Ranade and Bhandari [7], and references cited therein). Many such studies use and report pseudo-first order rate constants to describe performance of hydrodynamic cavitation process/device (for example, Rajoriya et al. [66], Capocelli et al. [67], and Saharan et al. [68]). Use of such a pseudo-rate constant which is a function of cavitation device configuration, operating conditions and number of passes through cavitation device is rather misleading and cannot be used to compare different cavitation devices. Some attempts have also been made to develop empirical correlation of performance with design and operating parameters (for example, see Sharma et al. [69]). However, such correlations have very limited applicability considering a wide variety of devices being used in practice. Detailed multi-scale or multi-layer computational fluid dynamics (CFD) models may potentially help to develop an ability to make quantitative simulations of flow and performance of hydrodynamic cavitation devices. The previously published studies on flow models may be broadly categorized into two groups: the first group focuses on detailed flow characteristics rather than performance of cavitation device/process (see for example, Ma et al. [70] and Hsiao et al. [71]). The second group focuses on simulating performance of cavitation devices using rather simplified CFD models (see for example, Capocelli et al. [72]). It is important to systematically present designs of hydrodynamic cavitation devices, experimental set-up required to evaluate performance of such devices and a suite of computational models useful for simulating hydrodynamic cavitation devices and processes based on those devices. In this book (Part 1), such an attempt is made. A brief introduction to applications of hydrodynamic cavitation is included in the following section.

1.3 Applications of Hydrodynamic Cavitation

As discussed earlier, hydrodynamic cavitation involves rapid phase change and generation of localized high pressure and temperature regions. Intense shear and high velocity jets as well as strongly oxidizing radicals are generated. These physico-chemical effects of cavitation can be harnessed for a variety of applications. Large number of studies reporting applications of hydrodynamic cavitation may be broadly classified into three groups as shown in Figure 1.6.

Several reviews listing these applications have been published. For example, see recent reviews by Tao et al. [73], Carpenter et al. [5], Albanese et al. [74], Holkar et al. [6], Panda et al. [75], Sun et al. [76] and Mancuso et al. [77]. Most of these reviews however primarily list various applications and provide only cursory information and therefore not very useful. Three major applications of hydrodynamic cavitation are wastewater treatment, water disinfection and pre-treatment of biomass. In this book, we have included dedicated chapters for these three applications. These primarily harness generation of oxidizing radicals as well as intense shear via cavity collapse. We have also summarized other applications. However, an attempt has been made to connect those applications of hydrodynamic cavitation to device design and wherever possible to modeling. The goal is not just to list the applications but provide adequate information to the reader so as to facilitate further work or implementation in practice. Each application, of course, will require corresponding domain knowledge and it is not possible to include all the intricacies with respect to numerous applications. However, an attempt is made here to present generic approach while discussing selected applications so that the approach may be adapted for applications not included in this book. The overall organisation of the book is briefly discussed in the following section.

1.4 Organization of the Book

The information in this book is organized to facilitate systematic understanding of hydrodynamic cavitation. The focus is on providing adequate framework for design of hydrodynamic cavitation devices, modeling of processes based on such

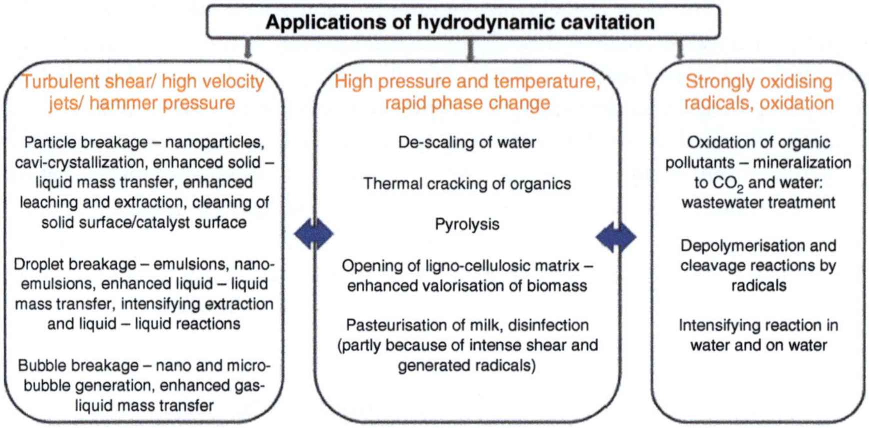

Figure 1.6 Applications of hydrodynamic cavitation.

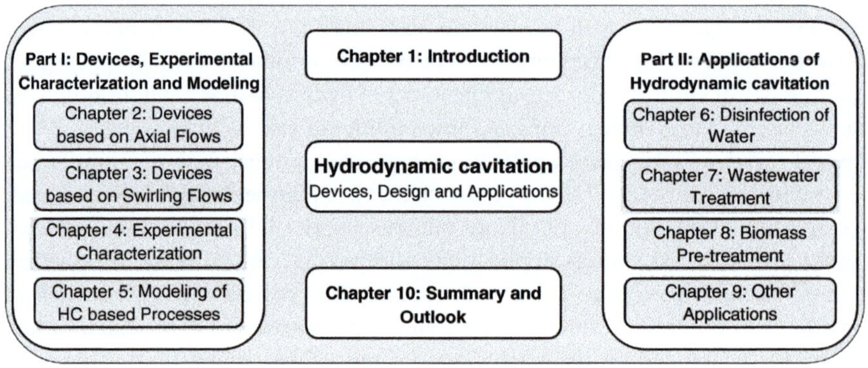

Figure 1.7 Organization of the book.

devices (experimentally as well as computationally) and developing applications of hydrodynamic cavitation. We have restricted the scope of this book to primarily discuss hydrodynamic cavitation realized by passive devices (without any moving parts). The material in this book is organized mainly in two parts (see the overall structure shown in Figure 1.7).

The first part presents design aspects of hydrodynamic cavitation and the second part presents various applications. These two parts precede with this introductory chapter. The last chapter on summary and outlook is included after the second part.

The first part comprises four chapters. Chapter 2 covers flow characteristics of hydrodynamic cavitation devices based on linear or axial flows. This primarily includes two most widely used cavitation devices namely orifice and venturi. Over the years, significant experimental data and information on these two devices have been accumulated. This chapter presents detailed computational fluid dynamics based models to provide useful insights and discuss various key design parameters of such cavitation devices. Some of the disadvantages of these linear flow based devices are highlighted. Hydrodynamic cavitation devices based on swirling flows are discussed in Chapter 3. The advantages realized by using swirling flows are highlighted. CFD models are used for bringing out key characteristics and key design parameters. Together these two chapters will provide adequate information on development and application of CFD models for simulating flow characteristics of hydrodynamic cavitation devices and clearly bring out the key design parameters. The presented discussion will not only be useful for further improvement and optimisation of devices discussed in this book, but will also be useful for developing new and better hydrodynamic cavitation devices.

Chapters 4 and 5 provide ways of experimentally characterizing various hydrodynamic cavitation devices and computationally modeling performance of processes based on cavitation devices, respectively. Key components of typical experimental set-up required to characterize and quantify performance of hydrodynamic cavitation devices are discussed in Chapter 4. Some of the common pitfalls in designing the experimental set-up are highlighted and recommendations to avoid these are included. The experimental investigations are mainly carried out to identify first inception of cavitation and then quantification of performance. Brief discussion on planning and executing such experiments on inception and quantification of performance is included. Specific recommendations are discussed. Chapter 4 and references cited therein will provide useful information for designing experimental set-up and carrying out experiments to characterize hydrodynamic cavitation devices.

Computational models for interpreting experimental data on performance of hydrodynamic cavitation devices and processes are discussed in Chapter 5. The processes based on hydrodynamic cavitation essentially use generated shear, localized hot spots, and hydroxyl radicals for realizing various physico-chemical transformations. Various approaches ranging from empirical models to rigorous multi-layer computational models for modeling hydrodynamic cavitation based transformation processes are discussed. Large number of studies use the empirical approaches – either based on effective rate constant or per-pass performance factor where these empirical parameters are obtained using the experimental data. These models are discussed in detail and compared with each other. A brief discussion on recently published studies about the possibility of developing Artificial Neural Networks (ANN's) based models is also included. After discussing these empirical models, physics based models starting with a single cavity to multiphase applications based on hydrodynamic cavitation are presented. A need for simultaneous application of multiple modeling approaches is highlighted. Some comments on future outlook and further advances in modeling of hydrodynamic cavitation based processes are also included.

The second part of the book presents various applications of hydrodynamic cavitation. Three major applications namely disinfection of water, wastewater treatment and pre-treatment of biomass are discussed in Chapters 6, 7, and 8, respectively. Chapter 6 is dedicated to discuss investigations of hydrodynamic cavitation based disinfection of water. Safe drinking water is important to mankind and quality of life, thus the disinfection of water by efficient processes becomes pertinent. This chapter discusses the advantages that cavitation could offer over conventional techniques.

Here the possibility of the technology as a standalone process to break down harmful microorganisms is discussed. Studies which have investigated the possibilities to enhance the performance of the cavitation process with hydrogen peroxide, ozone and aeration are discussed. Recent investigations to enhance the cavitation based disinfection with natural oils are discussed. The discussion will be useful for harnessing hydrodynamic cavitation for the disinfection of water with no or minimal use of other chemicals.

Chapter 7 discusses the wastewater treatment application. Water serves many functions in manufacturing industries and eventually ends up as wastewater. The contaminated and pollutant containing wastewater may have significantly higher chemical oxygen demand (COD) than permissible limits. It is therefore essential to treat such wastewater for oxidizing pollutants/species causing higher COD. Application of hydrodynamic cavitation which generates strongly oxidizing radicals for treating wastewater is discussed in Chapter 7. Published studies are critically reviewed to identify key process parameters and their influence on overall performance. These process parameters include operating temperature, pH, pressure drop across cavitation devices and initial COD levels. Possible ways of augmenting performance of hydrodynamic cavitation including use of aeration, addition of hydrogen peroxide or ozone and photocatalysis have also been briefly discussed. The discussion will be useful for developing intensified wastewater treatment processes based on hydrodynamic cavitation.

Application of hydrodynamic cavitation for enhancing valorisation of lignocellulosic biomass is discussed in Chapter 8. The key step in any such valorisation process of biomass is pre-treatment, which opens up complex lignocellulosic matrix and makes the cellulose and hemicellulose accessible for further chemical or biochemical process steps. Hydrodynamic cavitation can be used as a stand alone pre-treatment method or as a tool for intensifying other chemical based pre-treatments. Use of hydrodynamic cavitation for two key application areas namely bioethanol and biogas production is discussed. Key aspects are reviewed and influence of important process parameters on overall performance are brought out. The discussion will be useful for exploiting hydrodynamic cavitation for valorisation of a variety of complex biomass resources. Some comments on how hydrodynamic cavitation based bio-refineries may be developed are also included.

It is impossible to include every application of hydrodynamic cavitation in any book. Attempt is made here to present a generic approach towards development and optimisation of applications using hydrodynamic cavitation. Beyond these three major applications, additional applications are grouped by fundamental types such as gas–liquid, liquid–liquid and solid–liquid in Chapter 9. These applications namely floatation, oxidative desulphurization, emulsification, biodiesel production, cavi-crystallization and nano-particle generation are briefly reviewed. The key issues in the conventional processing are identified, the role of hydrodynamic cavitation is suggested, and the key studies in these application areas are provided as a directive to the reader.

In the last chapter of the book (Chapter 10), the current state of art on design of hydrodynamic cavitation devices and processes is summarized. A need for developing better multi-scale models and multi-layer models is emphasized. Need to develop appropriate verification, validation and model calibration practices is discussed. The lessons learnt from our experience of applying computational flow modeling to hydrodynamic cavitation based processes are briefly summarized. Some comments on selecting right level of complexity and modeling approach are included. Availability of non-invasive sensors such as acoustic sensors, real-time data processing, advances in multiphase flow modeling and machine learning may open up opportunities for developing high-fidelity computational models of hydrodynamic cavitation based processes. Besides comments on design and modeling, some discussion on current trends and outlook on hydrodynamic cavitation based applications are included. We hope that this book provides a framework and useful guidelines for harnessing hydrodynamic cavitation for a wide range of applications.

References

1 Brennen, C.E. (2014). *Cavitation and Bubble Dynamics*. Cambridge University Press.

2 Shah, Y.T., Pandit, A.B., and Moholkar, V.S. (1999). *Cavitation Reaction Engineering*. Springer Science & Business Media.

3 Suslick, K.S., Mdleleni, M.M., and Ries, J.T. (1997). Chemistry induced by hydrodynamic cavitation. *Journal of the American Chemical Society* 119 (39): 9303–9304.

4 Li, S. (2015). Tiny bubbles challenge giant turbines: three gorges puzzle. *Interface Focus* 5 (5): 20150020.

5 Carpenter, J., Badve, M., Rajoriya, S. et al. (2017). Hydrodynamic cavitation: an emerging technology for the intensification of various chemical and physical processes in a chemical process industry. *Reviews in Chemical Engineering* 33 (5): 433–468.

6 Holkar, C.R., Jadhav, A.J., Pinjari, D.V., and Pandit, A.B. (2019). Cavitationally driven transformations: a technique of process intensification. *Industrial and Engineering Chemistry Research* 58 (15): 5797–5819.

7 Ranade, V.V. and Bhandari, V.M. (2014). *Industrial Wastewater Treatment, Recycling and Reuse*. Butterworth-Heinemann.

8 Gągol, M., Przyjazny, A., and Boczkaj, G. (2018). Wastewater treatment by means of advanced oxidation processes based on cavitation – a review. *Chemical Engineering Journal* 338: 599–627.

9 Sarvothaman, V.P., Nagarajan, S., and Ranade, V.V. (2018). Treatment of solvent-contaminated water using vortex-based cavitation: influence of operating pressure drop, temperature, aeration, and reactor scale. *Industrial and Engineering Chemistry Research* 57 (28): 9292–9304.

10 Sarvothaman, V.P., Simpson, A.T., and Ranade, V.V. (2019). Modelling of vortex based hydrodynamic cavitation reactors. *Chemical Engineering Journal* 377: 119639.

11 Suryawanshi, P.G., Bhandari, V.M., Sorokhaibam, L.G. et al. (2018). Solvent degradation studies using hydrodynamic cavitation. *Environmental Progress & Sustainable Energy* 37 (1): 295–304.

12 Patil, P.B., Bhandari, V.M., and Ranadec, V.V. (2021). Wastewater treatment and process intensification for degradation of solvents using hydrodynamic cavitation. *Chemical Engineering and Processing Process Intensification* 108485.

13 Gaikwad, V. and Ranade, V. (2016). Disinfection of water using vortex diode as hydrodynamic cavitation reactor. *Asian Journal of Chemistry* 28 (8): 1867.

14 Jain, P., Bhandari, V.M., Balapure, K. et al. (2019). Hydrodynamic cavitation using vortex diode: an efficient approach for elimination of pathogenic bacteria from water. *Journal of Environmental Management* 242: 210–219.

15 Mane, M.B., Bhandari, V.M., Balapure, K., and Ranade, V.V. (2020). A novel hybrid cavitation process for enhancing and altering rate of disinfection by use of natural oils derived from plants. *Ultrasonics Sonochemistry* 61: 104820.

16 Mane, M.B., Bhandari, V.M., and Ranade, V.V. (2021). Safe water and technology initiative for water disinfection: application of natural plant derived materials. *Journal of Water Process Engineering* 43: 102280.

17 Mane, M.B., Bhandari, V.M., Balapure, K., and Ranade, V.V. (2020). Destroying antimicrobial resistant bacteria (AMR) and difficult, opportunistic pathogen using cavitation and natural oils/plant extract. *Ultrasonics Sonochemistry* 69: 105272.

18 Suryawanshi, N.B., Bhandari, V.M., Sorokhaibam, L.G., and Ranade, V.V. (2016). A non-catalytic deep desulphurization process using hydrodynamic cavitation. *Scientific Reports* 6 (1): 1–8.

19 Suryawanshi, N.B., Bhandari, V.M., Sorokhaibam, L.G., and Ranade, V.V. (2017). Developing techno-economically sustainable methodologies for deep desulfurization using hydrodynamic cavitation. *Fuel* 210: 482–490.

20 Ucbc, J., Kryževičius, Ž., Janutėnienė, J. et al. (2021). Desulfurizing of pyrolysis oil of used tires using a 3D-printed vortex diode and modeling of process. *Journal of Marine Science and Engineering* 9 (8): 876.

21 Nakashima, K., Ebi, Y., Shibasaki-Kitakawa, N. et al. (2016). Hydrodynamic cavitation reactor for efficient pretreatment of lignocellulosic biomass. *Industrial and Engineering Chemistry Research* 55 (7): 1866–1871.

22 Terán Hilares, R., Ramos, L., da Silva, S.S. et al. (2018). Hydrodynamic cavitation as a strategy to enhance the efficiency of lignocellulosic biomass pretreatment. *Critical Reviews in Biotechnology* 38 (4): 483–493.

23 Nagarajan, S. and Ranade, V.V. (2019). Pretreatment of lignocellulosic biomass using vortex-based devices for cavitation: influence on biomethane potential. *Industrial and Engineering Chemistry Research* 58 (35): 15975–15988.

24 Maddikeri, G.L., Gogate, P.R., and Pandit, A.B. (2014). Intensified synthesis of biodiesel using hydrodynamic cavitation reactors based on the interesterification of waste cooking oil. *Fuel* 137: 285–292.

25 Malani, R.S., Shinde, V., Ayachit, S. et al. (2019). Ultrasound–assisted biodiesel production using heterogeneous base catalyst and mixed non–edible oils. *Ultrasonics Sonochemistry* 52: 232–243.

26 Albanese, L., Ciriminna, R., Meneguzzo, F., and Pagliaro, M. (2017). Beer-brewing powered by controlled hydrodynamic cavitation: theory and real-scale experiments. *Journal of Cleaner Production* 142: 1457–1470.

27 Pathania, S., Ho, Q.T., Hogan, S.A. et al. (2018). Applications of hydrodynamic cavitation for instant rehydration of high protein milk powders. *Journal of Food Engineering* 225: 18–25.

28 Díez, D., Urueña, A., Barrios, C., and Antolín, G. (2019). Valorization of rendering fats to produce biodiesel by single and multi orifice plate cavitation reactor. *Waste and Biomass Valorization* 10 (12): 3773–3780.

29 Didenko, Y.T., McNamara, W.B., and Suslick, K.S. (1999). Hot spot conditions during cavitation in water. *Journal of the American Chemical Society* 121 (24): 5817–5818.

30 Gogate, P.R. and Pandit, A.B. (2005). A review and assessment of hydrodynamic cavitation as a technology for the future. *Ultrasonics Sonochemistry* 12 (1–2): 21–27.

31 Rayleigh, L.V.I.I.I. (1917). On the pressure developed in a liquid during the collapse of a spherical cavity. *The London, Edinburgh, and Dublin Philosophical Magazine and Journal of Science* 34 (200): 94–98.

32 Pandit, A.V., Sarvothaman, V.P., and Ranade, V.V. (2021). Estimation of chemical and physical effects of cavitation by analysis of cavitating single bubble dynamics. *Ultrasonics Sonochemistry* 77: 105677.

33 Franc, J.-P. and Michel, J.-M. (2006). *Fundamentals of Cavitation*. Springer Science & Business Media.

34 Ozonek, J. (2012). *Application of Hydrodynamic Cavitation in Environmental Engineering*. CRC Press.

35 Gogate, P.R. and Pandit, A.B. (2004). A review of imperative technologies for wastewater treatment I: oxidation technologies at ambient conditions. *Advances in Environmental Research* 8 (3–4): 501–551.

36 Meijn, G.-J. (2015). *Physical Modeling of Cavitation Using an Enthalpy Based Model*. Delft, The Netherlands: Faculty of Mechanical, Maritime and Material Engineering; Delft University of Technology.

37 Sun, X., Liu, J., Ji, L. et al. (2020). A review on hydrodynamic cavitation disinfection: the current state of knowledge. *Science of the Total Environment* 737: 139606.

38 Simpson, A. and Ranade, V.V. (2019). 110th Anniversary: comparison of cavitation devices based on linear and swirling flows: hydrodynamic characteristics. *Industrial and Engineering Chemistry Research* 58 (31): 14488–14509.

39 Barnaby, S.W. (1902). Torpedo-boat destroyers. *Journal of the American Society for Naval Engineers* 14 (2): 396–408.

40 Barnaby, S.W. (ed.) (1906). *Note on the Cavitation of Screw Propulsion (Including Appendix)*. Thomas Telford-ICE Virtual Library.

41 Young, F.R. (1999). *Cavitation*. World Scientific.

42 Asaithambi, N., Singha, P., Dwivedi, M., and Singh, S.K. (2019). Hydrodynamic cavitation and its application in food and beverage industry: a review. *Journal of Food Process Engineering* 42 (5): e13144.

43 Gogate, P.R. and Kabadi, A.M. (2009). A review of applications of cavitation in biochemical engineering/biotechnology. *Biochemical Engineering Journal* 44 (1): 60–72.

44 Richards, W.T. and Loomis, A.L. (1927). The chemical effects of high frequency sound waves I. a preliminary survey. *Journal of the American Chemical Society* 49 (12): 3086–3100.

45 Vyas, S. and Ting, Y.-P. (2018). A review of the application of ultrasound in bioleaching and insights from sonication in (bio) chemical processes. *Resources* 7 (1): 3.

46 Sancheti, S.V. and Gogate, P.R. (2017). A review of engineering aspects of intensification of chemical synthesis using ultrasound. *Ultrasonics Sonochemistry* 36: 527–543.

47 Manickam, S., Sivakumar, K., and Pang, C.H. (2020). Investigations on the generation of oil-in-water (O/W) nanoemulsions through the combination of ultrasound and microchannel. *Ultrasonics Sonochemistry* 69: 105258.

48 Ojha, K.S., Mason, T.J., O'Donnell, C.P. et al. (2017). Ultrasound technology for food fermentation applications. *Ultrasonics Sonochemistry* 34: 410–417.

49 Wang, M. and Yuan, W. (2016). Modeling bubble dynamics and radical kinetics in ultrasound induced microalgal cell disruption. *Ultrasonics Sonochemistry* 28: 7–14.

50 Bystryak, S., Santockyte, R., and Peshkovsky, A.S. (2015). Cell disruption of S. cerevisiae by scalable high-intensity ultrasound. *Biochemical Engineering Journal* 99: 99–106.

51 Angelopoulos, C. (2018). Ultrasound in dentistry: toward a future of radiation-free imaging. *Dental Clinics of North America* 62 (3): 481–489.

52 Reissland, N., Francis, B., Kumarendran, K., and Mason, J. (2015). Ultrasound observations of subtle movements: a pilot study comparing foetuses of smoking and nonsmoking mothers. *Acta Paediatrica* 104 (6): 596–603.

53 Fernández-Lao, C., Cantarero-Villanueva, I., Galiano-Castillo, N. et al. (2016). The effectiveness of a mobile application for the development of palpation and ultrasound imaging skills to supplement the traditional learning of physiotherapy students. *BMC Medical Education* 16 (1): 1–7.

54 Ashokkumar, M. (2016). *Advantages, Disadvantages and Challenges of Ultrasonic Technology. Ultrasonic Synthesis of Functional Materials*, 41–42. Springer.

55 Sutariya, S., Sunkesula, V., Kumar, R., and Shah, K. (2018). Emerging applications of ultrasonication and cavitation in dairy industry: a review. *Cogent Food & Agriculture* 4 (1): 1549187.

56 Petkovšek, M., Mlakar, M., Levstek, M. et al. (2015). A novel rotation generator of hydrodynamic cavitation for waste-activated sludge disintegration. *Ultrasonics Sonochemistry* 26: 408–414.

57 Garlicka, A., Zubrowska-Sudol, M., Umiejewska, K. et al. (2020). Effects of thickened excess sludge pre-treatment using hydrodynamic cavitation for anaerobic digestion. *Energies* 13 (10): 2483.

58 Zhao, Z., Wang, L., Fan, J. et al. (2020). Degradation of indigo carmine by coupling Fe (II)-activated sodium persulfate and ozone in a rotor-stator reactor. *Chemical Engineering and Processing Process Intensification* 148: 107791.

59 Garuti, M., Langone, M., Fabbri, C., and Piccinini, S. (2018). Monitoring of full-scale hydrodynamic cavitation pretreatment in agricultural biogas plant. *Bioresource Technology* 247: 599–609.

60 Sarvothaman, V.P. (2020). *Hydrodynamic Cavitation for Effluent Treatment: Using Vortex-Based Cavitation Devices*. Queen's University Belfast.

61 Ranade, V.V., Kulkarni, A.A., and Bhandari, V.M. (2016). inventors Vortex diodes as effluent treatment devices. Patent US9422952B2.

62 Dynaflow. Dynajets cavitating jets 2021. http://www.dynaflow-inc.com/Products/Jets/Jets-Cavitating.htm.

63 Gireesan, S. and Pandit, A.B. (2017). Modeling the effect of carbon-dioxide gas on cavitation. *Ultrasonics Sonochemistry* 34: 721–728.

64 Pawar, S.K., Mahulkar, A.V., Pandit, A.B. et al. (2017). Sonochemical effect induced by hydrodynamic cavitation: comparison of venturi/orifice flow geometries. *AIChE Journal* 63 (10): 4705–4716.

65 Simpson, A. and Ranade, V.V. (2019). Flow characteristics of vortex based cavitation devices: computational investigation on influence of operating parameters and scale. *AIChE Journal* 65 (9): e16675.

66 Rajoriya, S., Bargole, S., George, S., and Saharan, V.K. (2018). Treatment of textile dyeing industry effluent using hydrodynamic cavitation in combination with advanced oxidation reagents. *Journal of Hazardous Materials* 344: 1109–1115.

67 Capocelli, M., Prisciandaro, M., Lancia, A., and Musmarra, D. (2014). Hydrodynamic cavitation of p-nitrophenol: a theoretical and experimental insight. *Chemical Engineering Journal* 254: 1–8.

68 Saharan, V.K., Badve, M.P., and Pandit, A.B. (2011). Degradation of Reactive Red 120 dye using hydrodynamic cavitation. *Chemical Engineering Journal* 178: 100–107.

69 Sharma, A., Gogate, P.R., Mahulkar, A., and Pandit, A.B. (2008). Modeling of hydrodynamic cavitation reactors based on orifice plates considering hydrodynamics and chemical reactions occurring in bubble. *Chemical Engineering Journal* 143 (1–3): 201–209.

70 Ma, J., Hsiao, C.-T., and Chahine, G.L. (2017). A physics based multiscale modeling of cavitating flows. *Computers and Fluids* 145: 68–84.

71 Hsiao, C.-T., Ma, J., and Chahine, G.L. (2017). Multiscale tow-phase flow modeling of sheet and cloud cavitation. *International Journal of Multiphase Flow* 90: 102–117.

72 Capocelli, M., Musmarra, D., Prisciandaro, M., and Lancia, A. (2014). Chemical effect of hydrodynamic cavitation: simulation and experimental comparison. *AIChE Journal* 60 (7): 2566–2572.

73 Tao, Y., Cai, J., Huai, X. et al. (2016). Application of hydrodynamic cavitation to wastewater treatment. *Chemical Engineering and Technology* 39 (8): 1363–1376.

74 Albanese, L. and Meneguzzo, F. (2019). *Hydrodynamic Cavitation-Assisted Processing of Vegetable Beverages: Review and the Case of Beer-Brewing*, 211–257. Production and Management of Beverages.

75 Panda, D., Saharan, V.K., and Manickam, S. (2020). Controlled hydrodynamic cavitation: a review of recent advances and perspectives for greener processing. *Processes* 8 (2): 220.

76 Sun, X., Chen, S., Liu, J. et al. (2020). Hydrodynamic cavitation: a promising technology for industrial-scale synthesis of nanomaterials. *Frontiers in Chemistry* 8: 259.

77 Mancuso, G., Langone, M., and Andreottola, G. (2020). A critical review of the current technologies in wastewater treatment plants by using hydrodynamic cavitation process: principles and applications. *Journal of Environmental Health Science and Engineering* 18 (1): 311–333.

Part II

Hydrodynamic Cavitation Devices

2

Hydrodynamic Cavitation Devices Based on Axial/Linear Flow

2.1 Introduction

Typical passive hydrodynamic cavitation devices based on linear/axial flows use small restrictions like orifice or venturi for generating low-pressure regions. As liquid is forced through the constriction, the velocity increases at the expense of the pressure. Beyond a certain critical velocity, the mean pressure at the throat, or the vena contracta, falls below the vapor pressure of the liquid, leading to cavitation. The typical pressure profile in such devices is shown schematically in Figure 1.4. The orifice is one the most commonly used hydrodynamic cavitation devices [1–3] of this type. The conditions required to initiate cavitation inception with orifices are typically characterized by the non-dimensional parameter σ, the cavitation number, given below in Eq. (2.1), which is calculated as the pressure drop required to reach the liquid vapor pressure normalized by a characteristic dynamic head:

$$\sigma = \frac{p - p_v}{{}^1\!/_2 \rho_l v_t^2} \tag{2.1}$$

where p is downstream pressure, p_v is vapor pressure of liquid at operating temperature, ρ_l is density of liquid, and v_t is velocity at the constriction (or throat of orifice). In principle, the cavitation number approaches the value of unity at cavitation inception. However, as comprehensively discussed by Šarc et al. [4], this is far from the case in reality. Published literature features a broad range of definitions of cavitation number, and a wide variation in values at cavitation inception can be found across different types of devices. The specific geometric details of the restriction play an important role, as these details determine the resulting velocity and pressure profiles in the cavitating region. Figure 2.1 presents a schematic of a simple sharp-edged orifice, highlighting how the velocity and pressure gradients evolve along the restriction. Calculation of σ is highly sensitive to the choice of characteristic velocity value, and this must therefore be systematically determined from realistic velocity fields. Assumed average values based on flow rate and restriction area are often insufficient as a basis to use this σ parameter to infer behavior across different device geometries. This specific topic is covered in detail in Chapter 3, Section 3.3.

Hydrodynamic cavitation may also be generated by other mechanisms – for example, vortex cavitation and the use of rotor-stator type devices – both of which

Hydrodynamic Cavitation: Devices, Design, and Applications, First Edition.
Vivek V. Ranade, Vinay M. Bhandari, Sanjay Nagarajan, Varaha P. Sarvothaman and Alister T. Simpson.

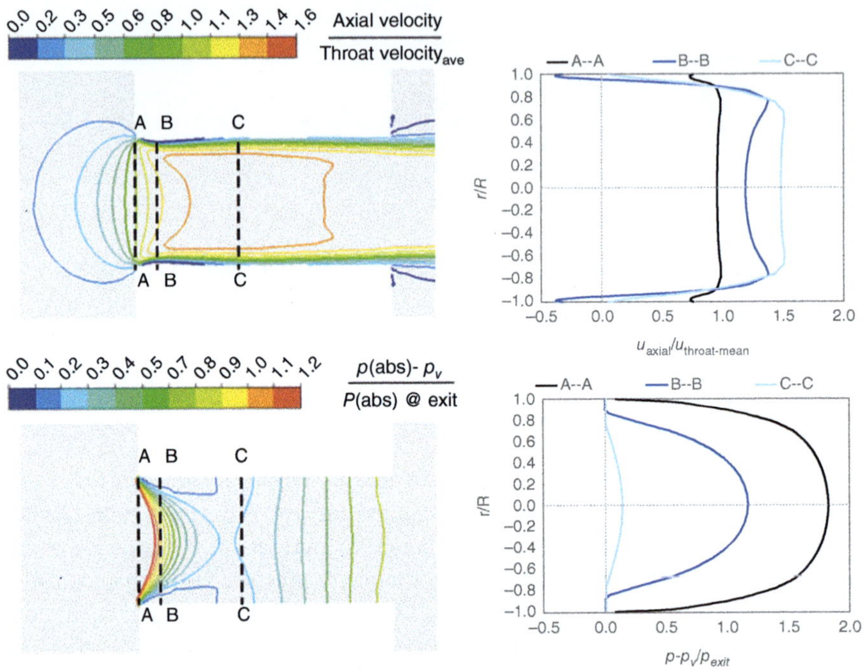

Figure 2.1 Cavitation inception in a sharp-edged orifice.

are covered in detail in Chapter 3. The scope of this chapter is restricted to axial-flow devices. The cavitation in a single axis-symmetric orifice is first introduced, with a focus on the influence of the key geometric parameters on the inception and evolution of cavitation. This is followed by a corresponding discussion of cavitation in single-hole, axis-symmetric venturi geometries, with the key geometric dependencies again presented in detail. The discussion then turns to alternative orifice and venturi designs, featuring multi-hole designs and different cross-section shapes, with a view to understanding potential ways to intensify cavitation performance in axial-flow geometries.

2.2 Orifice-Based Devices

Orifice-type devices are extensively used in pressurized fluid handling systems. The influence of geometry on both cavitating and non-cavitating flow behavior has been the subject of numerous studies. In fuel injection systems, for example, the orifice geometry plays a crucial role in the stability and uniformity of the spray generated as fuel is forced through the restriction, and ultimately on the emissions produced. Pearce and Lichtarowicz [5] presented experimental studies of the influence of geometry on the discharge coefficient C_d, for a range of submerged long orifice designs under both cavitating and non-cavitating conditions. Without cavitation, at Reynolds numbers of the order of 10 000, the discharge coefficient was found to remain constant for any given orifice. At these Reynolds numbers, they presented the following correlation for C_d with length-to-diameter ratio (l/d):

$$(C_d)_{max} = 0.827 - 0.0085(l/d) \tag{2.2}$$

where l is a depth or length of orifice and d is the diameter of orifice. The Eq. (2.2) indicates that with increase in (l/d) ratio, the discharge coefficient decreases. However, it should be noted that this study considered only long or deep orifice designs $(l/d > 2)$. Under cavitating conditions, an alternative equation is presented which gives the discharge coefficient as a function of a cavitation parameter, K:

$$C_d = C_c\sqrt{1+K} \tag{2.3}$$

where

$$K = \frac{P_2 - P_v}{P_1 - P_2} \tag{2.4}$$

and C_c is the contraction coefficient, which was stated to be equal to 0.61 for a sharp-edged inlet. P_1 and P_2 are upstream and downstream pressure respectively. The definition in Eq. (2.3) therefore describes a decreasing discharge coefficient with increasing pressure ratio, which tends to a minimum value equal to the value of C_c. The authors also reported that the pressure ratio required for cavitation inception tended to increase with increasing (l/d) ratio. At this point, it should be mentioned that Eq. (2.4) is one of a large number of definitions of cavitation parameter, or cavitation number, and as stated in the introduction to this chapter, the issues of consistent reporting of this parameter have been comprehensively discussed elsewhere [4]. For the duration of this text, unless explicitly stated otherwise for the purposes of comparison with published experimental values, the definition of cavitation number, σ, is as follows:

$$\sigma = \frac{p_d - p_v}{{}^1\!/\!_2\rho_l v_t^2} \tag{2.5}$$

where v_t is the velocity at the orifice throat and p_d is downstream pressure.

Nurick [6] presented a comprehensive study of a range of orifice designs with varying (l/d) ratios and different orifice inlet conditions. In the cavitating regime, Nurick proposed an alternative expression for C_d:

$$C_d = C_c\sqrt{\frac{P_1 - P_v}{P_1 - P_2}} \tag{2.6}$$

where

$$C_c = 0.62 + 0.38(A_1/A_0)^3 \tag{2.7}$$

Note that the relationships presented in Eqs. (2.3) and (2.6) are equivalent if a common value of C_c is imposed. The difference is therefore simply down to the higher values of C_c produced by Eq. (2.7), which result in higher overall calculated discharge coefficients. Nurick additionally observed that the orifice inlet conditions had a significant effect on the discharge coefficient and cavitation inception. The experimental measurements revealed that the critical pressure ratio required to initiate observable cavitation increased linearly with inlet roundness, and suggested the following relationship for the critical cavitation number, K_{cr}:

$$K_{cr} = \left(\frac{P_1 - P_v}{P_1 - P_2}\right) = -11.4(r/d) + 2.6 \tag{2.8}$$

where r is radius of inlet edge. More recently, Ohrn et al. [7] reported measured C_d values with varying (l/d) ratios, and found good agreement with the work presented in Pearce and Lichtarowicz [5] and Nurick [6]. They reported no appreciable influence on C_d with increasing (l/d) ratios above 2 for sharp-edged nozzles. However, inlet rounding was again found to have a significant effect. Measured discharge coefficients were found to increase from values around 0.68 at (r/d) ratios of 0.05 up to 0.98 at (r/d) ratios of 0.5. Ramamurthi et al. [8] also presented measurements of orifice atomizers with (l/d) ratios varying from 1 to 50, and also observed that cavitation inception was progressively delayed as aspect ratios increased beyond values of 5.

Despite these results demonstrating the significance of (l/d) and the sharpness of the orifice on cavitation behavior, many studies reporting the application of hydrodynamic cavitation focus only on the free area and the perimeter of orifice holes (see a recent review by Carpenter et al. [9] and Benito et al. [10]) and do not report these parameters. Many HC application studies use relatively thin orifices with (l/d) ratios typically less than 2. For example, Braeutigam et al. [11] studied a range of 25 orifice plates to determine the effect of area ratio on the decomposition of chloroform in water; of this series, and all but one of the 25 designs had (l/d) ratios of 2 or lower. There is little information in open literature on the behavior of such thin orifice plates under cavitating conditions and whether or not the relationships presented above can hold in such situations.

The key cavitating flow characteristics of orifices with varying thickness ratios (l/d) and inlet geometry have been presented previously by Simpson and Ranade [2]. Using numerical models validated against experimental data, a parametric study was carried out to investigate a wide range of (l/d) ratios from 5 down to a theoretical limiting case of 0 thickness. The baseline geometry considered in their work is presented in Figure 2.2, which featured an orifice diameter $d = 5.0$ mm and a length $l = 10$ mm, and the parametric variations studied are listed in Figure 2.3 with varying (l/d) ratio. The variation of cavitation inception with (l/d) ratio is shown in Figure 2.4, which clearly illustrates the significant influence of this parameter on inception. For the limiting zero thickness case, large pressure ratios of the order of 20 are required to induce cavitation, corresponding to $\sigma_i = 0.13$. The required operating pressures for inception show little variation above (l/d) ratios of 2, with cavitation inception numbers of the order of 1.5. Although the inception point is consistent

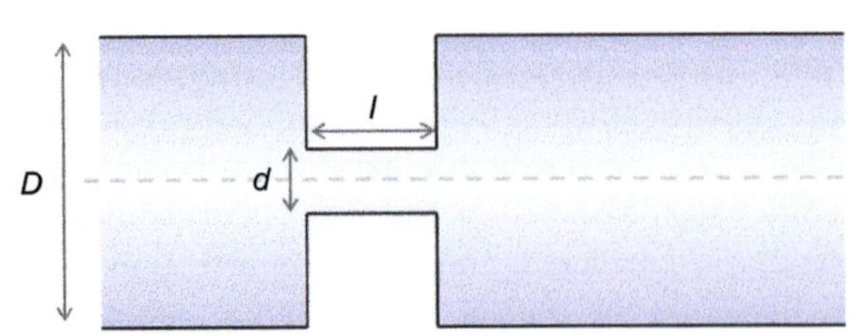

Figure 2.2 Baseline orifice geometry (Source: From [2]/with permission of Elsevier).

Parameter	Figure	Range
Thickness of orifice plate (l/d)	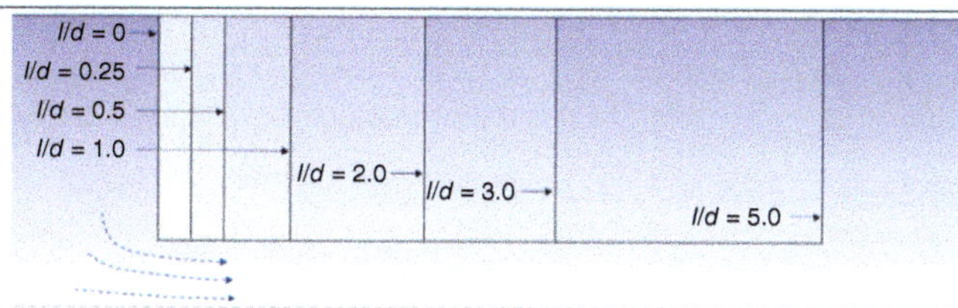	(l/d)= 0, 0.25, 0.50, 1.0

Figure 2.3 Orifice designs with varying (l/d) ratio (Source: From [2]/with permission of Elsevier).

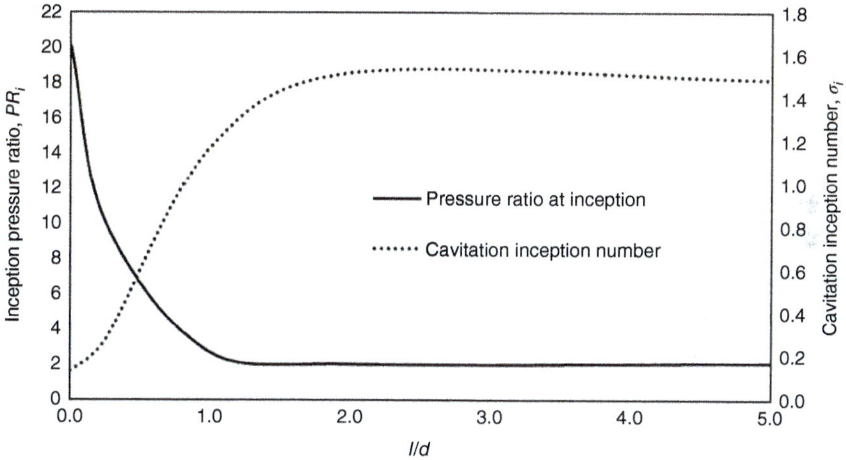

Figure 2.4 Cavitation inception number vs aspect ratio (Source: From [2]/with permission of Elsevier).

beyond values of 1.5, increasing (l/d) ratios beyond 2 has the effect of delaying the pressure recovery, and the low-pressure region is maintained through the length of the orifice. Subsequently, the predicted extent of the vapor cavity is shown to increase with increasing (l/d) ratio under matching operating pressure conditions (see Figure 2.5).

In Figure 2.6 the discharge coefficient is plotted as a function of pressure ratio for a range of different aspect ratios. It can be seen that for pressure ratios below 2, the discharge coefficient peaks at (l/d) ratio of 2. Beyond this, cavitation initiates and the vapor bubble at the inlet forms and grows. There is an initial fluctuation in the discharge coefficient, before a monotonic decrease is observed. If extrapolated, the value of discharge coefficient for each (l/d) ratio tends to values of the order of 0.61, which is in good agreement with the trends observed by Pearce and Licharowicz [5]. The plot of the correlation from Eq. (2.3), using a C_c value of 0.61 shows good agreement for the cavitating regime, with the trend closely following the curve for (l/d) ratio of 2.

Examining the maximum predicted discharge coefficients in the non-cavitating phase, the trends also show good agreement with experimental trends (see

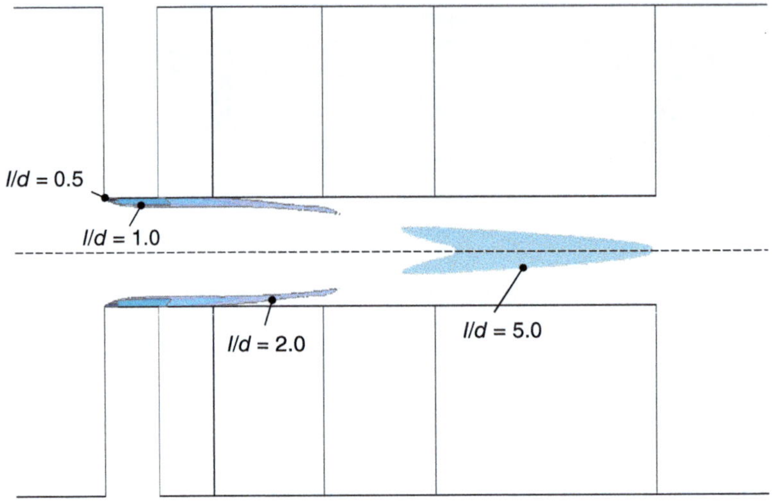

Figure 2.5 Contours of constant vapor volume fraction (=0.5) at varying aspect ratios (Pressure ratio, $Pr = P_1/P_2 = 5$) (Source: From [2]/with permission of Elsevier).

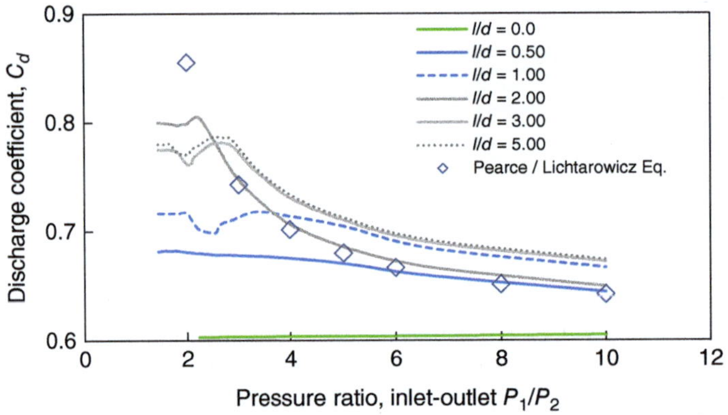

Figure 2.6 Orifice discharge coefficient vs pressure ratio for varying aspect ratios (Source: From [2]/with permission of Elsevier).

Figure 2.7). The calculated values of maximum discharge coefficient for different values of (l/d) ratio are shown in this figure along with the correlation from Pearce and Lichtarowicz [5] given in Eq. (2.2). This correlation was derived from experimental data for (l/d) ratios of 2 and above. Below these values, the results show significant deviation from the linear correlation. The majority of hydrodynamic cavitation studies based on orifice HC feature thin orifices of $(l/d) \leq 2$, and as such, the correlations suggested in Pearce and Lichtarowicz [5] and Nurick [6] would appear to be unsuitable at these lower values. Good agreement is however observed above $(l/d) = 2$, with predictions showing a reduction in C_d values.

The sharpness of orifice is also an important parameter that influences the characteristics of cavitation. Even a small rounding at the edge can dramatically change

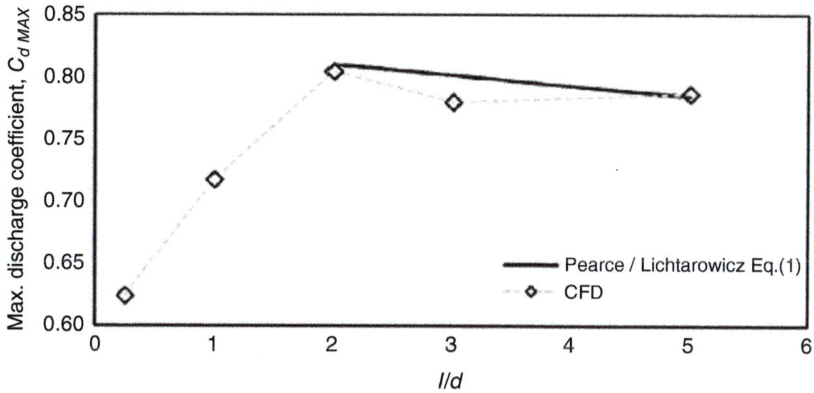

Figure 2.7 Maximum discharge coefficient as a function of (*l/d*) ratio for non-cavitating flows (Source: From [2]/with permission of Elsevier).

the flow field. A further parametric study on orifices with different rounded edges is presented in Simpson and Ranade [2], in which the inlet radius is varied from 0 mm (sharp) to an absolute value of 1.0 mm as detailed in Figure 2.8.

The predicted flow field results for the cases with and without an inlet radius are compared in Figure 2.8, at a pressure ratio of 1.5. It can be seen that different inlet radius sizes result in significant differences in flow behavior. At the operating pressure ratio of 1.5 ($\sigma = 2.2$), the inclusion of a small inlet radius of just 0.2 mm produces a very local acceleration at the inlet edge, which acts to suppress the separation bubble. This acceleration is sufficient to create a local reduction in pressure to the saturated vapor limit at the edge of the orifice. Elsewhere through the throat, however, flow acceleration is smoother than that shown for the sharp-edged orifice, and pressure losses are lower overall. The cavitation inception conditions are plotted as a function of inlet radius in Figure 2.9. At the larger inlet radii of 1 mm, the reduction in velocity gradient is sufficient to delay the predicted onset of cavitation significantly. The subsequent evolution of the vapor cavity at pressure ratios of 2.5 and 5 is shown in Figure 2.10. Although the predicted inception happens at lower pressure ratios for an inlet radius of 0.2 mm, the higher separation losses incurred by the sharp-edged orifice lead to the development of a larger vapor cavity along the outer wall as pressure ratios are subsequently increased. For the larger inlet radius, this smoothing out of the flow gradients is shown to inhibit the development of cavity formation.

Nurick [6] had also discussed the influence of orifice sharpness on cavitation inception. The cavitation inception predicted from the simulated results are compared with the correlation given by Nurick (Eq. (2.8)) in Figure 2.11. The CFD predictions presented in Simpson and Ranade [2] show an initial inflection with the introduction of a very small 0.2 mm radius ($(r/d) = 0.03$), due to the local acceleration produced at the edge. Beyond $(r/d) = 0.05$, the predictions qualitatively follow the trend indicated by the Nurick correlation, with all of the evidence pointing toward a requirement for higher pressure ratios to generate cavitation as the inlet radius is increased. The discharge coefficients of orifices with different sharpness as a function of inlet-to-outlet pressure ratio are shown in Figure 2.12.

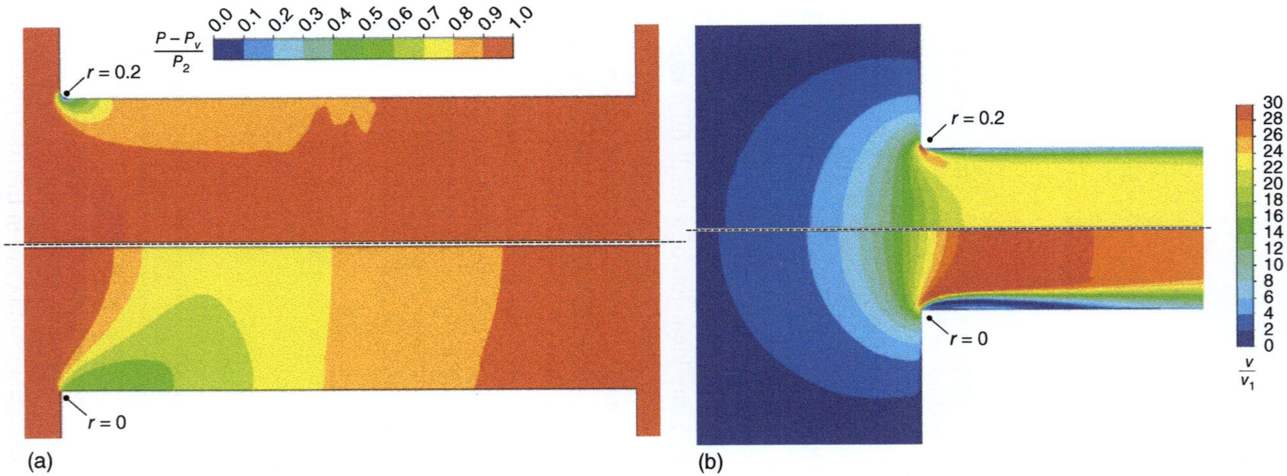

Figure 2.8 Comparison of pressure and velocity distributions in orifice with different inlet radii (Pressure ratio = 1.5). (a) Static pressure. (b) Velocity (Source: From [2]/with permission of Elsevier).

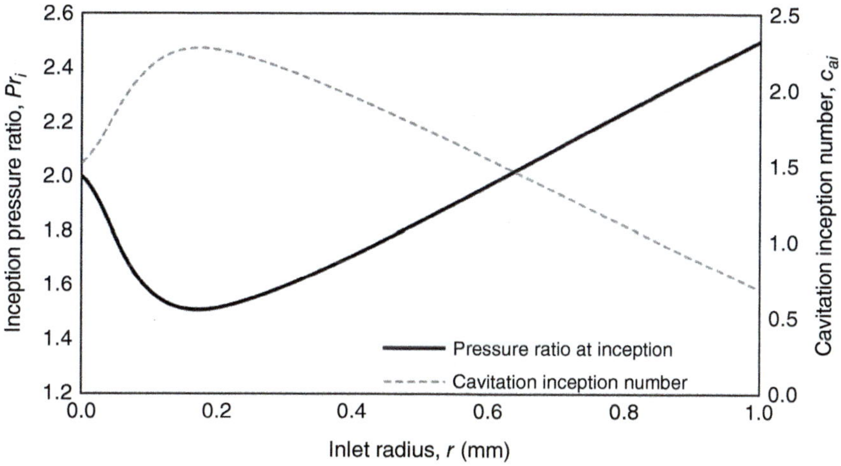

Figure 2.9 Cavitation inception number and inception pressure ratio vs. orifice inlet radius (Source: From [2]/with permission of Elsevier).

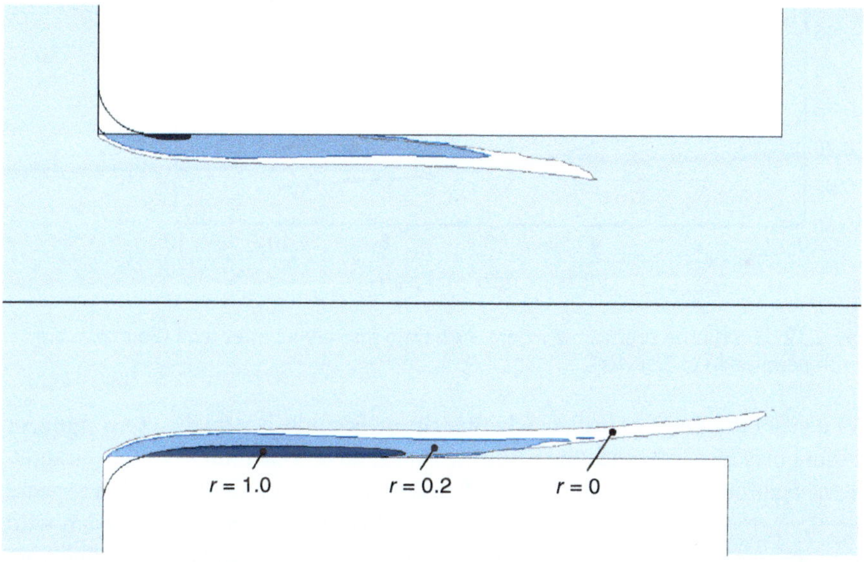

Figure 2.10 Iso-surfaces of constant vapor volume fraction ($\varepsilon = 0.5$) for different inlet radii (Top: pressure ratio = 2.5, bottom: pressure ratio = 5.0) (Source: From [2]/with permission of Elsevier).

The results show an inverse relationship between the hydrodynamic efficiency and the extent of vapor mass transfer, which again agrees with the results of previous experimental studies. This highlights the crucial influence of the inlet losses on the generated cavitation activity.

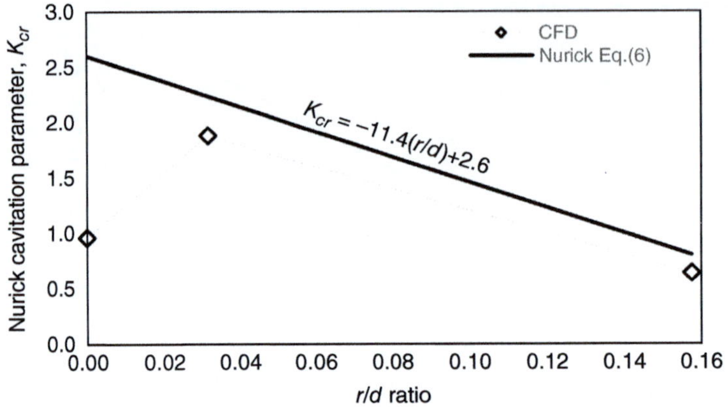

Figure 2.11 Comparison with predicted cavitation inception trend with correlation of Nurick (Source: From [6]).

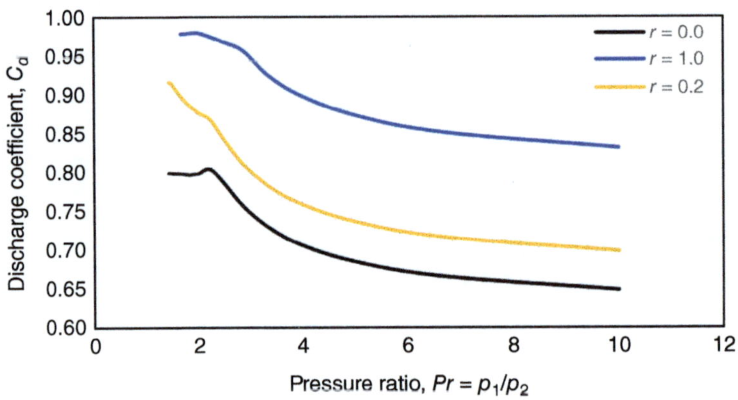

Figure 2.12 Discharge coefficient vs. pressure ratio for varying inlet radii (Source: From [2]/with permission of Elsevier).

For a given thickness of orifice, whether the orifice hole is straight, converging, or diverging may also influence the resulting flow field and cavitation. In the parametric investigation [2], three different orifice designs were compared with a straight restriction, converging restriction (+45°), and diverging section (−45°), each with (l/d) ratio = 0.5 (see Figure 2.13). The corresponding velocity and pressure fields are presented in Figure 2.14. As may be expected, the converging section has the effect of minimizing the separation bubble in comparison to both the straight and diverging section geometries. However, the pressure ratios required for cavitation inception are significantly higher than that required for the straight section. Figure 2.15 shows the minimum predicted pressure in the device versus operating pressure ratio. The corresponding cavitation inception numbers, determined from the conditions at which the minimum predicted pressure reaches the saturated vapor pressure, were found to be 0.66, 0.91, and 1.16 for the +45°, −45°, and straight sections respectively. Both angle profile sections require pressure ratios that are much higher (30–40%),

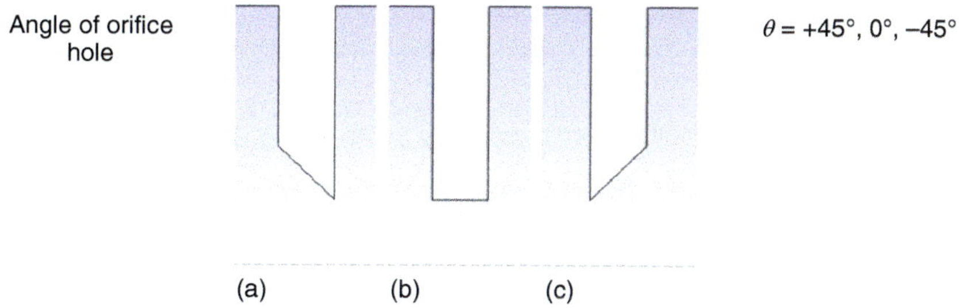

Figure 2.13 Orifice with varying hole angle (a) +45°, (b) 0°, (c) −45° (Source: From [2]/with permission of Elsevier).

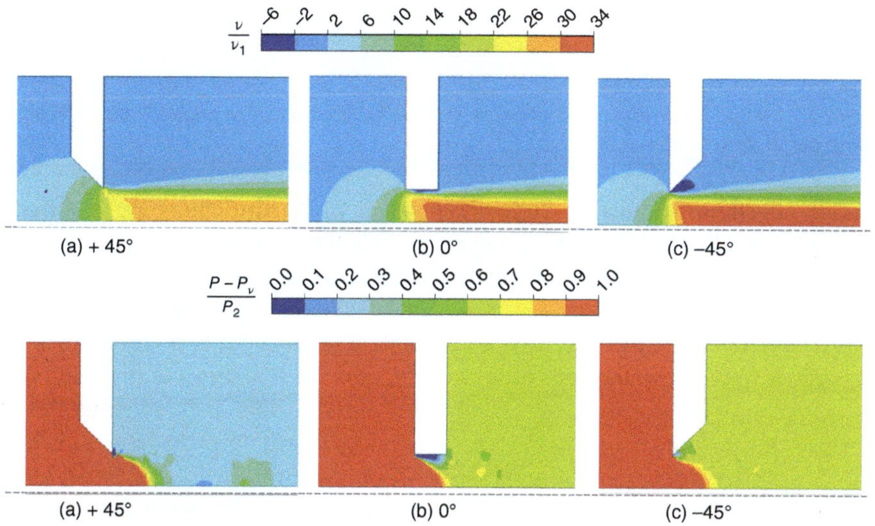

Figure 2.14 Velocity and pressure contours with varying orifice angle (pressure ratio = 5; (a) 45°, (b) 0°, (c) −45°) (Source: From [2]/with permission of Elsevier).

to drive the minimum pressure toward the saturated vapor pressure. This would suggest a significant impact on overall energy efficiency and consumption.

Up until this point, the flow fields and cavitation behavior have been discussed in terms of time-averaged properties. However, hydrodynamic cavitation is an inherently unsteady phenomenon, where the highly turbulent and multiphase flow features complex unsteady behavior. This is illustrated in Figure 2.16a, which shows unsteady periodic cavitation shedding across a range of operating conditions [12]. As the pressure differential increases across the orifice section, the cavitating region grows along the walls, eventually collapsing in a highly unsteady, complex, three-dimensional manner downstream of the orifice exit. Computational fluid dynamic (CFD) simulations using a RANS (Reynolds Averaged Navier Stokes) approach are effective at capturing the time-averaged properties, as shown in Figure 2.16b. The book by Ranade [13] may be referred to for more details on CFD simulations of turbulent flows. However, in order to resolve the details of these

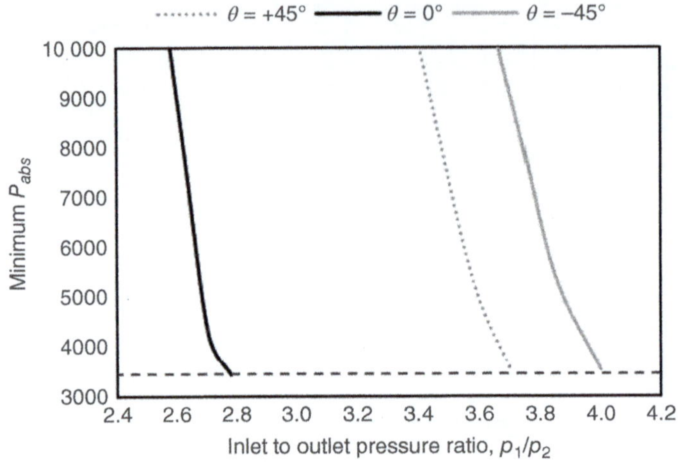

Figure 2.15 Minimum pressure vs. pressure ratio with varying orifice angles (Source: From [2]/with permission of Elsevier).

unsteady effects, full transient three-dimensional CFD simulations are required along with some level of scale, resolving the turbulence closure model.

Bai et al. [14] have presented results from full 3D Large Eddy Simulation (LES) models of cavitating orifice flow, revealing some detailed insights into the nature of the unsteady cavitation processes. As an example, the transient time–pressure history produced by the simulations is shown in Figure 2.17. A series of pronounced peaks in pressure of the order 1 MPa can be seen as the cavitation cloud grows from the orifice entrance and collapses into smaller cavitation structures downstream. Bai et al. [14] further interrogated the transient shedding of the cavitation cloud, and found a linear relationship between the dominant shedding frequency and cavitation number (see Figure 2.18). Detailed numerical simulations such as this provide an important step toward quantifying the performance of hydrodynamic cavitation devices for specific applications. Understanding and quantifying the key properties of these highly transient and turbulent multiphase flow processes is key to determining the resulting collapse intensities in terms of collapse pressures and temperatures, and the resulting reactions. Chapter 5 presents different approaches to combine such detailed CFD studies to cavity dynamics models, to establish modeling approaches that predict the performance of cavitating devices for specific applications.

2.3 Venturi-Based Devices

Similar to orifice-based hydrodynamic cavitation, the venturi is another one of the commonly used hydrodynamic cavitation devices [3, 15, 16]. The cavitating venturi (CV) is a relatively simple flow device, widely used in passive flow control and metering applications. One characteristic is that above a specific pressure ratio, a cavitating venturi becomes choked, beyond which the mass flow rate saturates at a certain

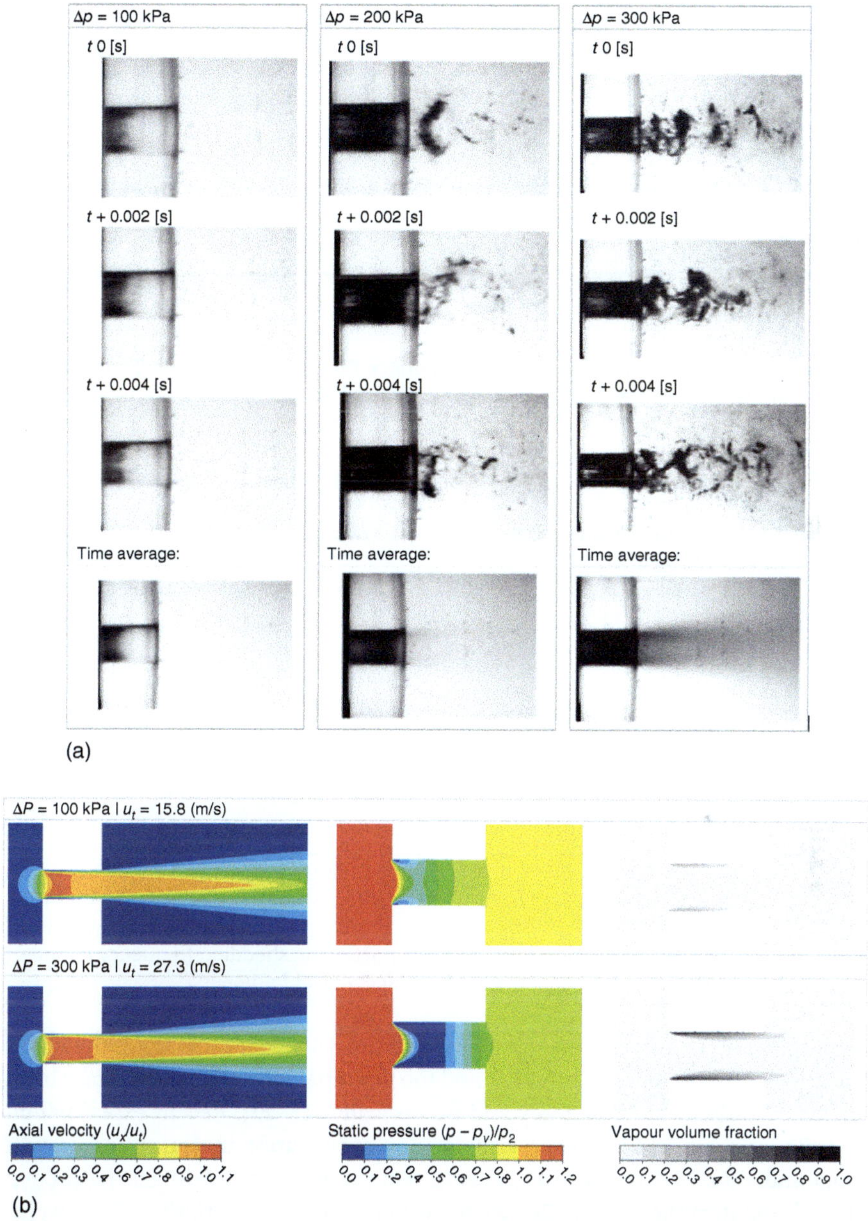

Figure 2.16 Transient cavitating flow in an orifice and time-averaged CFD predictions. (a) High-speed flow visualization of transient cavitating orifice flow. (b) Time-averaged RANS predictions of cavitating orifice (Source: From [12]/with permission of American Chemical Society).

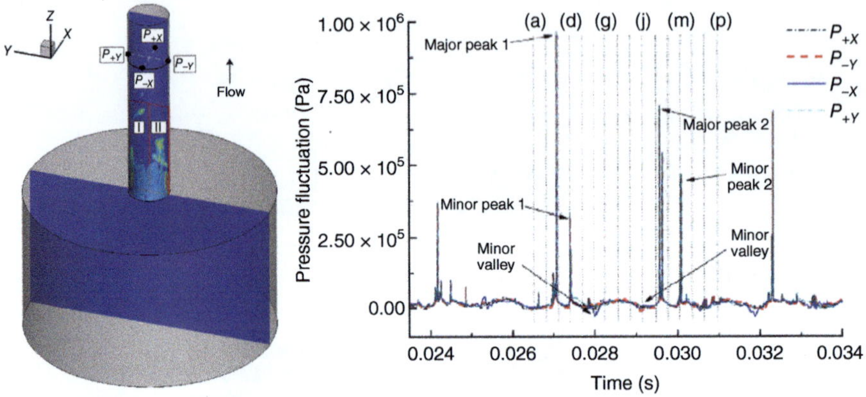

Figure 2.17 Pressure fluctuations in cavitating orifice flow. LES simulations (Source: From [14]/with permission of American Society of Mechanical Engineers).

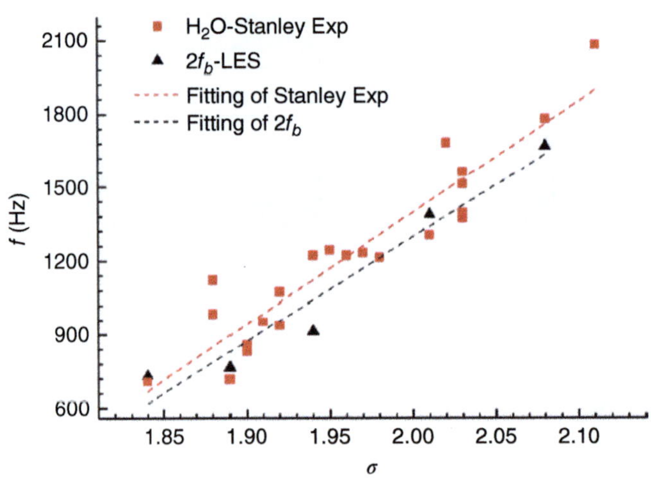

Figure 2.18 Influence of cavitation number on characteristic shedding frequency (Source: From [14]/with permission of American Society of Mechanical Engineers).

level despite any increase in overall pressure ratio. As a result, the CV found early application in rocket fuel systems [17], and the flow characteristics of CVs have been extensively studied experimentally and numerically [18–23]. Recent experimental studies have presented high-speed visualization of cavitation inception and growth through individual venturi devices, such as that presented by Abdulaziz [24], Long et al. [25], and Brinkhorst et al. [26], each presenting a clear description of the evolution of cavitation and its effect on overall device operation in terms of flow rate and pressure drop.

Cavitating venturi devices have been the subject of a number of recent experimental studies, highlighting their promising effectiveness in a range of wastewater treatment [27–29] and other process-intensification applications [30, 31]. However, the role that the numerous interacting design parameters play in overall device

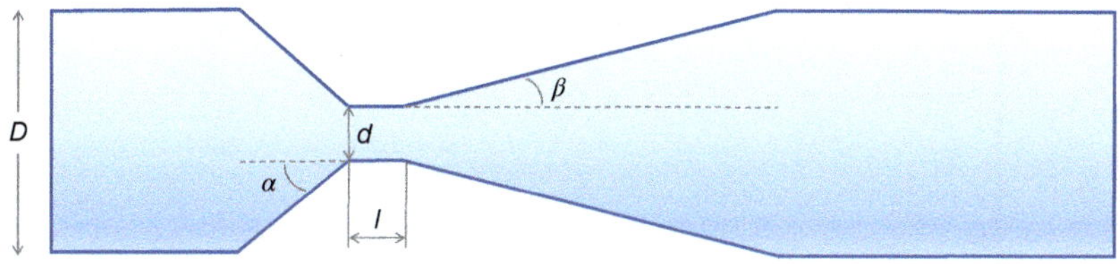

	Baseline	l/d = 1	l/d = 2	l/d = 3	β = 7.5°	β = 12.5°	β = 15°
D [mm]	12	12	12	12	12	12	12
d [mm]	2	2	2	2	2	2	2
l [mm]	0	2	4	6	0	0	0
α [°]	22.5	22.5	22.5	22.5	22.5	22.5	22.5
β [°]	9.25	9.25	9.25	9.25	7.5	12.5	15

Figure 2.19 Venturi design parameters (Source: From [32]/with permission of John Wiley & Sons).

performance is still not fully understood, and the reported experimental studies may employ non-optimized operating conditions and venturi geometries. To provide some guidance on these aspects, the following section details a parametric study of a typical cylindrical venturi device first presented by Simpson and Ranade [32]. The baseline design is based on the geometry and dimensions used by Musmarra et al. [28, 33], shown in Figure 2.19.

Simpson and Ranade [32] presented a CFD model that was shown to capture reported experimental data with cavitating venturi quite well, including the prediction of the cavitation inception point. As an illustration, the simulated flow rate at different pressure ratios was compared with the reported experimental data in

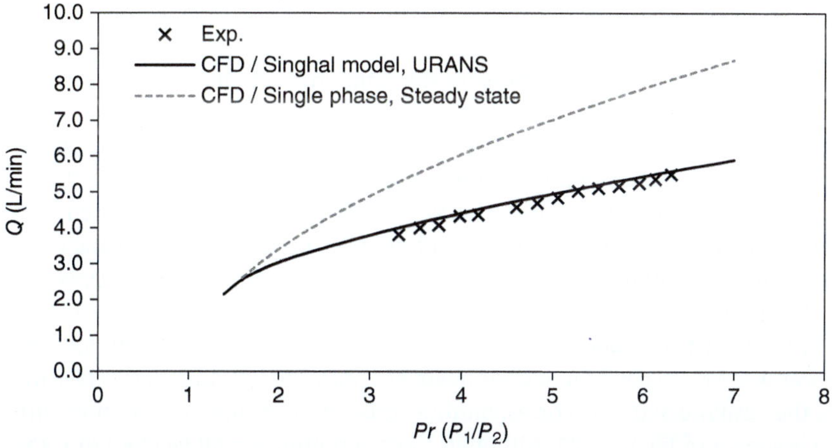

Figure 2.20 Baseline geometry: measured and predicted flow rate vs pressure ratio (Source: From [32]/with permission of John Wiley & Sons).

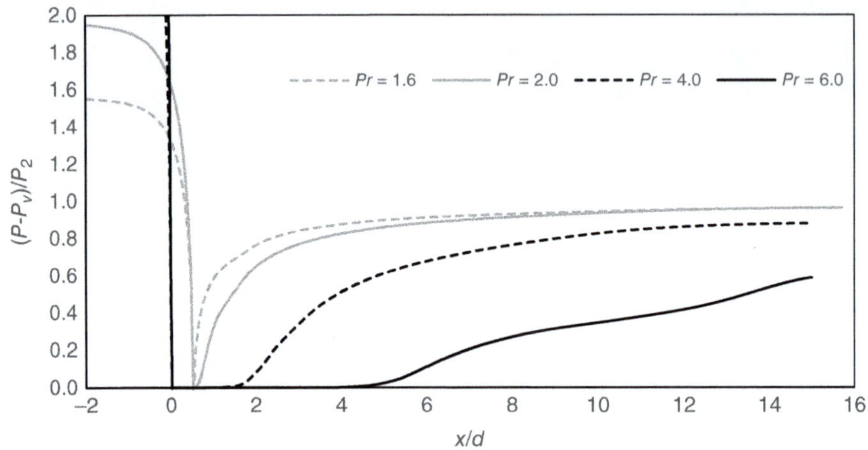

Figure 2.21 Predicted pressure vs. distance along outer wall (Source: From [32]/with permission of John Wiley & Sons).

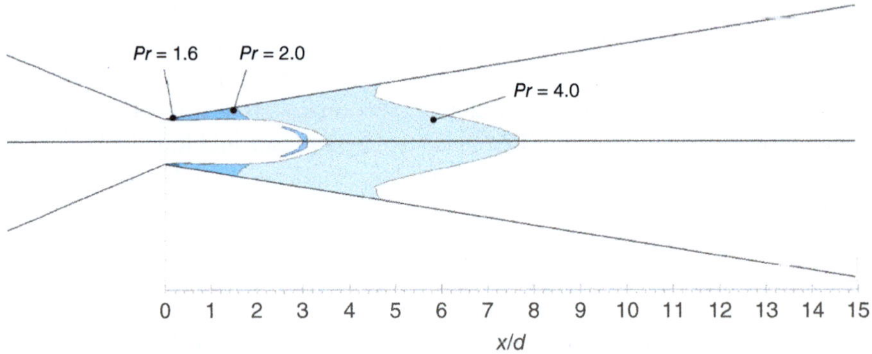

Figure 2.22 Contours of constant vapor volume fraction (0.5) (Source: From [32]/with permission of John Wiley & Sons).

Figure 2.20. The predicted static pressure distribution, both along the outer wall and the axis for one of the considered venturi, is shown in Figure 2.21. Cavitation inception is predicted to occur at a pressure ratio of 1.6, or a cavitation inception number σ_i of 0.96, which is in line with the experiment. Inception is initially localized around the outer wall of the venturi throat as a separation bubble is formed. The low-pressure region then extends axially and radially inward with increasing pressure ratio. Simulated enhancement of the cavity region with increase in pressure ratio is shown in Figure 2.22.

Similar to the orifice parametric study, the effect of throat section length for venturi devices was investigated for designs having l/d ratios of 1, 2, and 3, as well as the value at the baseline of 0. The corresponding variation of cavitation inception with l/d ratio is shown in Figure 2.23. Although inception numbers all lie close to 1, the aspect ratio has some influence on the inception point, with a maximum pressure

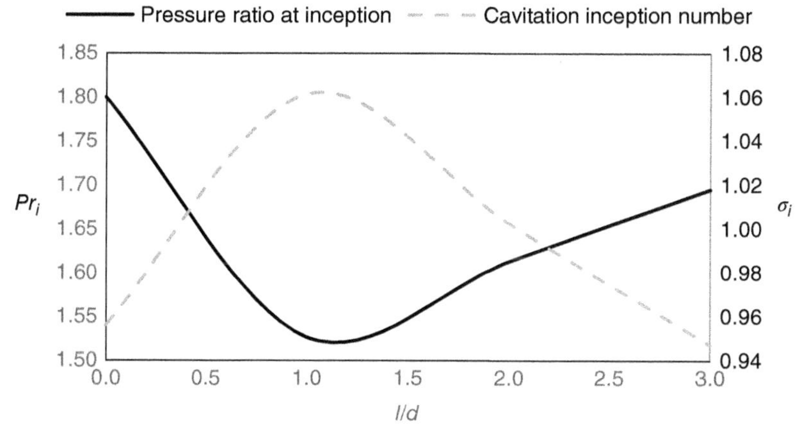

Figure 2.23 Cavitation inception against pressure ratio (Source: From [32]/with permission of John Wiley & Sons).

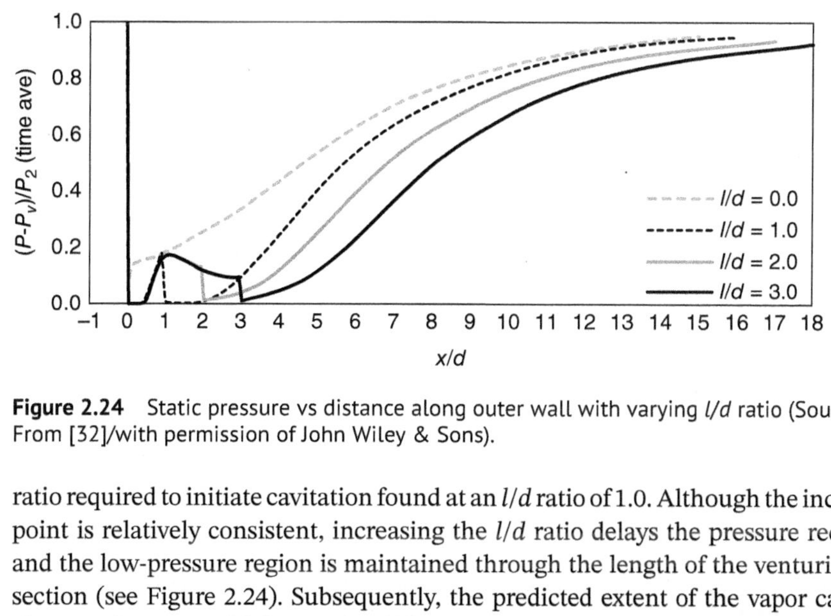

Figure 2.24 Static pressure vs distance along outer wall with varying l/d ratio (Source: From [32]/with permission of John Wiley & Sons).

ratio required to initiate cavitation found at an l/d ratio of 1.0. Although the inception point is relatively consistent, increasing the l/d ratio delays the pressure recovery, and the low-pressure region is maintained through the length of the venturi throat section (see Figure 2.24). Subsequently, the predicted extent of the vapor cavity is shown to increase with increasing l/d ratio under matching operating pressure conditions (Figure 2.25). Generally, with l/d ratios >0, there are two initiation sites for cavity formation: the initial edge of the throat entry, and a second site at the throat exit/diffuser entry where a second, larger cavity forms.

The effect of diffuser angle has also been shown to affect the inception and evolution of cavitation in a venturi significantly. Figure 2.26 shows the cavitation inception point plotted for diffuser angles, β, of 7.25, 9.2, 12.5, and 15°. The wall angle is shown to influence the inception point significantly. Larger wall angles were found to lead to delayed onset and growth of the predicted vapor cavity, with inception

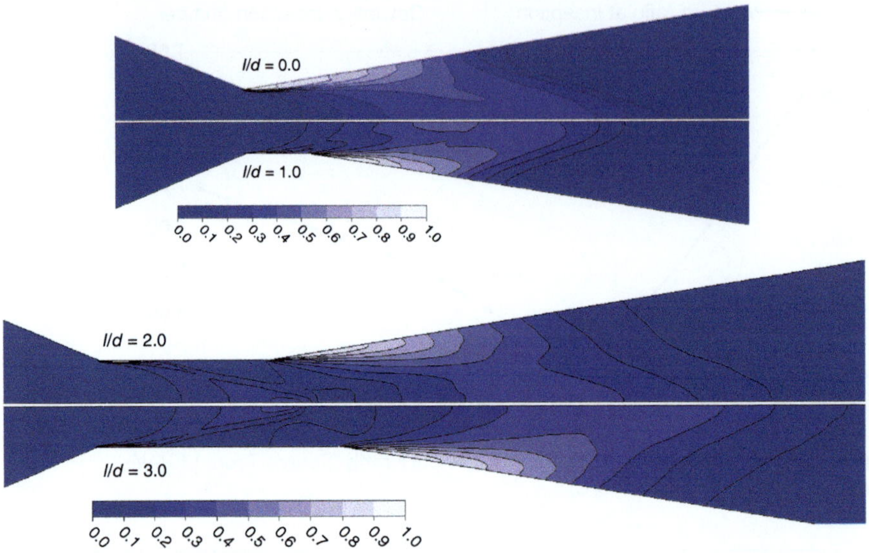

Figure 2.25 Variation in vapor cavity extent and shape with l/d ratio (Pr = 5.0) (Source: From [32]/with permission of John Wiley & Sons).

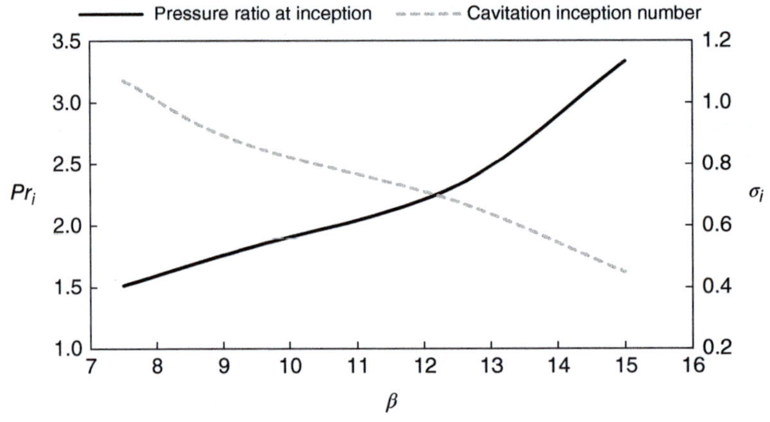

Figure 2.26 Cavitation inception versus diffuser angle (Source: From [32]/with permission of John Wiley & Sons).

number reducing to 0.45 with a wall half-angle of 15°. The pressure distribution along the outer wall is plotted in Figure 2.27 for each diffuser angle, which shows that smaller divergence angles act to delay pressure recovery. The corresponding variation in predicted vapor cavity extent and shape is presented in Figure 2.28.

The discussion on influence of various design parameters of orifice- and venturi-based hydrodynamic cavitation devices will be useful as a guide for the appropriate selection of device configuration. The scope of the discussion was restricted to single-throat devices. The following section briefly discusses the possibility of using multiple and non-circular throats.

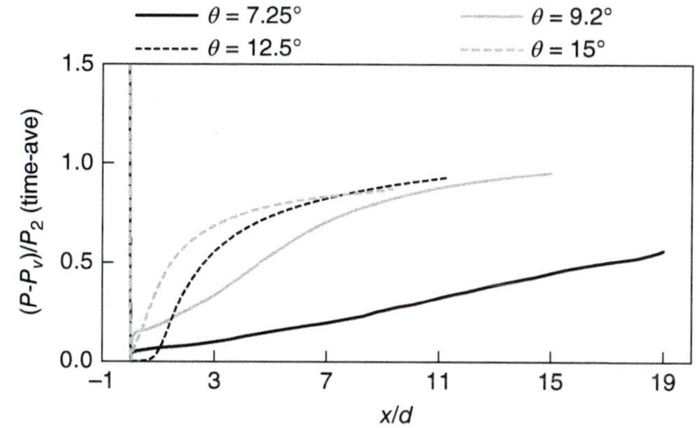

Figure 2.27 Static pressure vs. distance along outer wall with varying diffuser angles (Source: From [32]/with permission of John Wiley & Sons).

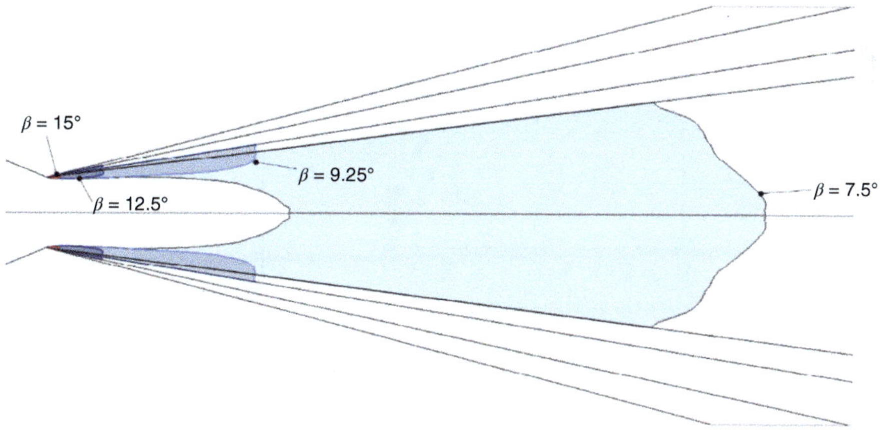

Figure 2.28 Predicted cavity shape and size with varying diffuser wall angles (Pr = 5.0) (Source: From [32]/with permission of John Wiley & Sons).

2.4 Enhancing Performance of Orifice/Venturi-Based Hydrodynamic Cavitation Devices

Beyond investigating the effects of the key geometric features of axial flow devices, a number of researchers have proposed alternative arrangements of cavitation systems to intensify the cavitation performance for use in different applications. One such example is the work carried out by Sivakumar and Pandit [1], who compared the performance of a series of orifice plate designs of differing hole numbers, hole diameters, and layouts, in degrading Rhodamine B. Although not all orifice plate designs were directly comparable in terms of flow rates and pressure drop, a significant variation in degradation performance was measured, as shown in Figure 2.29.

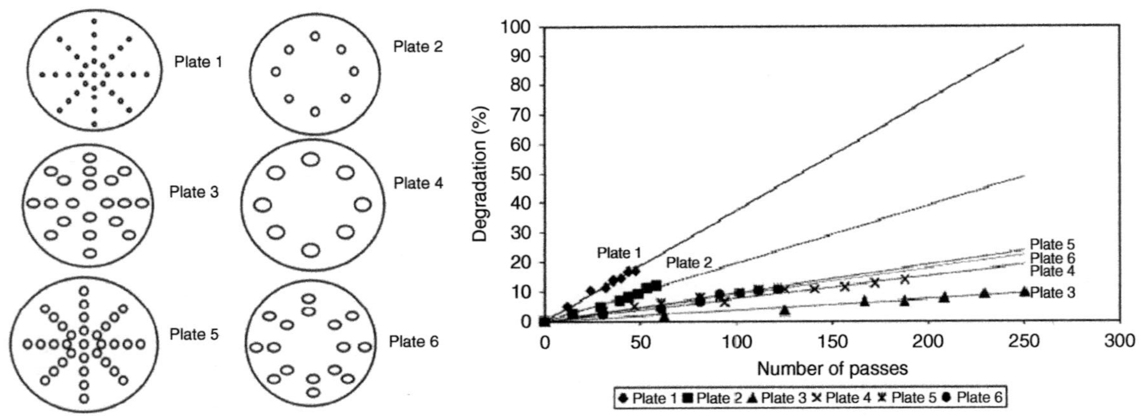

Figure 2.29 Performance comparison of different orifice plate arrangements (Source: From [1]/with permission of Elsevier).

It is particularly instructive to compare the significantly improved performance of plate 1 (33 holes, each of 1 mm diameter) over plate 2 (8 holes of 2 mm diameter), as the total flow area is closely matched (25.92 mm^2 vs. 25.1 mm^2). Similar observations have been recorded by Braeutigam et al. [34], who measured the degradation of different aqueous solutions of benzene and toluene with both single-hole and 32-hole orifice plates, each having the same total geometric cross-sectional area. The multi-hole orifice plate was found to offer improved degradation performance.

In addition to the number and size of holes making up the restriction, the cross-sectional shape has been shown to influence the performance of axial flow devices. Saharan et al. have investigated the influence of venturi cross-sectional shape on the degradation of orange G dye [35]. Both rectangular slot cross-section and circular cross-section venturis were constructed and tested, with a comparison carried out using a single-hole orifice. Again, although the flow characteristics in terms of pressure drop and flow rate were not consistent across the three devices, the measured degradation performance showed a significant difference between the geometries (Figure 2.30). The slit venturi was ranked first in terms of percentage of decolorization, with the orifice plate faring the worst out of the three devices.

Rajoriya et al. subsequently extended the study of different device geometries to compare slot and circular venturi devices to a series of six single-hole orifice plates, featuring both rectangular slots and round holes of different sizes [36]. Measuring the degradation of RB13 dye for each device, the authors showed a wide variation in performance. In general, the slot geometries produced improvements over the

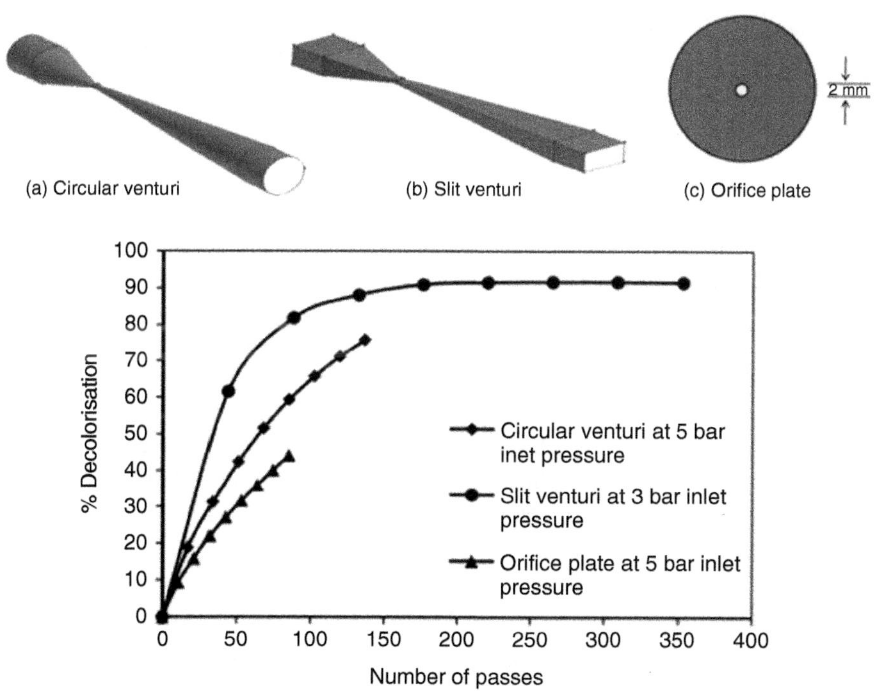

Figure 2.30 Performance comparison of different venturi and orifice cross-sections (Source: [35]/with permission of Elsevier).

corresponding circular geometry, and as before, the slot venturi design showed the best performance out of all of the geometries.

These results hint at the importance of two interdependent flow characteristics on the resulting cavitation intensity. Firstly, the effective perimeter, and hence the fluid shear, have an obvious influence on performance. This in turn leads to higher turbulence levels and turbulence frequencies, and the combined effect appears to produce higher intensity of cavitation for these studied applications. This suggests a potential route to hydrodynamic cavitation device optimization, in which the turbulence characteristics can be tuned to maximize performance for a given application. The key to achieving this is the ability to model the device cavitation characteristics accurately, alongside further systematic experimentation to quantify the influence of different geometric parameters on device performance. In particular, design optimization requires the ability to obtain a predictive measure of turbulence kinetic energy and frequencies for a given device, and the mechanisms by which the generated turbulence interacts with the vapor phase. An in-depth discussion of numerical modeling techniques and subsequent methodologies and techniques for device scale-up and optimization are presented in Chapter 5.

2.5 Summary and Outlook

Axial/linear flow devices such as orifice plates and venturi are considered the simplest hydrodynamic cavitation devices in terms of construction and operation. However, their cavitation characteristics and the influence of geometric variations are full of complexity. In terms of cavitation inception and evolution, the thickness of the orifice plate has a pronounced effect, as do the inlet radius and wall profile. Similarly, varying venturi restriction length and diffuser design has a significant effect on the resulting cavitation characteristics. For single-hole restrictions, the trends in cavitation characteristics with variation in key geometric parameters can be reasonably well defined, providing a sound basis for initial device design. In terms of subsequently optimizing device geometry for given HC applications, a review of published experimental studies shows that the interaction between the generated cavitation and the turbulence intensity within the device has a significant influence on device performance. Beyond axial and linear flows, other means of generating hydrodynamic cavitation can be used which have significantly different flow and cavitation characteristics, such as vortex-induced cavitation and rotor-stator type devices. Chapter 3 discusses these types of hydrodynamic cavitation devices based on rotational and swirling flows.

Notations

A_0	Orifice area
A_1	Plenum area
C_c	Contraction coefficient (–)

C_d	Discharge coefficient (–)
d	Diameter of linear cavitating device – orifice/venturi (mm)
D	Venturi internal pipe diameter (mm)
K_{cr}	Critical cavitation number (–)
K	Cavitation parameter (–)
l	Depth or length of orifice (mm)
P_1	Upstream pressure (kPa)
P_2	Downstream pressure (kPa)
p, p_d	Downstream pressure (kPa)
Pr	Pressure ratio (inlet/outlet)
p_v	Vapor pressure (kPa)
r	Radius of inlet edge (mm)
v_t	Velocity at the constriction or throat (m/s)

Greek Symbols

σ	Cavitation number (–)
β	Diffuser angle (°)
σ_i	Cavitation number at inception (–)
ρ_l	Density of liquid (kg/m^3)
θ	Orifice wall angle
v_t	Throat velocity (m/s)

References

1 Sivakumar, M. and Pandit, A.B. (2002). Wastewater treatment: a novel energy efficient hydrodynamic cavitational technique. *Ultrasonics Sonochemistry* 9 (3): 123–131.

2 Simpson, A. and Ranade, V.V. (2018). Modelling of hydrodynamic cavitation with orifice: influence of different orifice designs. *Chemical Engineering Research and Design* 136: 698–711.

3 Holkar, C.R., Jadhav, A.J., Pinjari, D.V., and Pandit, A.B. (2019). Cavitationally driven transformations: a technique of process intensification. *Industrial and Engineering Chemistry Research* 58 (15): 5797–5819.

4 Šarc, A., Stepišnik-Perdih, T., Petkovšek, M., and Dular, M. (2017). The issue of cavitation number value in studies of water treatment by hydrodynamic cavitation. *Ultrasonics Sonochemistry* 34: 51–59.

5 Pearce, I.D. and Lichtarowicz, A. (1971). Discharge performance of long orifices with cavitating flow. *Proceedings of Second Fluid Power Symposium*, pp. D2-13–D2-35, Guildford, Paper D2, University of Surrey.

6 Nurick, W.H. (1976). Orifice cavitation and its effect on spray mixing. *Journal of Fluids Engineering* 98 (4): 681–687.

7 Ohrn, T.R., Senser, D.W., and Lefebvre, A.H. (1991). Geometrical effects on discharge coefficients for plain-orifice atomizers. *Atomization and Sprays* 1 (2): 137–154.

8 Ramamurthi, K. and Nandakumar, K. (1999). Characteristics of flow through small sharp-edged cylindrical orifices. *Flow Measurement and Instrumentation* 10 (3): 133–143.

9 Carpenter, J., Badve, M., Rajoriya, S. et al. (2017). Hydrodynamic cavitation: an emerging technology for the intensification of various chemical and physical processes in a chemical process industry. *Reviews in Chemical Engineering* 33 (5): 433–468.

10 Benito, Y., Arrojo, S., Hauke, G., and Vidal, P. (2005). Hydrodynamic cavitation as a low-cost AOP for wastewater treatment: preliminary results and a new design approach. *WIT Transactions on Ecology and the Environment* 80: https://doi.org/10.2495/WRM050501.

11 Braeutigam, P., Franke, M., Wu, Z.L., and Ondruschka, B. (2010). Role of different parameters in the optimization of hydrodynamic cavitation. *Chemical Engineering and Technology: Industrial Chemistry-Plant Equipment-Process Engineering-Biotechnology* 33 (6): 932–940.

12 Simpson, A. and Ranade, V.V. (2019). 110th Anniversary: comparison of cavitation devices based on linear and swirling flows: hydrodynamic characteristics. *Industrial and Engineering Chemistry Research* 58 (31): 14488–14509.

13 Ranade, V.V. (2002). *Computational Flow Modeling for Chemical Reactor Engineering*. Academic Press.

14 Bai, W., Tijsseling, A.S., Wang, J. et al. (2021). Large eddy simulation investigations of periodic cavitation shedding with special emphasis on three-dimensional asymmetry in a scaled-up nozzle orifice. *Journal of Fluids Engineering* 143 (7): 071401.

15 Capocelli, M., Prisciandaro, M., Lancia, A., and Musmarra, D. (2014). Hydrodynamic cavitation of p-nitrophenol: a theoretical and experimental insight. *Chemical Engineering Journal* 254: 1–8.

16 Pawar, S.K., Mahulkar, A.V., Pandit, A.B. et al. (2017). Sonochemical effect induced by hydrodynamic cavitation: comparison of venturi/orifice flow geometries. *AIChE Journal* 63 (10): 4705–4716.

17 Randall, L.N. (1952). Rocket applications of the cavitating venturi. *Journal of the American Rocket Society* 22 (1): 28–38.

18 Stutz, B. and Reboud, J.L. (1997). Experiments on unsteady cavitation. *Experiments in Fluids* 22 (3): 191–198.

19 Xu, C., Heister, S.D., and Field, R. (2002). Modeling cavitating venturi flows. *Journal of Propulsion and Power* 18 (6): 1227–1234.

20 Ulas, A. (2006). Passive flow control in liquid-propellant rocket engines with cavitating venturi. *Flow Measurement and Instrumentation* 17 (2): 93–97.

21 Sayyaadi, H. (2010). Instability of the cavitating flow in a venturi reactor. *Fluid Dynamics Research* 42 (5): 055503.

22 Zhu, J., Xie, H., Feng, K. et al. (2017). Unsteady cavitation characteristics of liquid nitrogen flows through venturi tube. *International Journal of Heat and Mass Transfer* 112: 544–552.

23 Zhao, L., Mo, Z., Sun, L. et al. (2017). A visualized study of the motion of individual bubbles in a venturi-type bubble generator. *Progress in Nuclear Energy* 97: 74–89.

24 Abdulaziz, A.M. (2014). Performance and image analysis of a cavitating process in a small type venturi. *Experimental Thermal and Fluid Science* 53: 40–48.

25 Long, X., Zhang, J., Wang, J. et al. (2017). Experimental investigation of the global cavitation dynamic behavior in a venturi tube with special emphasis on the cavity length variation. *International Journal of Multiphase Flow* 89: 290–298.

26 Brinkhorst, S., Von Lavante, E., and Wendt, G. (2017). Experimental and numerical investigation of the cavitation-induced choked flow in a herschel venturi-tube. *Flow Measurement and Instrumentation* 54: 56–67.

27 Arrojo, S., Benito, Y., and Tarifa, A.M. (2008). A parametrical study of disinfection with hydrodynamic cavitation. *Ultrasonics Sonochemistry* 15 (5): 903–908.

28 Musmarra, D., Prisciandaro, M., Capocelli, M. et al. (2016). Degradation of ibuprofen by hydrodynamic cavitation: reaction pathways and effect of operational parameters. *Ultrasonics Sonochemistry* 29: 76–83.

29 Patil, P.N., Bote, S.D., and Gogate, P.R. (2014). Degradation of imidacloprid using combined advanced oxidation processes based on hydrodynamic cavitation. *Ultrasonics Sonochemistry* 21 (5): 1770–1777.

30 Nakashima, K., Ebi, Y., Shibasaki-Kitakawa, N. et al. (2016). Hydrodynamic cavitation reactor for efficient pretreatment of lignocellulosic biomass. *Industrial and Engineering Chemistry Research* 55 (7): 1866–1871.

31 Madison, M.J., Coward-Kelly, G., Liang, C. et al. (2017). Mechanical pretreatment of biomass–part I: acoustic and hydrodynamic cavitation. *Biomass and Bioenergy* 98: 135–141.

32 Simpson, A. and Ranade, V.V. (2019). Modeling hydrodynamic cavitation in venturi: influence of venturi configuration on inception and extent of cavitation. *AIChE Journal* 65 (1): 421–433.

33 Capocelli, M., Prisciandaro, M., Lancia, A., and Musmarra, D. (2015). Cavitational reactor for advanced treatment of contaminated water: the effect of recovery pressure. *Desalination and Water Treatment* 55 (12): 3172–3177.

34 Braeutigam, P., Wu, Z.L., Stark, A., and Ondruschka, B. (2009). Degradation of BTEX in aqueous solution by hydrodynamic cavitation. *Chemical Engineering and Technology* 32 (5): 745–753.

35 Saharan, V.K., Rizwani, M.A., Malani, A.A., and Pandit, A.B. (2013). Effect of geometry of hydrodynamically cavitating device on degradation of orange-G. *Ultrasonics Sonochemistry* 20 (1): 345–353.

36 Rajoriya, S., Bargole, S., and Saharan, V.K. (2017). Degradation of reactive blue 13 using hydrodynamic cavitation: effect of geometrical parameters and different oxidizing additives. *Ultrasonics Sonochemistry* 37: 192–202.

3

Hydrodynamic Cavitation Devices Based on Rotational/Swirling Flows

Hydrodynamic cavitation devices based on rotational or swirling flows may either be based on moving parts – typically involving a rotor and a stator with a small clearance between rotor and stator, or based on geometric configuration, to realize swirling/vortex flows. The latter type of passive hydrodynamic cavitation devices may be based on a combination of linear/axial flows and rotational/swirling flows. These devices involving rotational/swirling flows are discussed in this chapter. The discussion is mainly organized according to the device types, starting with the rotor-stator-based devices

3.1 Rotor-Stator Hydrodynamic Cavitation Devices

Cavitation in rotor-stator type devices or in rotor-rotor type devices is realized by spinning specially designed impellers (or rotors) rotating at sufficiently high rotational speeds in a confined space (stators with small clearances). This causes local reduction in pressure, which leads to cavitation when the reduced pressure approaches the vapor pressure limit. As the components rotate past one another, high levels of shear are generated as well. An example of one such device design is that reported by Zupanc et al. [1, 2], which consists of a pair of counter-rotating stainless steel rotors, each having a diameter of 90 mm with a small 0.8 mm gap in between (Figure 3.1).

As shown in Figure 3.1, Zupanc et al. observed by high-speed video [2] that there are three distinct cavitation regions. The leading edge of the rotating "teeth" shows an attached traveling cavitation bubble [1], whereas additional cavitation zones are also formed in the gap between the teeth [2] and also in the small clearance space between the teeth on adjacent rotors as they pass. The latter two cavitation zones are uniquely different to those formed in linear flow-type devices; they are intermittent, forming and collapsing as individual teeth rotate past each other. This is accompanied by relatively large, rapid pressure oscillations within the device of the order of 100 kPa [3].

This device type has been found to yield promising results for a wide range of applications. Using the device depicted in Figure 3.1 in combination with H_2O_2, Zupanc et al. reported more than 80% removal rates of pharmaceuticals (ibuprofen,

Hydrodynamic Cavitation: Devices, Design, and Applications, First Edition.
Vivek V. Ranade, Vinay M. Bhandari, Sanjay Nagarajan, Varaha P. Sarvothaman and Alister T. Simpson.

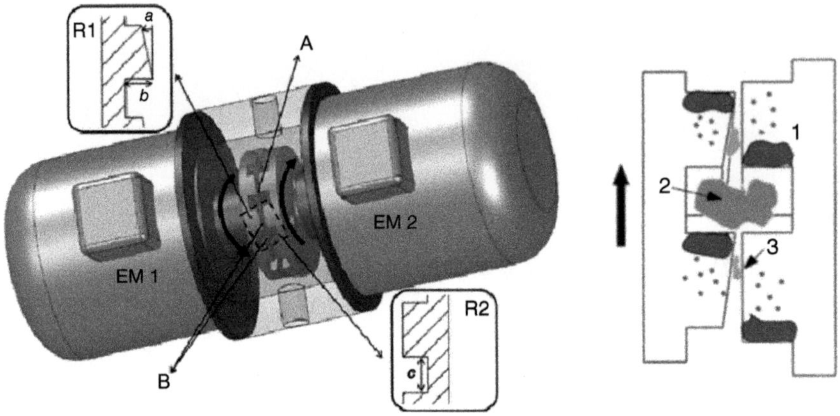

Figure 3.1 Rotor-rotor cavitation device (Source: Adapted from [1, 2]).

naproxen, ketoprofen, carbamazepine, and diclofenac) in wastewater [1]. The rotor-stator based cavitation devices have also been found to be effective in the treatment of waste activated sludge (WAS) – they produced better-digested sludge, increased the soluble chemical oxygen demand (SCOD) by a factor of 12, and improved biogas yields (12%) [3]. Elsewhere, similarly promising results have been reported for biogas production improvements using a similar rotor-stator device (see for example, Patil et al. [4]). Using a rotor-stator cavitation device to pre-treat wheat straw used in anaerobic digestion, they found that the pre-treated samples yielded double the quantity of biogas. Their study also highlighted some of the primary limitations of this type of device geometry, namely solids loading. Beyond wheat straw-to-water ratios of 1.5%, the authors reported that clogging became an issue. A number of variations of this device type and construction exist across the open literature. Recently, Kim et al. [5] have detailed the operation and performance of a rotor-stator type system for WAS disintegration (Figure 3.2). This device type incorporates a rotating plate and stationary cover, each incorporating a series of machined dimples at a constant radius. When tested, this device revealed a reduction in particle size of 77%, and a corresponding increase in soluble COD (sCOD) by a factor of 5.

Rotor-stator devices have also been successfully used in water disinfection studies. Cerecedo et al. [6] studied a series of prototype devices, shown schematically in Figure 3.3, each comprising a rotor with a number of semi-circular "vanes" which rotate within a stationary housing with a corresponding set of vanes. Measured bacterial disinfection for *Escherichia Coli* and *Enterococcus faecalis* showed that effective elimination could be achieved in a 250 ml sample in under 10 minutes.

In the rotor-stator type of cavitation devices, the cavitation is realized via accelerating flow through narrow gaps induced by rotation of the rotor (see Figure 3.4 taken from Gostisa et al. [7]). Recently, detailed CFD simulations of such rotor-stators has been published (for example, Sun et al. [8]). The typical simulated results are illustrated in Figure 3.5, clearly showing vortex as well as sheet cavitation generated in such devices.

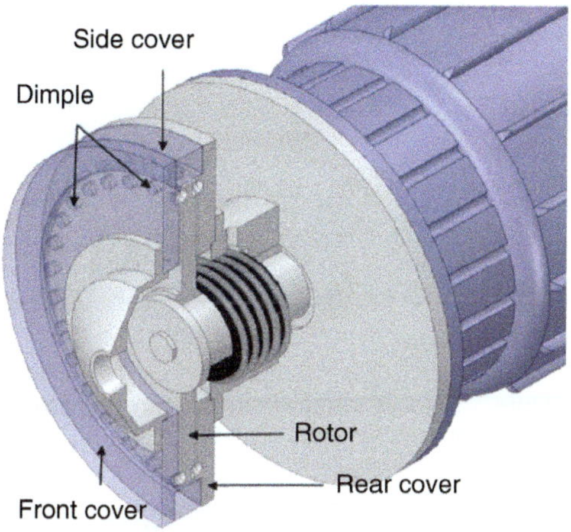

Figure 3.2 Rotor-rotor cavitation device (Source: From [5]/with permission of Elsevier).

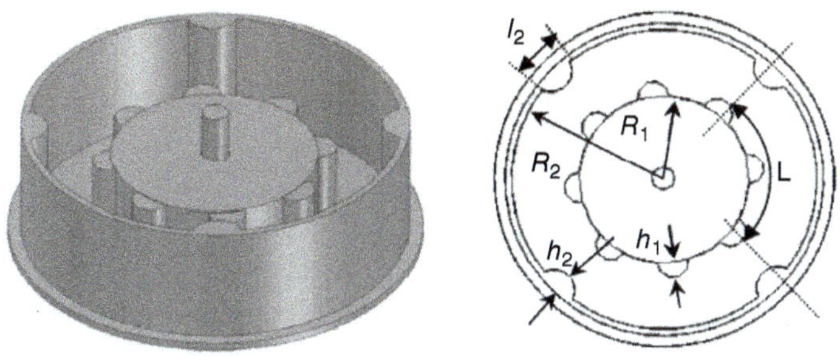

Figure 3.3 Rotor-stator cavitation device schematic (Source: From [6]/with permission of Elsevier).

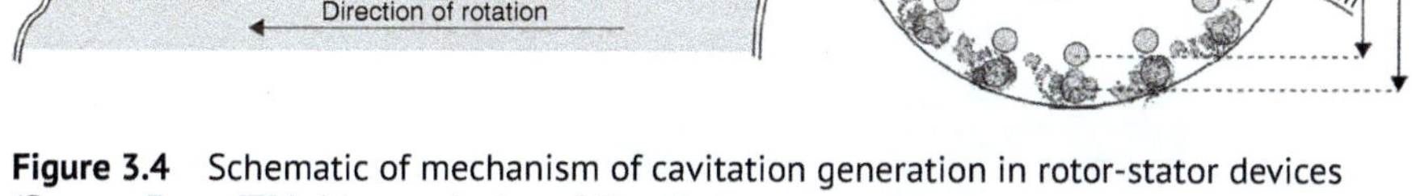

Figure 3.4 Schematic of mechanism of cavitation generation in rotor-stator devices (Source: From [7]/with permission of Elsevier).

The detailed CFD models are expected to lead to more insights and improvements in designing such rotor-stator devices for enhancing the effectiveness of realized cavitation. However, at the moment, the capital as well as the operating cost of rotor-stator- based devices are substantially higher than passive devices. The designs of passive devices are also continuously evolving and improving. Some of the recent designs of passive devices based on rotational/swirling flows are discussed in the following sections.

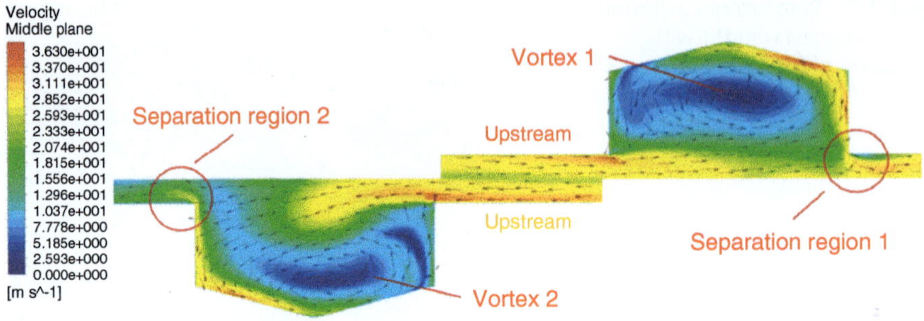

Figure 3.5 Illustrative simulated flow field in rotor-stator devices (Source: From [8]/with permission of Elsevier).

3.2 Vortex-Based Cavitation Devices

3.2.1 Vortex Cavitation

Vortex cavitation has been an extensively studied topic in marine engineering applications, both due to its influence on fluid drag and lift, as well as its erosion effects. Lifting surfaces such as hydrofoils and propeller blades generate tip vortices, often of sufficient strength such that a central cavitating core is formed. When such a tip vortex forms and impinges on adjacent surfaces, it is understood to be a major contributor to erosion in hydraulic machinery. It also leads to excessive noise and vibration. Arndt et al. have highlighted some of the complexities of such cavitating vortices, which exhibit a range of broadband pressure fluctuations and complex collapse characteristics [9, 10]. Couty in 2001 [11] conducted an experimental program to study this vortex cavitation mechanism in detail, using a cavitation vortex generator device (Figure 3.6).

The vortex generator constructed by Couty was installed in a test loop downstream of a rotating valve, which produced "slugging" type conditions in the upstream

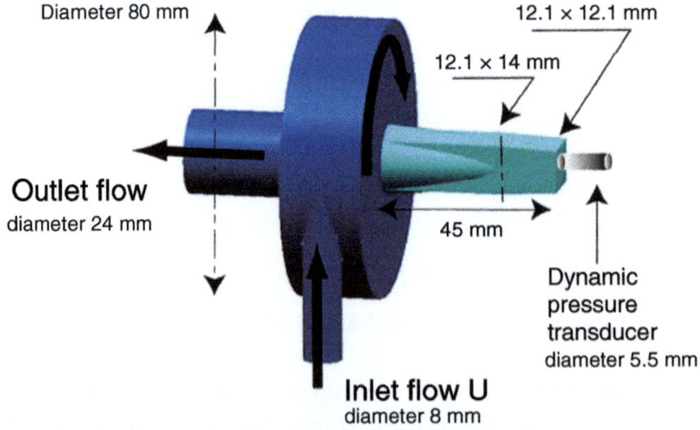

Figure 3.6 Experimental cavitation vortex generator (Source: From [11]. Reproduced from Dr Couty, EPFL 2001 thesis 2463).

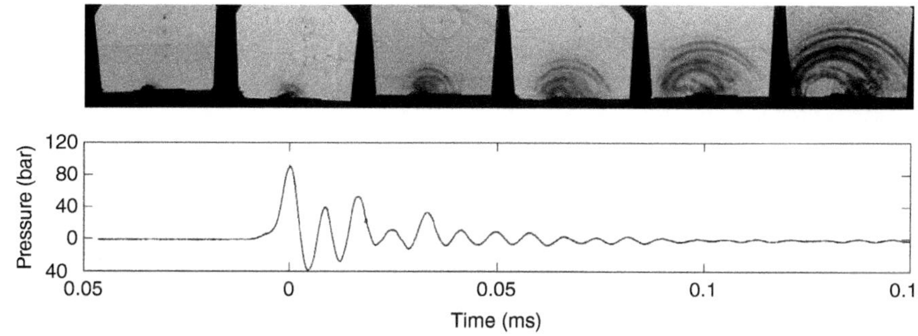

Figure 3.7 Multiple shock waves produced during vortex cavity collapse (Source: From [11]. Reproduced from Dr Couty, EPFL 2001 thesis 2463).

pipework, resulting in successive high- and low-pressure waves. The creation of the cavitation vortex was therefore cyclical, with the vortex forming and collapsing with each valve cycle. Using an intensified light camera, the authors were able to detect luminescence in the collapse phase. A shockwave was found to form, indicating the extremes of interfacial velocity present; the corresponding pressure transducer recordings captured an accompanying high- pressure oscillation (Figure 3.7).

These intense cavitation characteristics are of interest to engineers due to their potential role in enabling a wide range of physical and chemical transformations, from breaking up suspended particles to potentially generating oxidizing species. Consequently, there are numerous recent studies in the open literature that report using vortex cavitation for a variety of engineering applications. Examples include studies by Mancuso et al. on the degradation of waste dye (Rhodamine B) using a swirling jet cavitation reactor, the Ecowirl® [12]. Cavitation generated by an Ecowirl reactor was compared with a multi-hole orifice plate, and there were significant improvements in degradation performance using the swirling jet cavitation device. Swirling flow-based cavitation devices have also found use in bacterial disinfection applications; Loraine et al. have presented results on the reduction of both *E. coli* and *B. subtilis*, achieving reductions of log 7 and 4.5 respectively using a DynaJet® system [13]. On comparison with ultrasonic methods, they found an improvement in energy efficiency in excess of a factor of 10. These two devices are shown schematically in Figure 3.8. The use of a high-swirl vortex diode device has recently been presented by Jain et al. [15], and was found to be efficient in eliminating both *E. coli* and *Staphylococcus aureus*. The authors also compared their results to a linear flow orifice device, and found that the orifice required significantly higher energy input in order to obtain the same level of reactor performance. Considering better performance reported for vortex/swirling flow-based cavitation devices, some basic models of vortex are briefly discussed in the next section. Vortex-based cavitation devices are then discussed in Section 3.2.3.

3.2.2 Vortex Models

Regardless of the geometry adopted to generate a cavitating vortex, the central principle is the same: the local tangential velocity must be sufficient to lower the

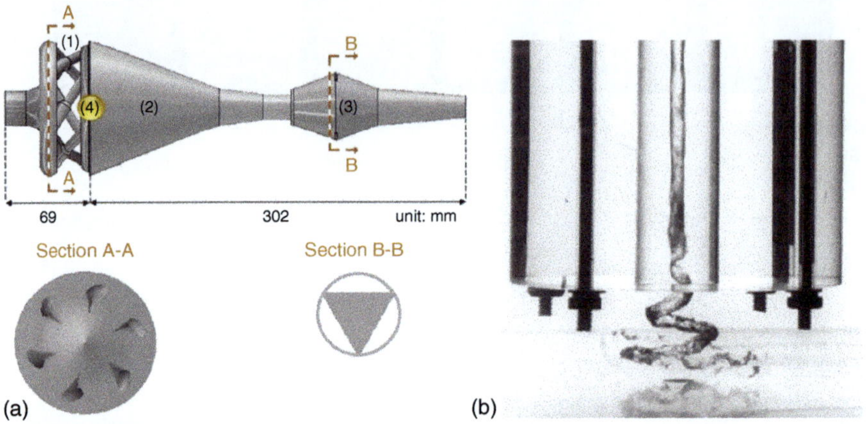

Figure 3.8 (a) Ecowirl reactor® (Source: From [14]/with permission of Elsevier) and (b) DynaJet reactor showing helical cavitation vortex breakdown (Source: Reprinted from [13], with permission from Elsevier).

static pressure to the vapor pressure limit of the liquid. This is just as in a linear flow device, in which the local axial velocity must drive down the local static pressure sufficiently for cavitation to initiate (Figure 3.9). There exist a number of empirical models to describe vortex flow in a viscous fluid, the simplest of which is the Rankine vortex model. The Rankine vortex model approximates the vortex as having a central forced vortex core, surrounded by free vortex. More physically representative models include the Lamb–Oseen vortex model [16] (which is summarized in Eqs. (3.1) and (3.2)).

$$u_z(r,0) = \frac{\Gamma_0}{4\pi \nu t} \exp\left(-\frac{r^2}{4\nu t}\right) \tag{3.1}$$

$$u_\theta = \frac{\Gamma_0}{2\pi r}\left(1 - \exp\left(-\frac{r^2}{4\nu t}\right)\right) \tag{3.2}$$

where $u_z(r, 0)$ is axial velocity at $z = 0$ and radius of r, u_θ is tangential velocity at a radius of r, t is time, and ν is kinematic viscosity. Γ_0 is the circulation strength of the vortex.

The Bachelor vortex model [17] similarly describes the velocity profiles in a viscous vortex:

$$u_\theta = \frac{\Gamma}{2\pi r}\left(1 - \exp\left(-\frac{Ur^2}{4\nu z}\right)\right) \tag{3.3}$$

$$u_z = \frac{BU^2}{8\nu z}\left(1 - \exp\left(-\frac{Ur^2}{4\nu z}\right)\right) \tag{3.4}$$

where U is the free stream velocity, B is a constant coefficient, and z is the axial distance.

Such models provide a useful starting point in the design of vortex systems, and typically reproduce real vortex flow properties with a reasonable degree of accuracy [17, 18]. Further, to determine a cavitation inception number, these velocity profiles,

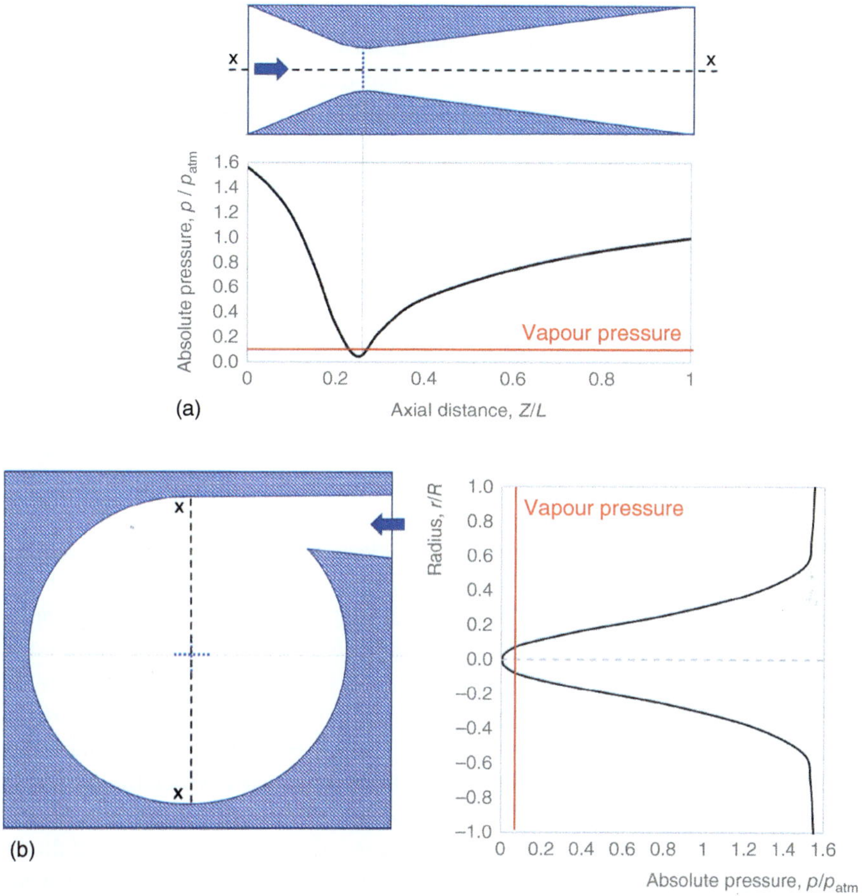

Figure 3.9 Schematic of cavitation devices (a) based on linear flow (b) based on swirling flows.

particularly the calculated maximum tangential velocity, must be related to the minimum pressure within the vortex. Arndt [19] proposed a relatively simple expression for a vortex of strength Γ and a viscous core of radius *a:*

$$\frac{p_{min} - p_0}{\frac{1}{2}\rho\left(\frac{\Gamma}{2\pi a}\right)} = C_{p\ min} \tag{3.5}$$

where $C_{p\ min} = -2.0$ if a Rankine model is used, and $C_{p\ min} = -1.74$ if a Lamb vortex is modeled.

For complex, 3D swirling flows, such analytical approaches are limited by two key uncertainties: first, the accurate determination of the maximum tangential velocities in the vortex, and second, the fundamental uncertainties over water quality and gas nuclei content. The latter is widely variable, and also widely understood to have a significant influence on vortex cavitation inception [20, 21]. Simpson and Ranade compared the cavitation inception in a vortex-based cavitation device to that in linear flow devices, both experimentally and numerically using CFD [22]. Using the classic definition of cavitation inception number (Eq. (3.6)), where C is used as a correction factor on the throat velocity, they determined that cavitation inception numbers of unity can be defined across different device geometries, provided the maximum velocities can be accurately determined. The maximum velocities within the device are difficult to determine by simple analytic relationships. However, accurate calculation of this key variable can rely on CFD simulations.

$$\sigma = \frac{p_d - p_v}{{}^1\!/_2\rho_l(Cu_t)^2} \tag{3.6}$$

Specific details of vortex-based cavitation devices are discussed in the following section.

3.2.3 Vortex Cavitation Devices

Confined vortex flow devices have found extensive use in a wide variety of industrial applications, most notably in cyclonic separation systems [23, 24]. Recent novel examples of vortex flow devices in chemical reaction engineering include rotating fluidized bed reactors [25–27], in which particle bed fluidization is achieved by generating highly swirling flow within a fixed geometry reactor. For hydrodynamic cavitation applications, a novel example is the use of a vortex device based on a vortex throttle, or vortex diode, shown schematically in Figure 3.10. Vortex diodes are bi-directional fluid flow devices, characterized by a higher flow resistance in one direction relative to the other. The basic construction of a vortex diode features a tangential inlet, swirl chamber, and an axial outlet port. When flow enters the tangential inlet, it is dominated by a confined, turbulent vortex, which breaks down as it travels along the axial port. When operating in the reverse direction as a fluidic diode, the axial port geometry is designed to create a smooth transition to the diode chamber in order to minimize losses. Fluidic diodes of different constructions have been in use since the early part of the twentieth century [28, 29], and the absence of moving parts and low maintenance overhead make them particularly suited to operation in harsh environments. This has led to their application in the nuclear industry as a type of non-return valve [30], and as part of pumping systems [31].

The complex internal flow characteristics of vortex devices have been studied in detail experimentally and numerically. Escudier et al. in 1980 presented detailed

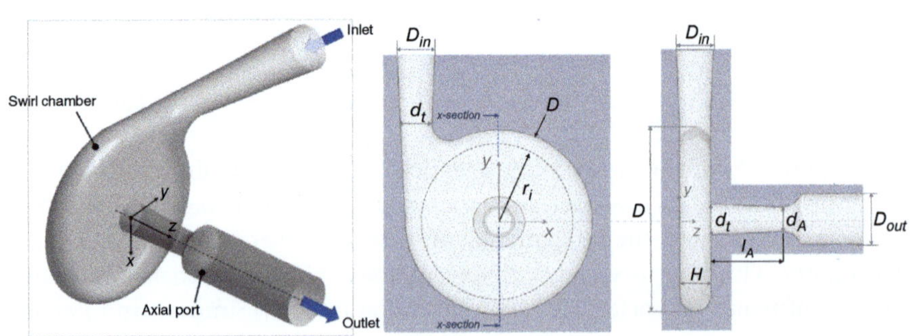

Figure 3.10 Vortex diode schematic.

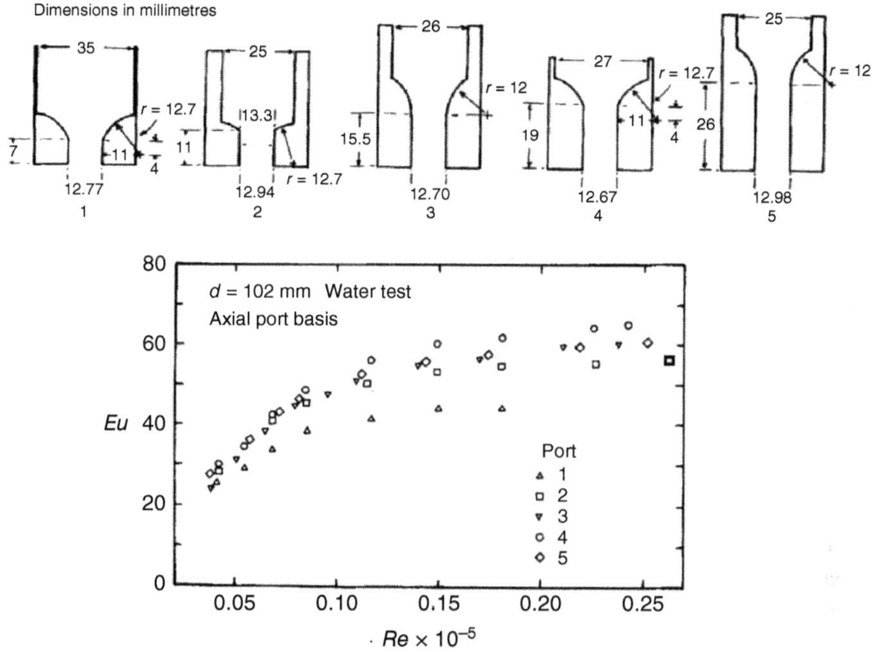

Figure 3.11 Vortex throttle exit geometry loss characteristics (Source: Adapted from [33]).

LDA (laser Doppler anemometer) measurements of the internal vortex structure in such a device having varying axial port geometries [32]. It was observed that the characteristics of the vortex structure and its breakdown in the axial port were strongly influenced by the axial port diameter. Larger diameters were found to produce a thick vortex core with significant levels of reverse axial flow. Smaller diameters, on the other hand, led to a thin annular core with peak axial velocities higher by an order of magnitude. Priestman et al. in 1987 [33] further tested a range of vortex "throttles," with a view to understanding the influence of the axial exit port geometry on pressure loss. Tests were performed on a wide range of geometries, having both square-edge and round-edge axial ports. Figure 3.11 presents a sample of round-edge geometries and their corresponding measured Euler number, indicating the strong influence of the radius of curvature.

As with cavitation in linear devices, cavitating vortex flow is highly complex and highly unsteady. In "The singing vortex," Arndt details the range of frequencies associated with vortex cavitation of the type generated in hydrofoil flows, which extends well into the kHz range [10]. Specifically relating to vortex cavitation devices, Pandare et al. subsequently presented a detailed internal flow investigation of vortex diode devices using CFD, revealing some further details of this complex unsteady vortex behavior [34]. The authors showed the formation of an inherently unsteady, precessing vortex core (PVC) along the axis of the device, predicted to oscillate at frequencies of the order of 60 Hz. This precessing vortex behavior has also been widely reported in vortex separation units [24, 35]. Along the axis of the

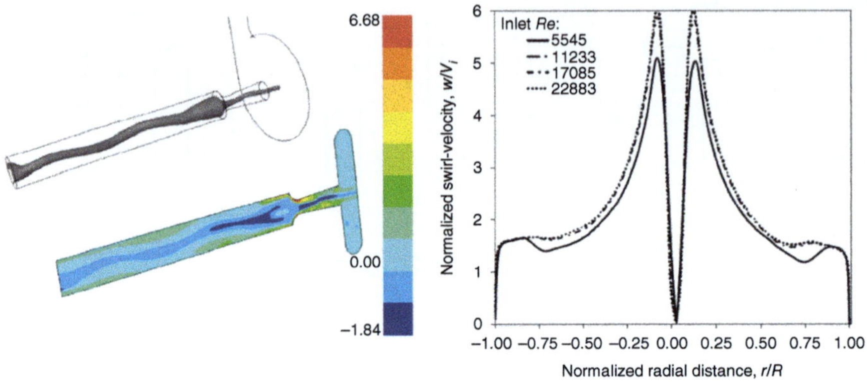

Figure 3.12 Left: Iso-surface of negative axial velocity (reverse flow) and contours of normalized axial velocity, right: normalized swirl velocity on center plane of diode (Source: From [34]/with permission of Elsevier).

vortex unit, Pandare and Ranade also reported a significant core of reverse flow in the axial port, which persists for up to 30 diameters into the downstream pipework. A sample of the results presented by Pandare and Ranade is shown in Figure 3.12, which shows persistent rotational flow in the axial outlet. The maximum tangential (swirling) velocity in the vortex diode was found to be about six times the velocity at the throat (the correction factor, C in Eq. (3.6) may be assumed as 6).

Simpson and Ranade presented further studies of multiphase cavitating flow in vortex diodes of various scales using a combination of CFD and high-speed flow visualization [22, 36]. The strong swirling flows produce a cavitating vortex core, which proceeds to grow along the axis of the device up to the extent of the exit of the axial port, where the coherent vapor core breaks down into a highly turbulent cloud of cavitation bubbles (Figure 3.13). The trajectories of individual cavitation bubbles were also predicted using a Discrete Phase Model, showing the typical time pressure histories of bubbles as they form and are then convected into the downstream pipework. The predictions reveal a unique feature of confined vortex cavitation: due to the presence of a strong radial pressure gradient, individual bubbles experience rapid changes in the surrounding liquid pressure as they are transported from the axis to the outer walls by the highly turbulent flow. Smaller cavities in particular can experience rapid fluctuations between the vapor pressure, p_v, and values in excess of the device outlet pressure, p_2.

The axial and radial pressure gradients have been compared in Figure 3.14. It can be seen that the radial pressure gradient in the vortex diode remains relatively constant up to the exit of the axial port. Comparison of this to the maximum axial pressure gradient shows the radial gradient to be higher by a factor of 3. Contrastingly, in linear flow devices, the maximum pressure gradient experienced by individual cavities is limited to the difference between vapor pressure and the device outlet pressure (p_2). Other research in this field has found that increasing the rate of pressure recovery has a positive effect on device performance in terms of increasing the intensity of cavitation bubble collapse. Particularly relevant is a study by Soyama

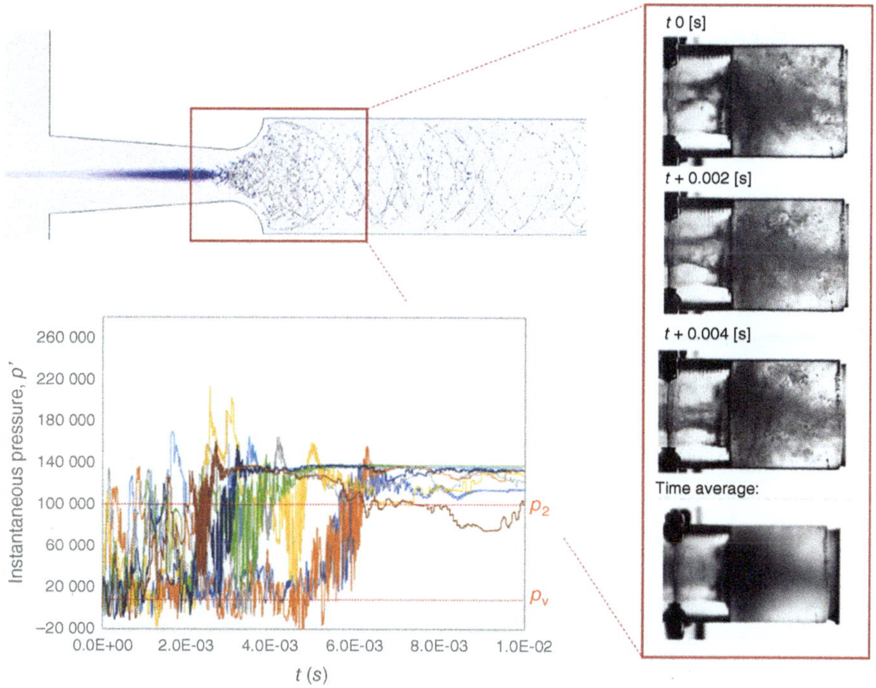

Figure 3.13 Left: Multiphase CFD with cavity trajectory predictions (Source: From [36]/with permission of John Wiley & Sons) and right, recorded high-speed flow visualization (Source: Reprinted from [22], with permission from American Chemical Society).

and Hoshino [37], who found that increasing the back pressure in a venturi cavitation reactor by a factor of 2 resulted in a recorded increase in acoustic power of two orders of magnitude. Capocelli et al. [38] have also studied the effect of increasing back pressure in a venturi cavitation reactor on the degradation of *p*-Nitrophenol in water. For the same overall device pressure drop, an increase in back pressure of the order of 0.4 bar was reported to deliver an appreciable improvement in degradation performance, with recorded concentration ratio found to reduce by 15%. Therefore, in principle, the rapid pressure recovery attained in the vortex unit axial port may contribute to achieving more intense cavity collapse, and by utilizing the radial pressure gradient, the performance of the vortex device may be optimized for practical applications. A possibility of combining linear and swirling flow-based devices for realizing cavitation is discussed in the following section.

3.3 Devices Based on Combinations of Linear and Swirling Flows

Generating cavitation in a flow with a sufficient degree of swirl has a number of advantages. Firstly, the position of cavitation inception is shifted to the interior of

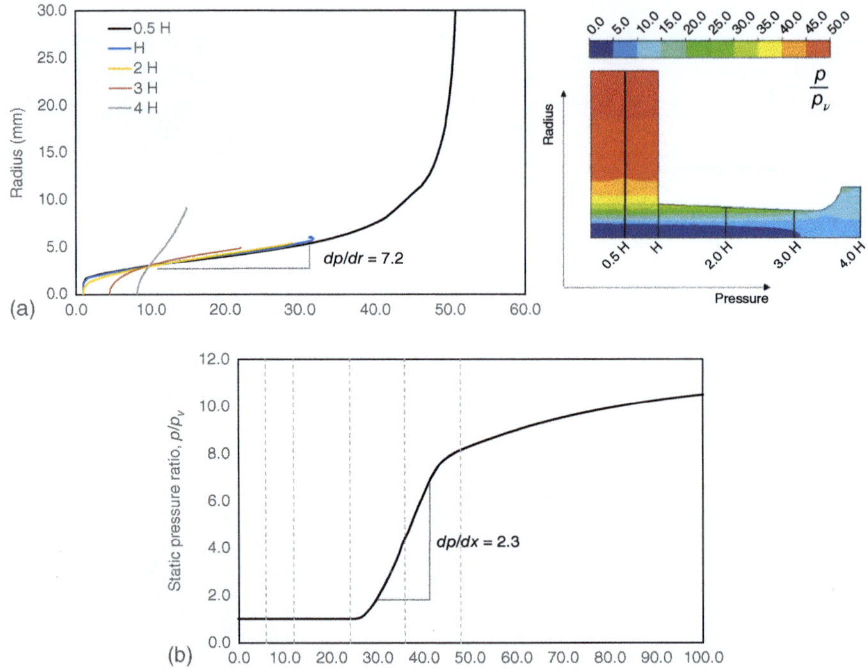

Figure 3.14 Predicted static pressure distributions (Diode with 6 mm throat diameter, $u_t = 3.3$ m/s). (a) X-axis: Static pressure ratio, p/p_v. (b) X-axis: Distance along axis, z (mm) (Source: From [36]/with permission of John Wiley & Sons).

the device, away from solid surfaces, in contrast to the case with only an axial flow. In addition, the increased radial pressure gradient can be designed such that individual cavities experience large pressure oscillations, in excess of the axial pressure recovery experienced downstream of the cavitation region. Vortex diode or vortex throttle-type devices are one such way to generate cavitation in this method. However, the design of simple orifice or venturi-type devices can also be modified to impart significant amounts of swirl upstream of the cavitation region. One method of combining axial and swirling flows is to install a swirl generator-type component upstream of the cavitation region. A good example of such a configuration has been studied by Kozák et al. [39]. The authors used a set of guide vanes upstream of the throat of a venturi tube, generating a swirl number of the order of 0.2. High-speed camera recordings and the accompanying numerical results showed two main cavitation regions: one occurring close to the walls at the exit of the venturi throat, and a second cavitating vortex core at the center of the device extending into the diffuser (Figure 3.15). In this case, the swirler design was successful in significantly modifying the cavitation characteristics, moving a significant portion of the cavitating flow to the center of the device. The addition of swirl is therefore a potential means of alleviating cavitation erosion risks in devices which use a restriction as the primary means of generating cavitation.

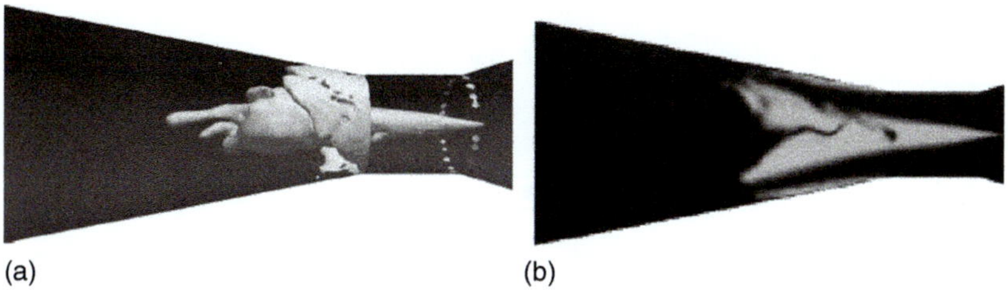

Figure 3.15 Vapor volume fraction in a cavitating venturi with swirl. (a) Numerical prediction. (b) High-speed camera recording (Source: Reprinted from [39], with permission from The American Society of Mechanical Engineers).

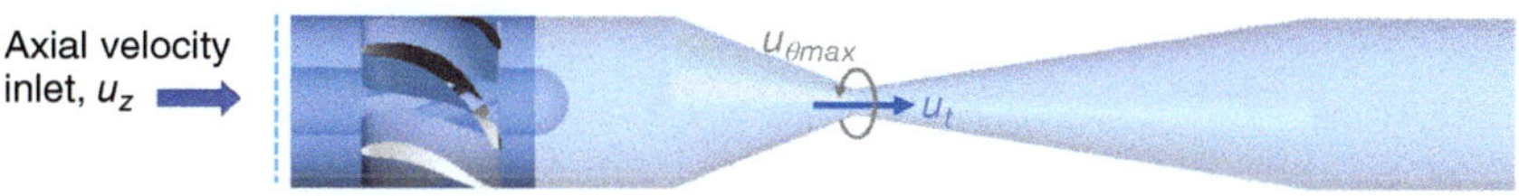

Figure 3.16 Venturi with in-line swirler (Source: From [22]/with permission of American Chemical Society).

A detailed comparison of the flow characteristics of devices which combine axial and swirling flows against a vortex device has been presented previously by Simpson and Ranade [22]. In this study, an in-line swirler was designed for installation upstream of either an orifice or a venturi device (see Figure 3.16). The swirler comprised five guide vanes, designed to generate swirl ratios of the order of 2 at the throat of the device (Figure 3.17a). This was 3D-printed from nylon, with the orifice, venturi, and vortex devices each machined from Perspex, in order to facilitate flow visualization using a high-speed camera system. Each device was designed such that the pressure drop versus flow rate characteristics were closely matched, to enable a like-for-like comparison (Figure 3.17b).

$$Re = \frac{\rho \overline{u_t} d_t}{\mu} \tag{3.7}$$

$$S = \frac{u_{\theta\ max}}{\overline{u_t}} \tag{3.8}$$

The investigation showed that with sufficient attention to swirler design, it is possible to generate confined vortex cavitation in conventional venturi/orifice-type devices, characterized by a cavitating vortex core at the axis of the device. Figure 3.18 shows the influence of generating pre-swirl in the inlet flow approaching a venturi and an orifice under cavitating conditions. In both geometries, the primary cavitation region moves from the edges of the restriction to the axis. In each case, pre-swirl leads to the formation of a distinct vortex core formed at the axis of the device, which breaks down into a series of bubble clouds as the pressure recovers in the turbulent flow downstream of the restriction. This behavior is successfully reproduced in corresponding CFD predictions (Figure 3.19).

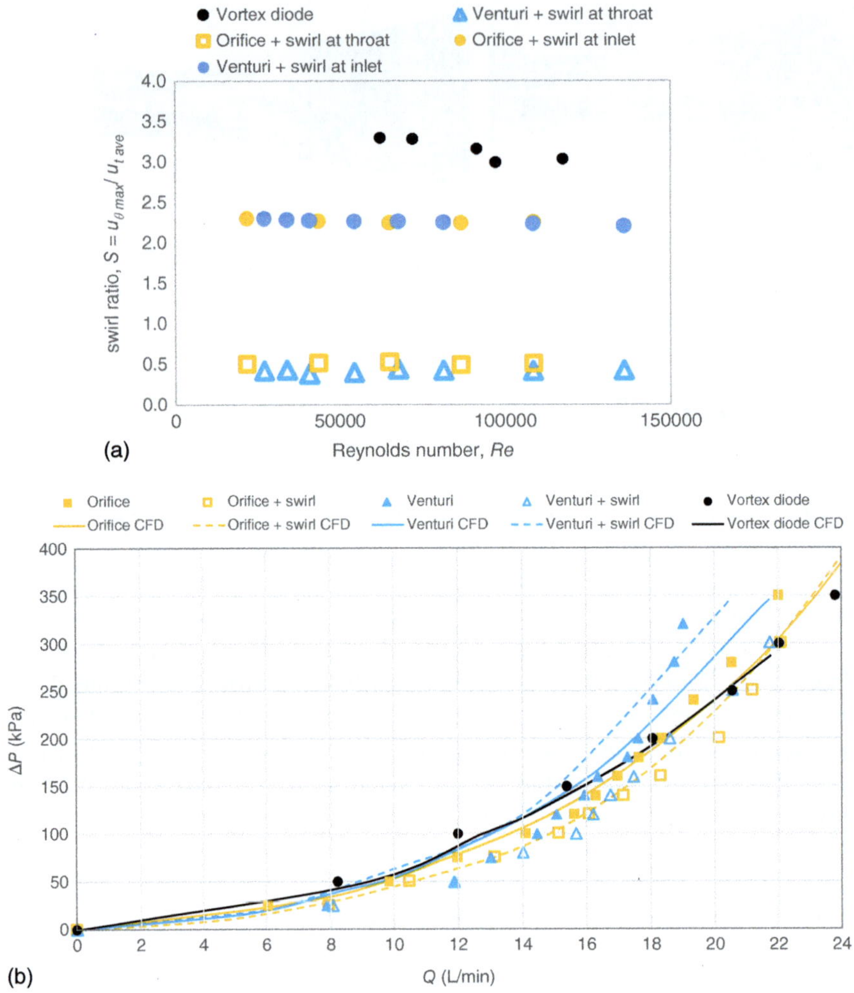

Figure 3.17 Comparison of cavitation devices with and without swirl. (a) Device swirl characteristics. (b) Pressure drop vs flow rate (Source: From [22]/with permission of American Chemical Society).

These results show that is possible to manipulate the region of cavitating flow to constrain it to the interior of the device, away from solid surfaces. At inlet swirl ratios of around 2, this can also be achieved with negligible increases in device pressure loss, as quantified in Figure 3.17. This study was also extended to investigate higher inlet swirl ratios, with the trends in device characteristics shown in Figure 3.20. Predictions show that higher swirl ratios – beyond values of 2 – incur additional pressure loss. Accompanying flow field predictions, summarized in Figure 3.21, show some key trends in behavior as inlet swirl is increased. Firstly, a region of reverse flow is formed downstream of the restriction, which grows in extent as swirl ratio is increased. Secondly, as the strength of the vortex core is increased, so too is the predicted gas hold-up. These trends in behavior suggest that for given applications

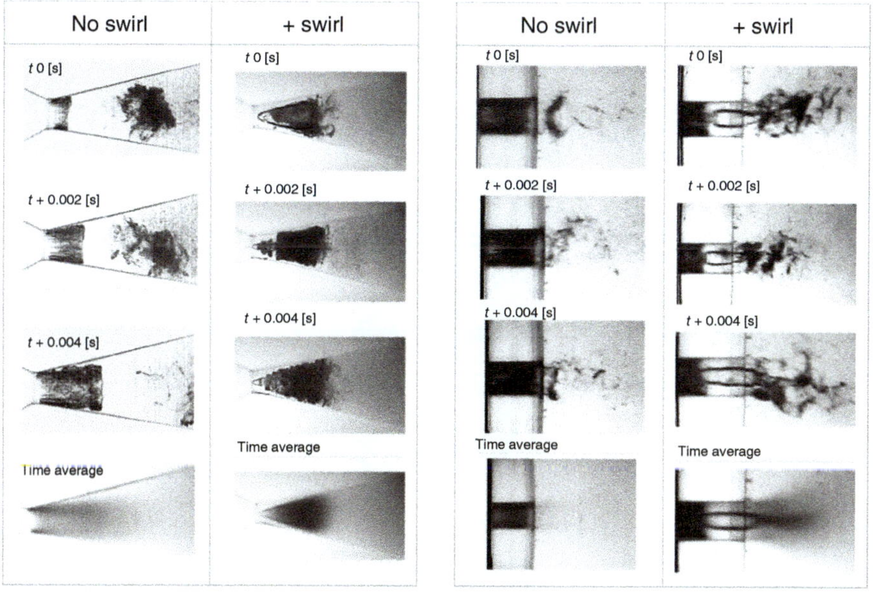

Figure 3.18 High-speed camera captures, venturi (left) and orifice (right), with and without swirl (Source: Reprinted from [22], with permission from American Chemical Society).

and flow requirements, an optimum geometry can be achieved which removes the risks associated with cavitation adjacent to solid surfaces, while also incurring relatively small additional pressure loss and therefore negligible additional energy input beyond a conventional linear flow device. This offers a promising route forward for reliable and robust industrial-scale hydrodynamic cavitation device designs.

3.4 Summary and Outlook

Cavitation devices which utilize swirling or rotational flows offer a number of advantages over conventional linear flow devices, and show exciting potential for a number of industrial applications. In particular, numerous studies have shown that with vortex-based devices, it is possible to generate cavitation in the bulk fluid, sufficiently removed from adjacent solid surfaces. This has the potential to minimize the risk of erosion, which is one of the primary limitations restricting the adoption of hydrodynamic cavitation technology more widely in industrial applications. Generating cavitation using high swirl/vortex- generating devices has the additional advantage in that cavitation generation does not rely on forcing flow through a small restriction. As a consequence, this opens up the possibility of handling working fluids with higher solids loading, and larger solids particle sizes, without the risk of device clogging. For applications such as biomass treatment and wastewater sludge processing, this offers a promising route towards reliable robust reactors at industrial scales. Achieving the full potential of such devices is dependent on the development of computational modeling strategies that can capture the complex physics, alongside high-fidelity experimental studies. Therein lies the challenge ahead: the complex,

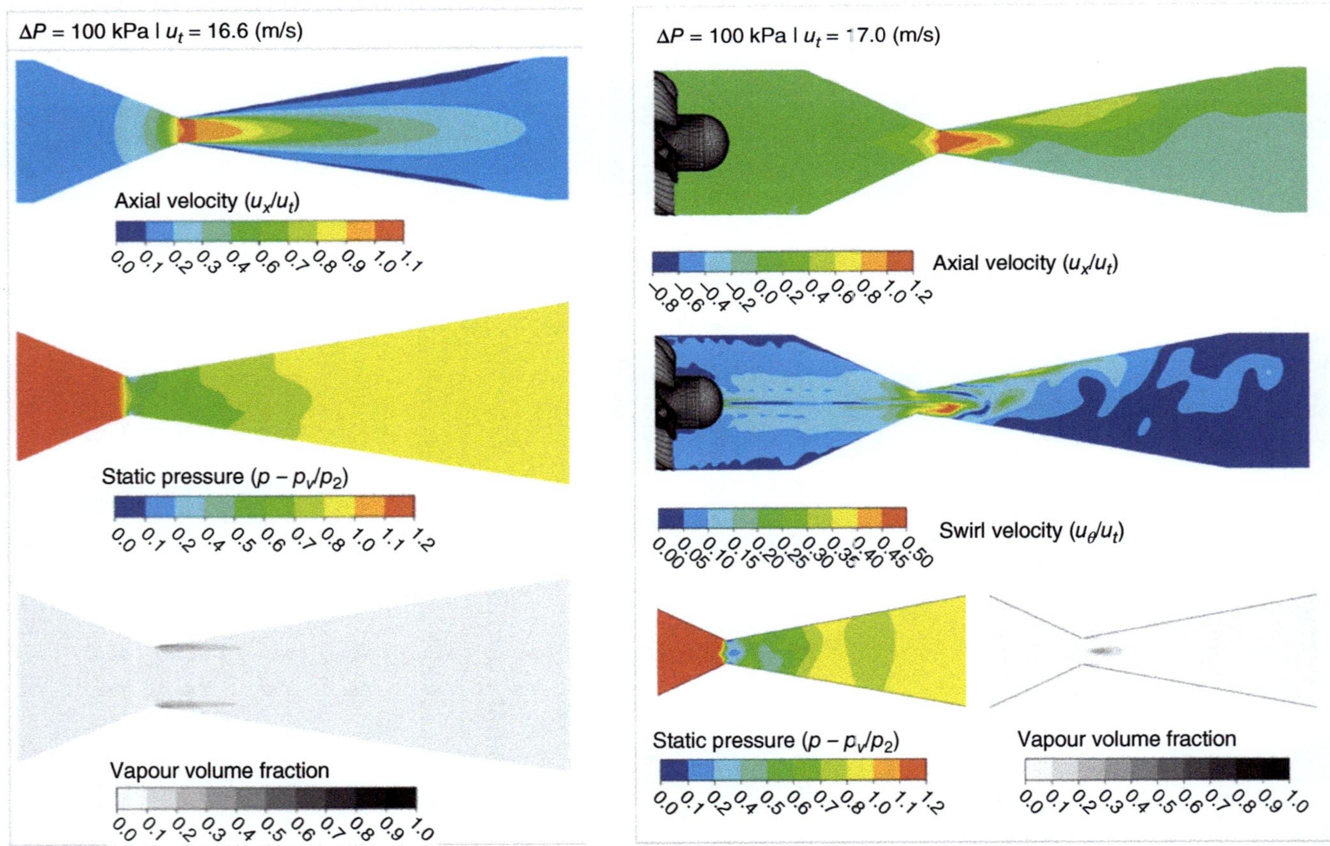

Figure 3.19 Computed flow fields, venturi without (left) and with swirl (right) (Source: From [22]/with permission of American Chemical Society).

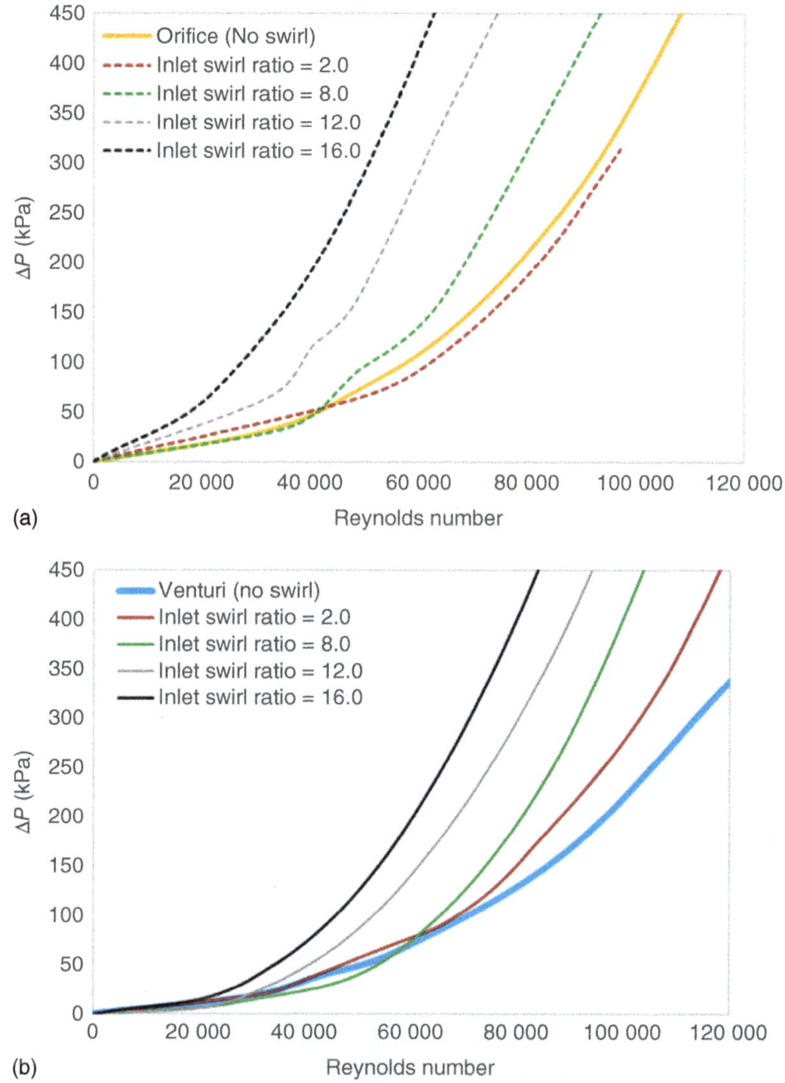

Figure 3.20 Pressure drop vs. Re number at high swirl ratios. (a) Orifice. (b) Venturi (Source: From [22]/with permission of American Chemical Society).

transient, multiphase behavior of cavitating flows features a wide range of lengths and timescales, and the interactions and mass transfer mechanisms between the liquid, vapor, and dissolved gases is far from well understood. Developing methods to quantify the intensity of this highly complex cavitation behavior, and understanding how this relates to the performance of the devices in real applications is therefore key to realizing the full potential of these technologies. Chapter 5 outlines the current state-of-the art in cavitation modeling, and sets out frameworks for integrating advanced cavitation models with bubble dynamics calculations and chemical reaction engineering models to describe device performance.

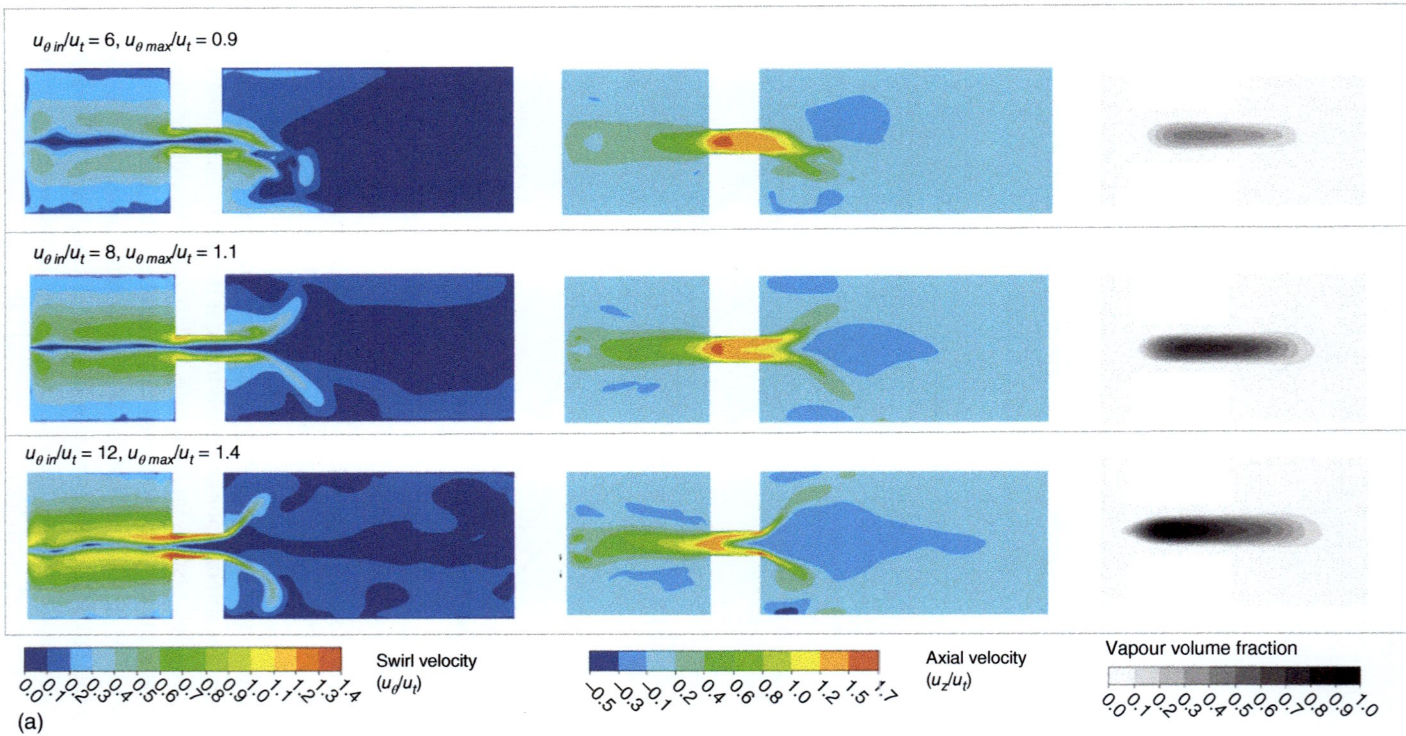

Figure 3.21 Influence of high swirl ratios on cavitation characteristics. (a) CFD predictions; orifice with high swirl ratios. (b) CFD predictions; venturi with high swirl ratios (Source: From [22]/with permission of American Chemical Society).

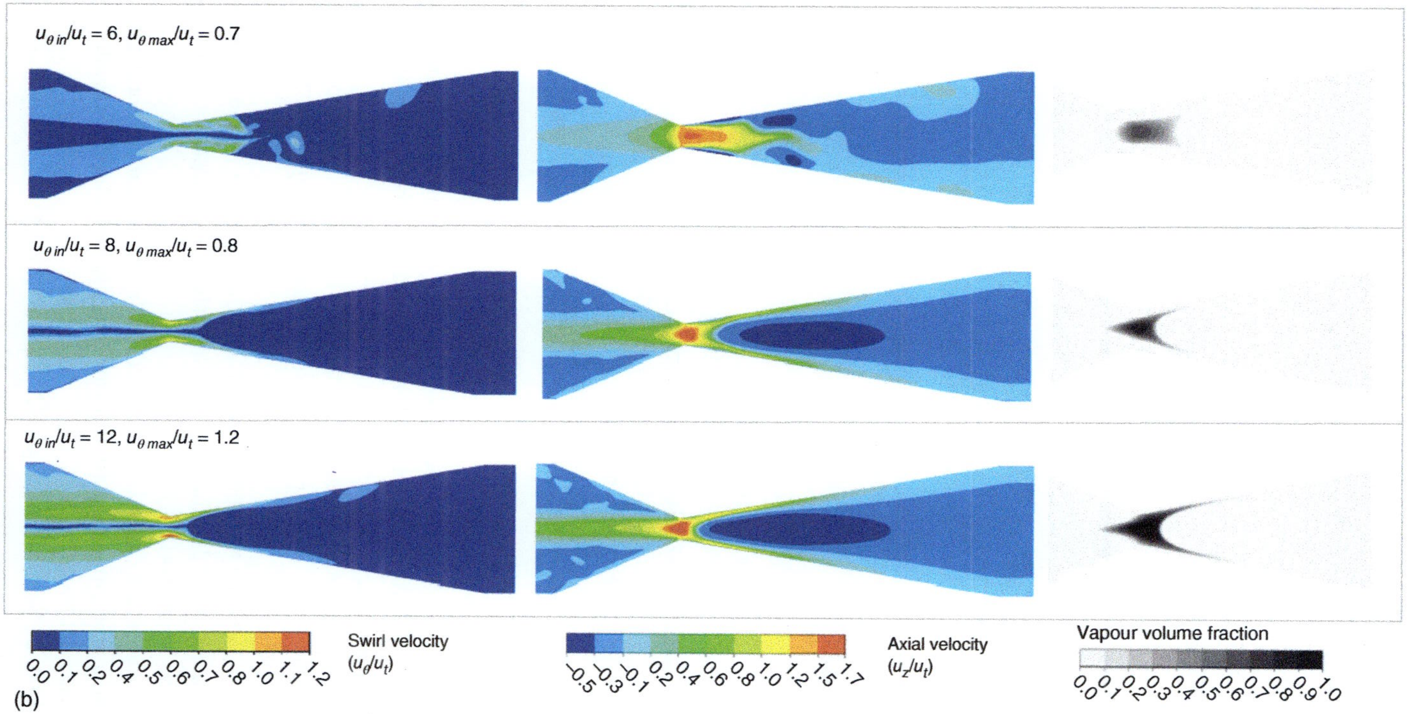

Figure 3.21 (*Continued*)

Notations

a	Radius of viscous core (m)
B	Constant coefficient
C_p	Pressure coefficient
C	Correction factor on throat velocity
d_t	Throat diameter (mm)
p_0	Atmospheric pressure (kPa)
p_d	Discharge pressure (kPa)
p_{min}	Minimum pressure within vortex (kPa)
p_v	Vapor pressure (kPa)
Re	Reynolds number (–)
u_t	Throat velocity (m/s)
u_z	Axial velocity (m/s)
u_θ	Tangential velocity (m/s)
U	Free stream velocity (m/s)
t	Time (s)
z	Axial distance (m)

Greek Symbols

μ	Viscosity (N s/m^2)
ρ	Density (kg/m^3)
ρ_l	Density of liquid (kg/m^3)
σ	Cavitation number (–)
Γ_0	Circulation strength of vortex
Γ	Strength of vortex
ν	Kinematic viscosity (m^2/s)

References

1 Zupanc, M., Kosjek, T., Petkovšek, M. et al. (2014). Shear-induced hydrodynamic cavitation as a tool for pharmaceutical micropollutants removal from urban wastewater. *Ultrasonics Sonochemistry* 21 (3): 1213–1221.

2 Petkovšek, M., Zupanc, M., Dular, M. et al. (2013). Rotation generator of hydrodynamic cavitation for water treatment. *Separation and Purification Technology* 118: 415–423.

3 Petkovšek, M., Mlakar, M., Levstek, M. et al. (2015). A novel rotation generator of hydrodynamic cavitation for waste-activated sludge disintegration. *Ultrasonics Sonochemistry* 26: 408–414.

4 Patil, P.N., Gogate, P.R., Csoka, L. et al. (2016). Intensification of biogas production using pretreatment based on hydrodynamic cavitation. *Ultrasonics Sonochemistry* 30: 79–86.

5 Kim, H., Koo, B., Sun, X., and Yoon, J.Y. (2020). Investigation of sludge disintegration using rotor-stator type hydrodynamic cavitation reactor. *Separation and Purification Technology* 240: 116636.

6 Cerecedo, L.M., Dopazo, C., and Gomez-Lus, R. (2018). Water disinfection by hydrodynamic cavitation in a rotor-stator device. *Ultrasonics Sonochemistry* 48: 71–78.

7 Gostiša, J., Širok, B., Repinc, S.K. et al. (2021). Performance evaluation of a novel pilot-scale pinned disc rotating generator of hydrodynamic cavitation. *Ultrasonics Sonochemistry* 72: 105431.

8 Sun, X., Xuan, X., Song, Y. et al. (2021). Experimental and numerical studies on the cavitation in an advanced rotational hydrodynamic cavitation reactor for water treatment. *Ultrasonics Sonochemistry* 70: 105311.

9 Arndt, R.E.A., Arakeri, V.H., and Higuchi, H. (1991). Some observations of tip-vortex cavitation. *Journal of Fluid Mechanics* 229: 269–289.

10 Arndt, R., Pennings, P., Bosschers, J., and Van Terwisga, T. (2015). The singing vortex. *Interface Focus* 5 (5): 20150025.

11 Couty, P. (2002). *Physical Investigation of Cavitation Vortex Collapse*. EPFL. PhD Thesis.

12 Mancuso, G., Langone, M., Laezza, M., and Andreottola, G. (2016). Decolourization of Rhodamine B: a swirling jet-induced cavitation combined with NaOCl. *Ultrasonics Sonochemistry* 32: 18–30.

13 Loraine, G., Chahine, G., Hsiao, C.-T. et al. (2012). Disinfection of gram-negative and gram-positive bacteria using DynaJets® hydrodynamic cavitating jets. *Ultrasonics Sonochemistry* 19 (3): 710–717.

14 Mancuso, G. (2018). Experimental and numerical investigation on performance of a swirling jet reactor. *Ultrasonics Sonochemistry* 49: 241–248.

15 Jain, P., Bhandari, V.M., Balapure, K. et al. (2019). Hydrodynamic cavitation using vortex diode: an efficient approach for elimination of pathogenic bacteria from water. *Journal of Environmental Management* 242: 210–219.

16 Wu, J.-Z., Ma, H.-Y., and Zhou, M.-D. (2015). *Vortical Flows*. Springer.

17 Liu, Z., Fox, R.O., Hill, J.C., and Olsen, M.G. (2015). A batchelor vortex model for mean velocity of turbulent swirling flow in a macroscale multi-inlet vortex reactor. *Journal of Fluids Engineering* 137 (4): https://doi.org/10.1115/1.4028784.

18 Wood, V.T. and White, L.W. (2011). A new parametric model of vortex tangential-wind profiles: development, testing, and verification. *Journal of the Atmospheric Sciences* 68 (5): 990–1006.

19 Arndt, R.E.A. (2002). Cavitation in vortical flows. *Annual Review of Fluid Mechanics* 34 (1): 143–175.

20 Chen, L., Zhang, L., Peng, X., and Shao, X. (2019). Influence of water quality on the tip vortex cavitation inception. *Physics of Fluids* 31 (2): 023303.

21 Arndt, R.E.A. and Keller, A.P. (1992). Water quality effects on cavitation inception in a trailing vortex. *Journal of Fluids Engineering* 114 (3): 430–438.

22 Simpson, A. and Ranade, V.V. (2019). 110th Anniversary: comparison of cavitation devices based on linear and swirling flows: hydrodynamic characteristics. *Industrial and Engineering Chemistry Research* 58 (31): 14488–14509.

23 Hreiz, R., Gentric, C., Midoux, N. et al. (2014). Hydrodynamics and velocity measurements in gas–liquid swirling flows in cylindrical cyclones. *Chemical Engineering Research and Design* 92 (11): 2231–2246.

24 Hreiz, R., Gentric, C., and Midoux, N. (2011). Numerical investigation of swirling flow in cylindrical cyclones. *Chemical Engineering Research and Design* 89 (12): 2521–2539.

25 De Wilde, J. and de Broqueville, A. (2007). Rotating fluidized beds in a static geometry: experimental proof of concept. *AIChE Journal* 53 (4): 793–810.

26 Volchkov, E.P., Dvornikov, N.A., Lukashov, V.V., and Abdrakhmanov, R.K. (2013). Investigation of the flow in the vortex chamber with centrifugal fluidizing bed with and without combustion. *Thermophysics and Aeromechanics* 20 (6): 663–668.

27 Niyogi, K., Torregrosa, M.M., Marin, G.B. et al. (2018). On the mechanisms of secondary flows in a gas vortex unit. *AIChE Journal* 64 (5): 1859–1873.

28 De Vries, S.F., Florea, D., Homburg, F.G.A., and Frijns, A.J.H. (2017). Design and operation of a Tesla-type valve for pulsating heat pipes. *International Journal of Heat and Mass Transfer* 105: 1–11.

29 Zobel, R. (1936). Experiments on a hydraulic reversing elbow. *Mitteilungen des Hydraulischen Instituts der Technischen Hochschule München* 8: 1–47.

30 Yoder, G.L., Elkassabgi, Y., DeLeon, G. et al. (2011). Vortex diode analysis and testing for fluoride salt-cooled high-temperature reactors. ORNL/TM-2011/425, ORNL, Oak Ridge, TN.

31 Tippetts, J.R. and Priestman, G.H. (1983). The flow states of liquid through two series-connected diffusers subject to cavitation. *Journal of Fluids Engineering* 105 (3): 350–355.

32 Escudier, M.P., Bornstein, J., and Zehnder, N. (1980). Observations and LDA measurements of confined turbulent vortex flow. *Journal of Fluid Mechanics* 98 (1): 49–63.

33 Priestman, G.H. (1987). A study of vortex throttles part 2: viscid flow analysis. *Proceedings of the Institution of Mechanical Engineers, Part C: Journal of Mechanical Engineering Science* 201 (5): 337–343.

34 Pandare, A. and Ranade, V.V. (2015). Flow in vortex diodes. *Chemical Engineering Research and Design* 102: 274–285.

35 Yin, J., Li, J., Ma, Y. et al. (2015). Study on the air core formation of a gas–liquid separator. *Journal of Fluids Engineering* 137 (9): 091301.

36 Simpson, A. and Ranade, V.V. (2019). Flow characteristics of vortex based cavitation devices: computational investigation on influence of operating parameters and scale. *AIChE Journal* 65 (9): e16675.

37 Soyama, H. and Hoshino, J. (2016). Enhancing the aggressive intensity of hydrodynamic cavitation through a venturi tube by increasing the pressure in the region where the bubbles collapse. *AIP Advances* 6 (4): 045113.

38 Capocelli, M., Prisciandaro, M., Lancia, A., and Musmarra, D. (2015). Cavitational reactor for advanced treatment of contaminated water: the effect of recovery pressure. *Desalination and Water Treatment* 55 (12): 3172–3177.

39 Kozák, J., Rudolf, P., Hudec, M. et al. (2019). Numerical and experimental investigation of the cavitating flow within venturi tube. *Journal of Fluids Engineering* 141 (4).

Part III

Characterizing and Modeling of Cavitation Devices

4

Experimental Characterization of Hydrodynamic Cavitation Devices

The previous two chapters discussed the key characteristics of hydrodynamic cavitation devices. To evaluate the performance of existing and newly developed cavitation devices, it is essential to carry out systematic experiments. Though a large number of experimental studies on hydrodynamic cavitation are published (Sarvothaman et al. [1]; Simpson and Ranade [2] and references cited therein), information on designing the required experimental set-up and on planning and execution of experiments is rather limited. We hope to bridge this gap in this chapter. Typically, batch experiments are carried out for characterizing hydrodynamic cavitation devices [3]. The three main tasks of characterizing hydrodynamic cavitation devices are to provide basic data on pressure drop as a function of flow rate [1], to identify inception of cavitation [4], and to quantify the overall performance (extent and efficacy of cavitation) of the hydrodynamic cavitation-based application/process [5]. In the following section, we describe and discuss key aspects of designing an experimental set-up. The subsequent sections discuss the use of such a set-up for identifying inception and quantifying overall performance.

4.1 Experimental Set-up for Characterization of Hydrodynamic Cavitation Devices

Hydrodynamic cavitation – either passive (without moving parts) or active (with moving parts) – needs to be experimentally characterized to provide useful data for design, optimization, and scale-up. A schematic of a typical set-up to characterize hydrodynamic cavitation devices is shown in Figure 4.1.

Such a typical cavitation-based experimental set-up consists mainly of five key components: the holding tank, pump, hydrodynamic cavitation device, piping network, and a suite of in-line sensors for detecting various parameters of interest. In addition, of course, the experimental characterization will require additional analytical facilities, which may provide additional information via off-line analysis of collected samples. A discussion on the key aspects of these components of experimental set-up follows.

Hydrodynamic Cavitation: Devices, Design, and Applications, First Edition.
Vivek V. Ranade, Vinay M. Bhandari, Sanjay Nagarajan, Varaha P. Sarvothaman and Alister T. Simpson.

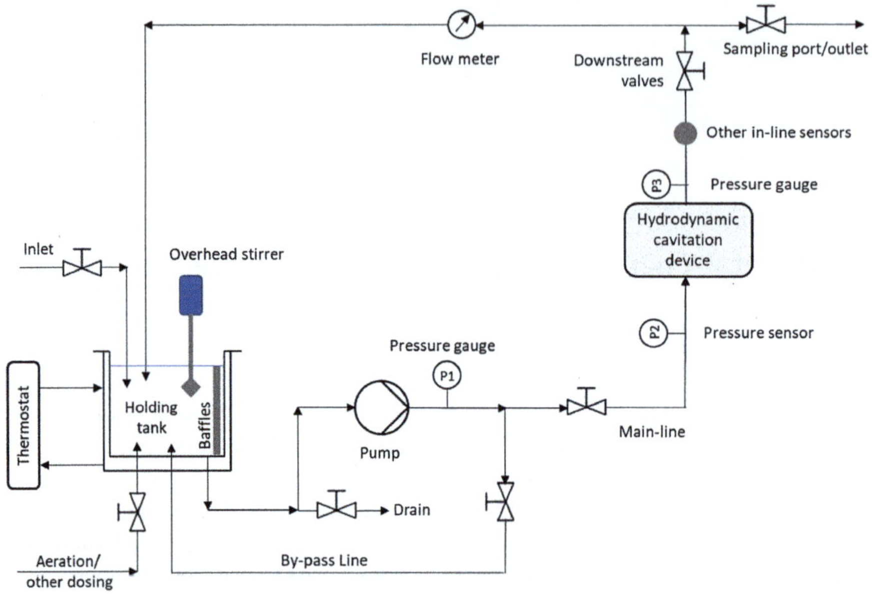

Figure 4.1 Schematic of overall experimental set-up.

4.1.1 Holding Tank

A quick review of numerous experimental studies published on characterizing the performance of hydrodynamic cavitation devices unfortunately reveals that the design of the holding tank has not received as much attention while designing the experimental set-up. The holding tank is a crucial component of any cavitation experimental set-up and it is important for adequate attention to be given for its design. Some of the key aspects related to design and configuration of the holding tank are briefly discussed in the following section.

Size of the holding tank: The minimum size is dictated by the type of pump/operating flow rates. The larger the size, the more the time required for completing a specific number of passes through the cavitation device. Typically, volume of the holding tank is selected to achieve ~100 passes through the cavitation devices per hour.

Shape of the holding tank: Typically, cylindrical or rectangular holding tanks have been used. If the tank is cylindrical, baffles are recommended, particularly when an overhead stirrer is employed. The width, number, and locations of baffles may be selected based on other internals and the extent of mixing required. The aspect ratio (ratio of height to diameter or width) is important to avoid gas entrainment, mixing, and suspension. The design may also require internal cooling coils for temperature control. The material of construction for both the holding tank and the internals, cooling coil, etc. needs to be considered appropriately depending upon the application area. Ease of tank cleaning also needs attention in cases where there is likelihood of deposits on the surfaces/walls, to ensure consistency in the results and avoid contamination in subsequent experiments.

Shape of the tank bottom: The shape of the tank bottom and the overall configuration of the holding tank and internals will control flow and mixing in the holding tank. Particularly when handling suspended solids, the shape of the tank bottom needs to be selected in such a way that there are no dead spots where solids may accumulate. Typically, flat, conical, or elliptical tank bottoms have been used.

Top cover of the holding tank: It is important to ensure that the top of the holding tank is appropriately covered. In many reported studies, open holding tanks have been used. The use of open tanks may lead to evaporation of water and possibly contamination by pollutants, which may skew observed results. The potential of evaporation loss needs to be quantified, and wherever appropriate, the top of the holding tank should be covered.

Overhead stirrer for the holding tank: Though flow from by-pass and returning main lines may be used to realize mixing in the holding tank, a separate overhead stirrer may be needed when mixing is critical – particularly when there are fluids with significant differences in density and viscosity. Usually, propeller or pitch-blade turbines like axial flow impellers will provide adequate flow and mixing in the holding tank. A typical thumb rule for ensuring complete mixing in the holding tank is to ensure that typical circulation time in the holding tank is at least three to five times smaller than the residence time in the holding tank.

Location of the suction line: The location of the suction line going to the pump needs to be such that there is no vortex formation because of suction and no gas entrainment in the pump suction line. For cases with solids suspension, it is also important to ensure that clogging risks are mitigated appropriately.

Location of by-pass return line: This is important, particularly when bypass flow is being used for mixing in the holding tank. Care needs to be taken to avoid short-circuiting. When bypass flow is used for mixing, jet-mixing principles are often used to decide the location and angle of nozzle used with the bypass return line.

Location of main return line: It is generally recommended to submerge the returning main line in the liquid contained in the holding tank. That will avoid gas induction from the top surface. Here again, it is important to ensure appropriate placement of the return line location and suction location to avoid any short-circuiting.

Maintaining temperature of holding tank: Multiple passes through cavitation devices will generally increase the temperature of liquid in the system. The energy consumed by the pump will be dissipated in the liquid. The expected temperature rise may be estimated by using a simple heat balance like:

$$MC_p\frac{dT}{dt} = \frac{\Delta P\, Q}{\text{Pump efficiency}} - \text{Loss to surrounding} \tag{4.1}$$

where M is the mass and C_p is the heat capacity of liquid in the holding tank and ΔP and Q are pressure drop across the pump and total flow rate through the pump, respectively. If we ignore losses to the surrounding and the pump efficiency (considering that the two effects are against each other, this assumption will not be too far from reality), we may estimate temperature rise as a function of the number of passes, n (which is a ratio of elapsed time and residence time) as:

$$\Delta T = \frac{\Delta P\, n}{\rho\, C_p} \tag{4.2}$$

If we consider pressure drop across the pump as 400 kPa, the density and heat capacity of water, we can estimate that temperature rise will be approximately 10 °C per 100 passes. Ignoring losses to the surrounding would be a reasonable assumption as long as the temperature difference between the liquid in the system and the surrounding is not high. Beyond a certain temperature rise, losses to the surrounding will be significant and eventually will be equal to the energy consumption rate, leading to an equilibrium temperature of the system even if there is no external cooling. Generally, this balance between the energy consumption and losses occurs around 30 °C above the ambient temperature as a first approximation. When there is an interest in carrying out experiments at specific temperatures, it is recommended to use either the external jacket or internal coils in the holding tank for maintaining the desired temperature of the considered liquid. The alternative is to use an in-line heat exchanger, either on the by-pass line or on the main return line, to avoid jacket or internal coil in the holding tank. The usual heat exchanger design methods may be used for sizing the required heat exchanger.

Aeration/dosing of other augmenting agents: In many experiments, hydrodynamic cavitation is augmented by using aeration or by using additives like steam [6], hydrogen peroxide [7, 8], ozone [7, 9], catalysts, etc. Aeration is used to maintain the concentration of dissolved oxygen in liquid, though it can also possibly contribute toward cavity inception, growth, and collapse, and thereby has an impact on the overall cavitation process. Aeration is typically achieved by sparging of air (or oxygen) in liquid – either using a conventional sparger in the holding tank or using an in-line sparger in the main line [10]. If the considered system contains solid particles, it is often useful to install spargers with holes facing the downward direction to avoid potential clogging of sparger holes. It may also be possible to use the low-pressure zone generated in a hydrodynamic cavitation device for sucking air from the surrounding. Additives like steam, hydrogen peroxide, or ozone are typically introduced either just before or in the hydrodynamic cavitation unit, to maximize their impact. Depending on the specific application and augmentation strategy, bespoke dosing arrangements may be designed [11].

4.1.2 Pump

A variety of pumps have been used in the published studies of hydrodynamic cavitation. It is not possible here to cover the detailed discussion and criteria for selecting appropriate pumps. Several excellent texts and monographs are available for the same [12–14]. Instead, we provide a list of commonly used pumps for the set-up being discussed, with brief comments on each type.

Peristaltic pumps: These pumps do not contain a mechanical seal and are suitable for handling corrosive fluids. They offer a compact and flexible solution suitable particularly for small-scale set-ups and devices [15]. However, the pumps are generally not able to generate discharge pressure above the 200 kPa gauge and generate oscillatory flow. The potential impact of oscillatory flow generated by peristaltic pumps on generated cavitation is not yet entirely clear and quantified. The results of Thaker and Ranade [15] on liquid–liquid emulsions indicate that this influence may not be significant.

Centrifugal pumps: This may be one of the most commonly used pumps in published studies on hydrodynamic cavitation [2, 10]. These are quite versatile and a wide range of options are available for selection. These pumps could be either single-stage or multi-stage, depending on the desired range of discharged pressure and flow. It must be noted that there is often very less clearance between the impeller and the pump casing. Use of such pumps is therefore restricted to rather clean liquids without any suspended solids. Compatibility of seals used in the pumps must be checked while selecting the pump for the specific system under consideration. Care should be taken to operate these pumps with a very low pressure drop – high flow under such conditions may result in pump cavitation. This will not only damage the pump but will also corrupt the experimental results aimed at cavitation in the device. It is generally recommended to use a variable frequency drive (VFD) with the centrifugal pump to expand its operating range and avoid the possibility of cavitation due to unfavorable operating conditions.

Progressive cavity pump: These pumps can handle slurries and complex liquids (viscous, shear-sensitive, corrosive, and abrasive). Generally, these pumps can handle solids up to 25 mm in diameter. These pumps also tolerate some amount of entrained air/gases.

Lobe pumps: These pumps are employed in the pulp and paper, pharmaceutical, and food industries. They are especially useful for food handling by virtue of their construction. Generally, these pumps can handle solids up to 100 mm in diameter.

Gear pumps: These pumps are capable of pumping high-viscosity fluids. These are suited for high-viscosity applications in the petrochemical (lube oil/crude oil), pulp and paper (acid/lye), and the food industries (chocolate/molasses). These pumps are not suitable for handling abrasives or solids.

Cutter pumps: As the name signifies, these pumps incorporate a "cutting" operation, with a cutting edge attached to the leading edge of the impeller vane. They can be used for handling wastewater containing fibrous solids and other heavy-duty industrial applications.

Submersible pumps: These pumps can be employed within the fluid to be pumped, with the main advantage being that pump cavitation can be avoided. They were installed first within an oil well, and subsequently in sewage pumping, general industrial plumbing, and water wells. The discharge pressure achievable by these pumps usually is not large and may be insufficient for many hydrodynamic cavitation applications.

Diaphragm pumps: These pumps are used for metering applications and are very accurate (flow rate is maintained within $\pm 1\%$). This can be specifically useful for metering food ingredients. These pumps can tolerate solids up to 25 mm in diameter and a wide range of liquids.

Torque pumps: These pumps are used for fluids with high solid content and fibrous particles and gas bubbles, which can be typical of contaminated fluids such as sewage and wastes [16].

While it is almost impossible to discuss all aspects of pump selection, the brief summary included here may provide some help in the selection of an appropriate type of pump for a hydrodynamic cavitation application. There are instances where a pneumatic pressure is used to realize flow, instead of a pump (see for example, Zupanc et al. [17]). However, this involves an air compressor, etc., and is generally not preferred unless there is no suitable pump available.

4.1.3 Hydrodynamic Cavitation Device

A variety of hydrodynamic cavitation devices are available commercially. However, most of the published studies use "homemade" devices rather than commercially available devices. This makes the comparison of results obtained by different research groups rather difficult. As discussed in Chapters 2 and 3, small changes in fabrication, particularly for laboratory-scale devices, can make large differences in the observed characteristics and performance of hydrodynamic cavitation devices. For example, the sharpness of edge of the orifice can make significant difference in cavitation performance. It should be remembered that "look-alike" devices will not "perform alike" because of minute differences in fabrication. It is therefore important to take utmost care in fabricating hydrodynamic cavitation devices and to specify clearly all the dimensions and tolerances. This will facilitate comparison of results among different research groups, which will stimulate further advances.

Apart from ensuring adequately small fabrication tolerances, it is also important to ensure that hydrodynamic cavitation devices are installed appropriately in the set-up. Bends and fittings close to the device inlet and outlet should be avoided. As far as possible, the piping arrangement should facilitate realization of fully developed flow at the inlet and from the outlet. Typically, a straight portion of at least 40 times the pipe diameter is recommended at both sides of the cavitation device. For the cases of swirling flow-based cavitation devices, an even longer development zone at the outlet is needed to ensure that swirling is subsided.

The material of construction and surface finish of cavitation devices is important as well. If the interest is to visualize flow field in the cavitation devices, transparent material needs to be used for fabricating cavitation devices (like acrylic). The polishing of surfaces after machining becomes really important for such visualization experiments. Even otherwise, surface finish may influence the performance characteristics of devices. There is an increasing trend to use additive manufacturing (3D printing) for fabricating cavitation devices. A variety of materials are used for additive manufacturing and careful selection is needed. Some of the commonly used materials with additive manufacturing may have limitations on the maximum pressure they can withstand. With high pressure, there may be some expansion (of polymeric materials). This may lead to changes in the dimensions of the device – leading to inappropriate comparison. The cleaning of 3D printed devices to remove any supports used while printing and the surface finish are both critically important to obtain reproducible results. Care needs to be taken to use appropriate connectors between 3D-printed devices and the rest of the piping, since such connections have generally been found to be the weakest link leading to leakages and other issues.

4.1.4 Piping Arrangements/Fittings

No details of piping arrangements and pipefittings are generally provided while describing experimental set-ups in the published research papers. However, it is an important component of the experimental set-up and needs to be designed appropriately. It is important to ensure that there are no dead zones in the entire piping network. Overall piping arrangements should be designed to facilitate in situ cleaning between the two experiments. The presence of dead zones often hinders such in situ cleaning. The drain location needs to be identified appropriately to ensure that it enables complete draining of liquid from the holding tank, pump casing, as well as the entire piping network. It is also essential to accurately measure volume of the piping network. To relate the number of passes with elapsed time of experiments, it is essential to include the total volume of the system, comprising the volume of piping network, liquid in the holding tank, pump casing, and cavitation device.

One of the important considerations of the piping arrangements and pipe fittings (valves, flow meters, etc.) used in the set-up is to ensure that cavitation does not occur anywhere else in the network except in the hydrodynamic cavitation device under investigation. This is especially important if the flow rate through the cavitation device is controlled by manipulating flow through the bypass line (instead of the variable frequency drive). In such cases, cavitation may occur in the bypass line. In some experiments, in order to generate higher pressure at the discharge of cavitation device, a valve is employed at the downstream of the cavitation device. Depending on the desired pressure at the outlet of the cavitation device and the extent of throttling needed at such a downstream valve, cavitation may occur in the valve. In both such cases, to reduce the extent of throttling needed to realize the desired pressure drop, it is advisable to use a series of valves for realizing the desired pressure drop instead of a single valve. This will reduce or eliminate chances of cavitation in the bypass line or in the downstream valve.

Another important point for consideration is the potential for adsorption of pollutants on the internal surfaces of pipes. This is particularly important in case of the commonly used plastic piping. Many organic pollutants may adsorb on PVC, and therefore incorrectly indicate the extent of degradation by the hydrodynamic cavitation. For example, coumarin, which is one of the commonly used dosimetric probes for quantifying generation of hydroxyl radicals [18], is adsorbed on plastic pipes [19]. In this lab report [19], the authors discuss adsorption of commonly used organic pollutants in hydrodynamic cavitation studies.

4.1.5 In-line Sensors

In addition to the components mentioned above, the experimental set-up may also comprise various in-line sensors. Pressure gauges and flow meters are the bare minimum to ensure correct operation of the test set-up. A wide range of flow meters and pressure sensors are available. The details of these sensors are not discussed here. Several excellent resources are available for selecting appropriate flow meters and pressure sensors. Depending on the application under consideration, several other in-line sensors may be used. For example, if identification of cavitation inception is of interest, acoustic sensors or hydrophones may be used to pick up the characteristic signature of cavitation (see Section 4.2). In-line pH, temperature, dissolved oxygen, and conductivity sensors may provide useful information that is relevant for understanding various applications of water treatment. Conventionally, COD (chemical oxygen demand) and BOD (biological oxygen demand) measurements are time consuming. However, in recent years, in-line sensors for these are available [20, 21]. For multiphase applications such as emulsions or crystallization, sophisticated in-line sensors like FBRM (focused beam reflectance measurement), PVM (particle vision and measurement) or Raman probes may also be used to characterize particles (shapes/size distribution) and concentrations. More information on these in-line process analytical tools may be obtained from Raman and Abbas [22], John et al. [23], and Wang et al. [24, 25]. The location of the in-line sensors needs to be carefully selected. Acoustic sensors need to be placed near the cavity collapse zone. The other in-line sensors may be placed at a sufficient distance away from the device outlet, to avoid potential damage because of cavitation.

The in-line sensors often need to be complemented by more thorough off-line analytical facilities. These may include conventional analytical equipment like gas chromatography [26], high-performance liquid chromatography [27], UV [1], GC–MS [28], and NMR [29]. For multiphase particulate systems, an additional particle sizer or image analysis may also be useful.

Together, all the components and the schematic shown in Figure 4.1 will offer an opportunity to carry out adequate experimental characterization of hydrodynamic cavitation devices. This will allow optimization of process conditions for a specific application under development. The experimental data may be interpreted using the mathematical models discussed in Chapter 5. Together, the models and experimental data can provide a pathway for process optimization and scale-up. The use of the experimental set-up discussed here for identifying inception of cavitation and for quantifying overall performance of cavitation-based processes is briefly described in Sections 4.2 and 4.3 respectively.

4.2 Identification of Inception of Hydrodynamic Cavitation

It is important to identify the operating conditions at which inception of hydrodynamic cavitation occurs. Such quantitative information about inception is useful for specifying nominal capacity of the considered hydrodynamic cavitation device and for identifying suitable conditions for operating any process based on such cavitation. Usually, cavitation inception is identified either in terms of the pressure drop across the device or in terms of the characteristic velocity through the device (usually velocity at the throat of the device). It is therefore necessary to first characterize the flow rate versus pressure drop relationship for any hydrodynamic cavitation device. For all passive cavitation devices, such a relationship is a unique function of device configuration, dimensions, and properties of working fluid. The experimental set-up shown schematically in Figure 4.1 can be used to generate the required data of pressure drop across the cavitation device as a function of flow rate or throat velocity. Typically, the observed pressure drop characteristics of a hydrodynamic cavitation device are represented in terms of Euler number, *Eu* as:

$$Eu = \frac{\Delta P}{\frac{1}{2}\rho V_C^2} \tag{4.3}$$

where ΔP is pressure drop, ρ is density of working fluid, and V_C is the characteristic velocity of the hydrodynamic cavitation device. If the characteristic velocity V_C is taken as the maximum velocity magnitude in the device, typical values of *Eu* are in the range of 2–3. For linear flow devices like orifice or venturi, the characteristic velocity is taken as throat velocity, which is close to maximum velocity in the device. For vortex diode-based cavitation devices (discussed in Chapter 3), the maximum velocity in the device is typically five times larger than the throat velocity. Therefore, if the Euler number is defined in terms of throat velocity, the typical range is of 50–75. Generally, value of Euler number will increase with the Reynolds number. In principle, it may be possible to identify inception of cavitation by observing a discontinuity in the pressure versus throat velocity data. However, the transient nature of the flow and turbulent pressure fluctuations often make this rather difficult.

Several alternative methods have been used for identifying inception of cavitation. It should be noted that as the low pressure generated in a cavitation device approaches vapor pressure of liquid, cavities emerge from single-phase flow. Therefore, visual methods, which can detect generation of cavities in a single-phase flow, can be used for identifying inception of cavitation. For example, Soyama and Hoshino [30] used the visualization method to detect inception occurring in a venturi (see Figure 4.2).

The visualization method of course requires that the cavitation devices and piping immediately downstream of the devices are transparent [30]. High-speed imaging [31], light scattering methods [32], and particle imaging velocimetry [33] have also been used for identifying inception. For opaque devices and piping, visual methods (either manual or with cameras) are not applicable. The dosimetric methods of identifying cavitation inception rely on detecting the presence of hydroxyl radicals generated during the cavity collapse. These methods use radical scavengers to quantify generated radicals (indirectly by analyzing the products of reaction between the selected scavenger and OH radicals). The commonly used scavengers for cavitation are salicylic acid [34], terephthalic acid [35], potassium iodide [36], and coumarin [18]. There are no generalized recommendations of one mentioned dosimetry method over the other and the choice is based on availability of the dosimeter and analytical technique. It is important to verify that the dosimeter does not produce any oxidized products under non-cavitating conditions, as this would

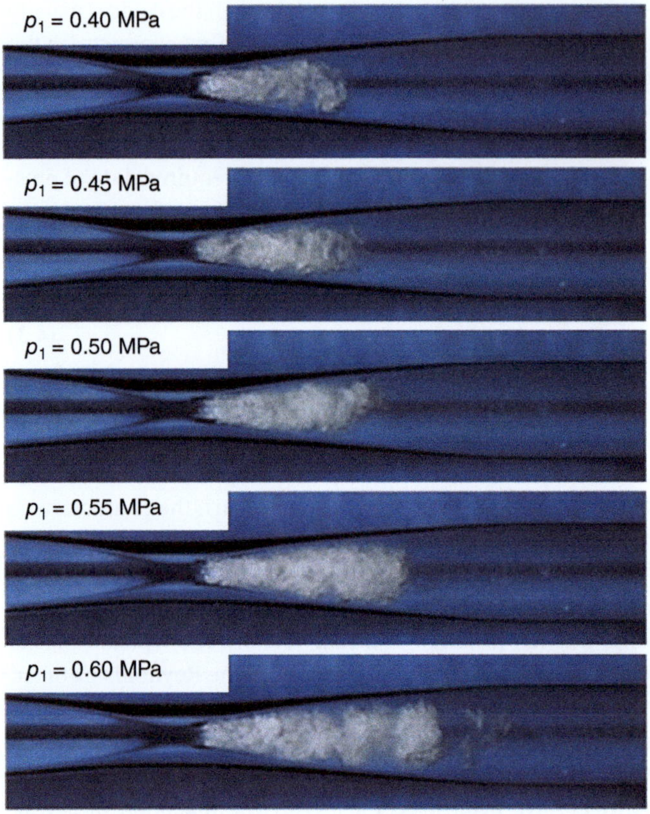

Figure 4.2 Photograph of venturi at different operating pressure drop (Source: [30], from AIP Publishing LLC).

lead to false interpretation of cavitation, as is the case with potassium iodide [37]. With the exception of potassium iodide, the other discussed possibilities – salicylic acid, terephthalic acid, Fricke dosimetry, and coumarin – do not have any known ambiguity. An experimental profile of salicylic hydroxylation at four operating conditions is presented in Figure 4.3. At the lowest operating condition (<100-psi), it was observed that no dihydroxybenzoic acid was formed. At higher operating flow (2000, 3000, and 4000 psi), the quantity of hydroxylation monotonically increases. These experiments for dosimetry can be performed with experimental set-up, as presented in Figure 4.1, measuring the concentration of the dosimeter at dedicated time intervals. The inception condition by this approach can be marked by operating the hydrodynamic cavitation device modifying the operating conditions with gradual increments, ranging from non-zero (non-cavitating conditions) to higher operating flow conditions.

Acoustic sensors have also been used for identifying inception of cavitation [38–41]. Such acoustic methods rely on the increased sound levels generated by the collapsing cavities. Several studies have described cavitation noise: for example, a tip vortex was reported as a "singing vortex" [42] because of the cavitation noise

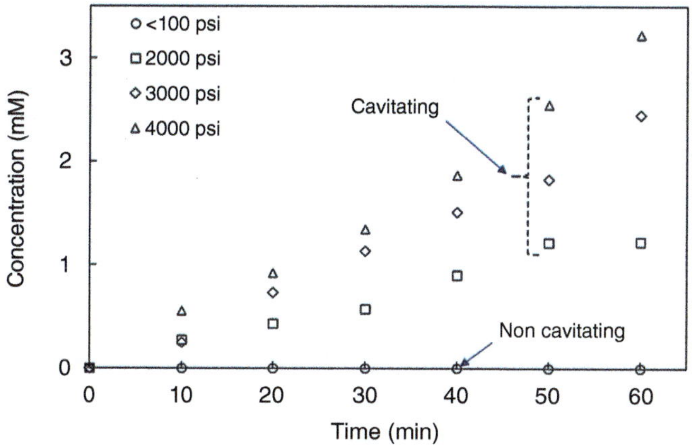

Figure 4.3 Effect of upstream pressure on the extent of hydroxyl radical production as measured by dihydroxybenzoic acid formation (Source: From [27]/with permission of Elsevier).

or the cavitation noise from an orifice was described as the sound of pebbles being rolled through the device [2]. The experimental procedure for identifying inception of cavitation comprises three parts: (i) construct a cavitation loop to operate the cavitation device under consideration at different flow rates; (ii) acquire acoustic or pressure fluctuation data; and (iii) process the acquired data for identifying inception. The construction of a typical cavitation loop is discussed in Section 4.1. A hydrophone type of instrument is commonly used for the acquisition of noise [39, 43]. Inexpensive microphones can also be used. For example, Ranade et al. [4] have used an ordinary mobile phone for acquiring acoustic signals from the cavitation device. The overall arrangement is shown schematically in Figure 4.4.

Different data processing techniques based on the time and frequency spectra are used to analyze the cavitation noise: for example, the counting of number of peaks larger than a certain threshold [44–46] and Fast Fourier Transform spectra [40, 41, 47–50] are used to identify the inception condition. Several studies have reported analysis of acoustic signals of cavitating flow. For example, orifice [51], tip vortex cavitation [52], and cavitation in valves [53]. Recently, Ranade et al. [4] have

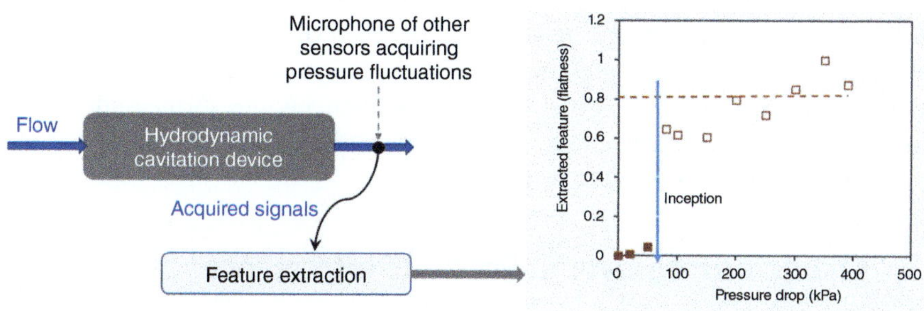

Figure 4.4 Schematic arrangement for using acoustic signals for identifying inception.

used sophisticated analysis of acoustic signals acquired via simple mobile-based sensors for identifying inception of cavitation. Typically, data acquisition frequency of 22 kHz was found to be adequate. Several standard toolboxes are available for conditioning and processing acoustic signals. For example, MATLAB R2020b has "Audio," "Signal Processing," and "DSP System" toolboxes for enabling extraction of several acoustic features. Similar tools are available from other sources (including open source codes) as well. The cited papers may be referred to for more details on sensors as well as data processing required for identifying inception of cavitation.

An example of the difference between cavitating and non-cavitating conditions for a vortex diode-based cavitation device is presented in Figure 4.5a. Here, the amplitude profiles at two operating conditions are presented. By choosing three pre-calibrated amplitudes with visual observations (represented here as solid horizontal lines), it can be seen that the black curve varies multiple times over and beyond these horizontal lines, whereas the orange curve does not cross the set amplitude thresholds. In the work [46], the number of times the black and orange curves crossed the threshold was counted in order to analyze different operating conditions and inspect onset of cavitation. In Figure 4.5b, three operating conditions for a spool valve from the work by Martin et al. [54] are presented. Two cavitating conditions can be differentiated from the non-cavitating condition by inspecting the FFT profiles.

Figure 4.6 shows a sample result from the study by Ranade et al. [4] in which they have extracted and used a variety of features from the acquired acoustic signals. In this figure, a feature called ZCR (zero crossing rate), which is a measure of noisiness of the data, is used to identify inception of cavitation. The paper by Ranade et al. [4] may be referred for more details.

Some commercial sensors for identifying inception of cavitation are available. For example, sensors marketed as i-Alert (Cran Engineering) and a Cavitation Detection System (Yokogawa Electric Corporation). These commercial products would gather time waveform or differential pressure to detect abnormalities and present them to the end-user for identifying cavitation in operations. These sensors may also be used for identifying inception of hydrodynamic cavitation.

Most of the published studies use acoustic signals only for identifying inception. Recently, Ranade et al. [4] have showed a possibility of using acoustic signals not only to identify inception of cavitation but also to extract information about the extent of cavitation, and thereby, about performance of the considered hydrodynamic cavitation devices. However, further work is needed for full development of an effective approach for establishing a relationship between the features extracted from the acquired acoustic signals and the extent of cavitation or performance of hydrodynamic cavitation device. The methods for quantitatively measuring and characterizing overall performance of hydrodynamic cavitation devices and processes are briefly discussed in the following section.

4.3 Characterizing Overall Process Performance

Hydrodynamic cavitation is being harnessed for numerous industrial applications such as water disinfection, water treatment, biomass pre-treatment, oxidative

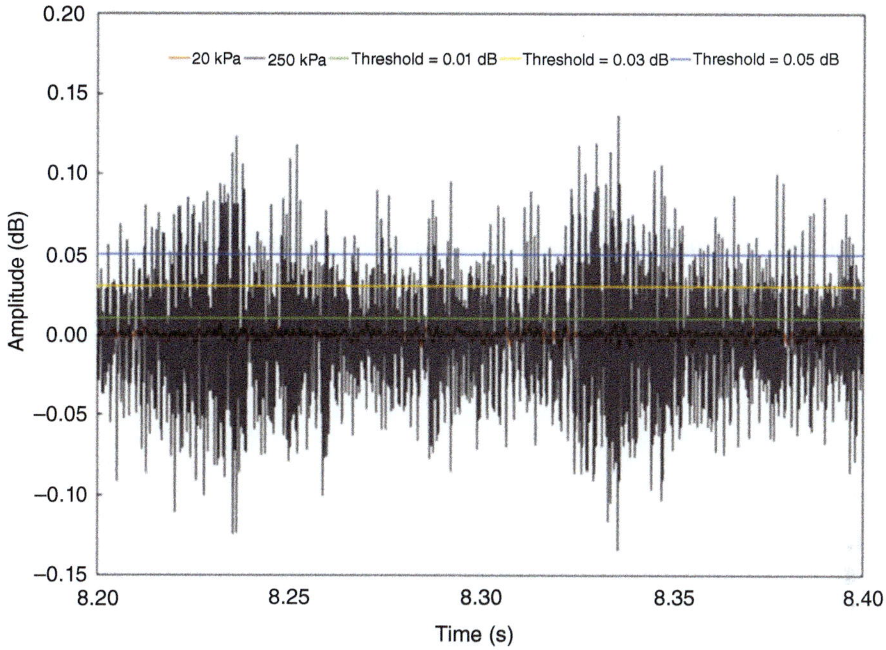

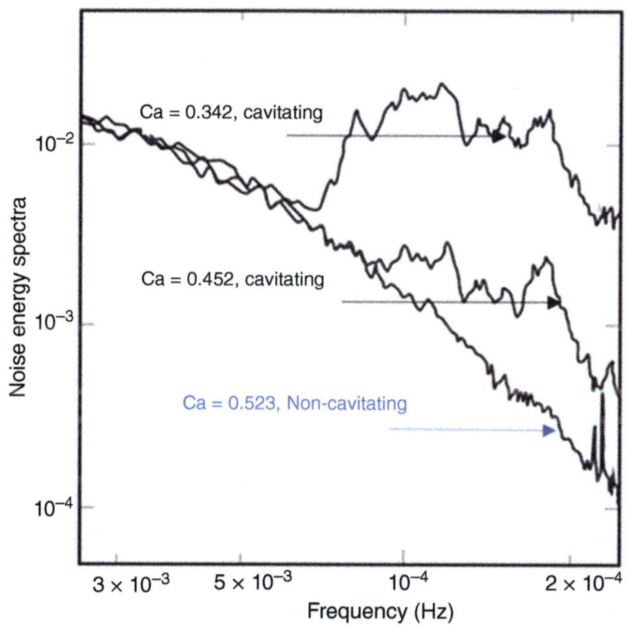

Figure 4.5 (a) Amplitude profiles for cavitating and non-cavitating conditions in a vortex diode-based cavitation device (Source: From [46]) and (b) frequency spectra for cavitating and non-cavitating conditions (Source: From [47]/with permission of Cambridge University Press).

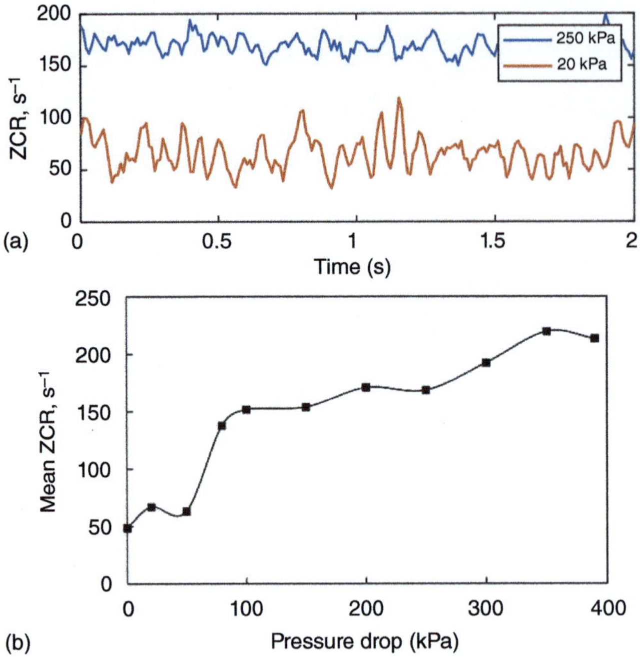

Figure 4.6 Identifying inception of cavitation based on value of ZCR (a) comparison of ZCR at 20 and 250 kPa, (b) influence of pressure drop on mean value of ZCR (Source: Adapted from [55]).

desulfurization, emulsions, dairy products, nano particles, and so on. Some examples of these applications have been cited in Chapter 1. The purpose of this brief section is not to discuss these applications. A focused discussion on some of the prominent applications is included in the later chapters of this book (for example, water disinfection in Chapter 6, wastewater treatment in Chapter 7, valorization of biomass via pre-treatment in Chapter 8, and other miscellaneous applications in Chapter 9). The discussion in this section is limited to some comments on experimental investigations of these applications. Obviously, depending on particular applications, the specific details will vary. Here, we have attempted to compile general recommendations and tips for characterizing and quantifying performance of hydrodynamic cavitation devices and the processes based on those.

Hydraulic testing: It is recommended to start the experimental program with a fully drained and cleaned experimental set-up (shown schematically in Figure 4.1). A variety of different methods may be used depending on specifics of the set-up (material of construction, fittings used, compatibility issues, requirements, and so on). This may include washes by detergents, mild acid, and higher-than-ambient temperature water (up to 40 °C). It is a good idea to carefully fill the set-up with water and run a few experiments with water as a working fluid. This will allow accurate measurements of system volume. It is best to prepare a chart of system volume as a function of height in the holding tank for a quick reference. It will be useful to open the drain valves and sample ports to ensure that there are no air pockets trapped in the set-up, and that the set-up is completely filled with liquid. After establishing that there are no leaks anywhere in the loop, further measurements may be carried out. Measurements of pressure drop across the hydrodynamic cavitation device as a function of flow rate through the device and comparison of it with the vendor-supplied device characteristics is a good starting point to ensure that the device is installed and working as per the specifications. If the device is homemade, it will be useful to compare the newly acquired data with the reference data set. It will also be useful to identify clearly the lower and upper boundaries of the operating conditions considering the specifications of the pump and the cavitation device [10].

Thermal testing: As mentioned earlier, running the hydrodynamic cavitation set-up in a batch mode will lead to rise in operating temperature. The extent of hydrodynamic cavitation and thereby overall performance will be influenced by the operating temperature. It is therefore essential to characterize the heat transfer coefficients and loss to surroundings by carrying out specific experiments. Monitoring the temperature of liquid in the holding tank and the saturation level attained after a prolonged experiment will be useful to estimate the effective heat transfer coefficient between tank contents and cooling medium (either through jacket or internal coil) and the overall losses to surroundings. These two quantities will be a function of the operating conditions of the cavitation device as well as flow rate and temperature of the cooling medium. It will be best to estimate the required heat transfer parameters over the range of interests for further use.

Ensuring no cavitation anywhere except the device under consideration: Before proceeding to execute experiments for quantifying performance of the considered hydrodynamic cavitation device, it is essential to ensure that cavitation occurs only in the device and nowhere else in the flow loop. The other potential locations where cavitation may occur are pumps or valves. The pump should be appropriately selected by considering the operating range of interest to avoid pump cavitation [10]. Pressure drop across any valve used in the set-up needs to be as small as possible to avoid cavitation in the valve. As indicated in Section 4.1, if a high pressure drop is desired, it is better to use multiple valves in series with the pressure drop across each valve limited to a small value. Presence of cavitation occurring in the set-up may be detected by audible cavitation noise (which can be mastered with some practice) or any other techniques used for detecting the presence of cavitation (as discussed in Section 4.2). Control experiments (see next point) often help as well.

Control experiment: It is important to carry out a control experiment at an operating condition when there is no cavitation in the considered cavitation device (typically at a lower-than-critical flow rate at which inception occurs). For example, Ranade et al. [55] conducted degradation of dichloroaniline in water using a vortex-based cavitation. Their data, obtained at control (non-cavitating) and cavitating conditions are shown in Figure 4.7. If such control experiments show non-negligible degradation, it will be worthwhile to investigate the potential reasons. It will also be useful to quantify net degradation because of the hydrodynamic cavitation device under investigation.

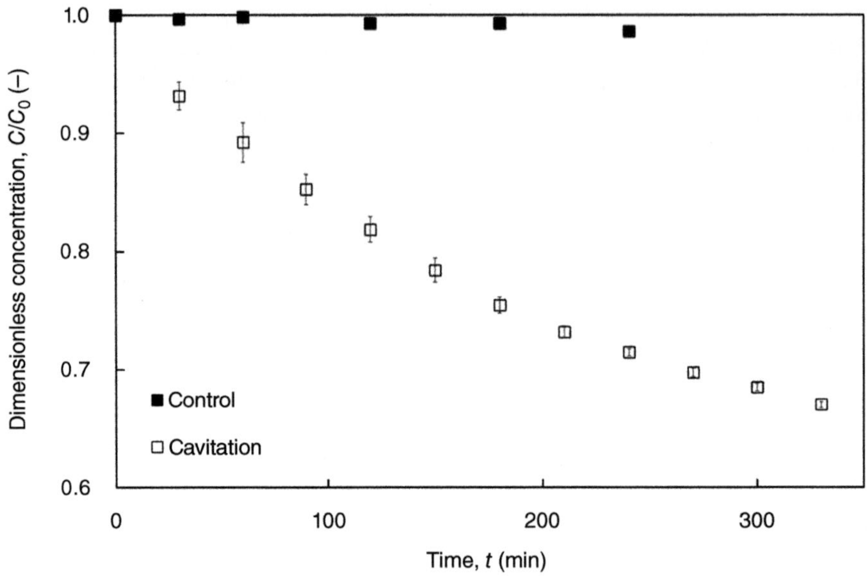

Figure 4.7 Degradation profile for dichloroaniline at cavitating and control conditions (Source: From [55]/with permission of Elsevier).

Mixing/eliminating dead zones: Prior to performing experiments, it is important to ensure that the sampling port is working fine and samples are a true representation of the system of interest. It is therefore essential to ensure adequate mixing in the holding tank and the absence of dead zones in the entire flow loop. Sampling is especially critical in the case of multiphase systems. For example, if the system under consideration is a slurry with certain specific solids loading (for example in a biomass pre-treatment application), it is not straightforward to ensure that the sampled portion will have same solids loading as in the main system. The bends and pipefittings used in the sampling port often will change the amount of solids in the sampled portion. Adequate care needs to be taken to minimize such skewed sampling and to account for such a possibility while interpreting the experimental data. If the interest is in measuring droplet (in case of say emulsion-related applications) or particle (in case of say crystallization or precipitation reactions) size distributions, it is important to ensure that the process of sampling does not alter droplet or particle size distributions significantly. In some cases, significant volume of samples may be necessary. For example, for biomass pre-treatment applications, the sampled portion may be of the order of ~100 ml. In such cases, it will be essential to account for changes in the system volume as multiple samples are withdrawn.

Executing experiments/troubleshooting: As in the case of any scientific experiments, careful design of experiments with clearly defined goals is very useful to minimize the number of experiments. It is always recommended to evaluate the performance of the device/process under investigation using multiple analytical techniques to cross-validate the data. Despite careful planning of experiments, some failures in the experimental set-up/experiments may be encountered. Similar to many new insights gained during the debugging phase of any complex computer code, troubleshooting of experimental set-up offers interesting learning opportunities. Simple problems like clogging shed new lights on motion of suspended particles in the flow loop and provide a pathway to design better systems and devices. Considering the immense variety in applications and possible designs of experimental set-up, it will be impossible to discuss all the potential failure scenarios and troubleshooting strategies. A watchful analysis of failure scenarios encountered in practice and following the general suggestions included in this chapter will be useful for systematic evaluation of hydrodynamic cavitation devices and processes.

When the objective of the experimental study is to evaluate different types of hydrodynamic cavitation devices, it is essential to develop a rational basis for their comparison. Several published studies have used a rather ad-hoc basis for comparing different devices. Considering the typical open designs of hydrodynamic cavitation devices, it is generally difficult to compare different devices based on their performance as a function of energy dissipation per unit mass (like many other process equipment). The boundaries of device volume are often difficult to delineate. It is therefore recommended to compare the performance of different devices based on total energy consumption per kg of processed material, which is $\left(\frac{n\Delta P}{\rho}\right)$ J/kg. This can be applied for batch as well as continuous systems. An even better comparison strategy is to design different configurations and sizes of devices in such a way that all the devices exhibit similar pressure drop versus flow rate characteristics. For example, Simpson and Ranade [2] designed five orifice, venturi, and vortex-based devices that exhibit similar pressure drop versus flow characteristics and compared their performance.

We hope that the discussion of experimental set-up, identifying inception of cavitation, and evaluating cavitation devices and the processes based on them will be useful for readers to plan and execute experiments. The interpretation of experimental data using the computational models and their use for further optimization and scale-up are discussed in the next chapter (Chapter 5).

4.4 Conclusions

This chapter discussed the experimental characterization of the different types of hydrodynamic cavitation devices and their applications. Key aspects of designing the experimental set-up and selecting its various components are discussed in Section 4.1. Specific recommendations included in this section will be useful for designing a new experimental set-up or improving an existing set-up. Sensors, experimental set-up, and methods for identifying inception of cavitation are discussed in Section 4.2. The general methodology for quantifying overall performance of hydrodynamic cavitation devices and processes based on those is discussed in Section 4.3. The cited references and specific recommendations included here will be useful for experimentally characterizing hydrodynamic cavitation devices.

Notations

C	Concentration (ppm)
C_0	Initial concentration (ppm)
C_p	Specific heat capacity (J/kg K)
Eu	Euler number (−)
M	Mass (kg)
n	Number of passes through cavitation device (−)
ΔP	Pressure drop (Pa)
Q	Total flow through pump (m^3/s)
t	Time (s)
T	Temperature (K)
V_c	Characteristic velocity (m/s)
ZCR	Zero-crossing rate (s^{-1})
ρ	Density of liquid (kg/m^3)

References

1 Sarvothaman, V.P., Simpson, A., and Ranade, V.V. (2020). Comparison of hydrodynamic cavitation devices based on linear and swirling flows: degradation of dichloroaniline in water. *Industrial and Engineering Chemistry Research* 59 (30): 13841–13847.

2 Simpson, A. and Ranade, V.V. (2019). 110th Anniversary: comparison of cavitation devices based on linear and swirling flows: hydrodynamic characteristics. *Industrial and Engineering Chemistry Research* 58 (31): 14488–14509.

3 Ranade, V.V. and Bhandari, V.M. (2014). *Industrial Wastewater Treatment, Recycling and Reuse*. Butterworth-Heinemann.

4 Ranade, N.V., Sarvothaman, V., and Ranade, V.V. (2021). Acoustic analysis of vortex-based cavitation devices: inception and extent of cavitation. *Industrial and Engineering Chemistry Research* 60: 8255–8268.

5 Nagarajan, S. and Ranade, V.V. (2019). Pretreatment of lignocellulosic biomass using vortex-based devices for cavitation: influence on biomethane potential. *Industrial and Engineering Chemistry Research* 58 (35): 15975–15988.

6 Mahulkar, A.V., Bapat, P.S., Pandit, A.B., and Lewis, F.M. (2008). Steam bubble cavitation. *AIChE Journal* 54 (7): 1711–1724.

7 Patil, P.N. and Gogate, P.R. (2015). Degradation of dichlorvos using hybrid advanced oxidation processes based on ultrasound. *Journal of Water Process Engineering* 8: e58–e65.

8 Kumar, M.S., Sonawane, S.H., Bhanvase, B.A., and Bethi, B. (2018). Treatment of ternary dye wastewater by hydrodynamic cavitation combined with other advanced oxidation processes (AOP's). *Journal of Water Process Engineering* 23: 250–256.

9 Thanekar, P. and Gogate, P.R. (2020). Improved processes involving hydrodynamic cavitation and oxidants for treatment of real industrial effluent. *Separation and Purification Technology* 239: 116563.

10 Sarvothaman, V.P., Nagarajan, S., and Ranade, V.V. (2018). Treatment of solvent-contaminated water using vortex-based cavitation: influence of operating pressure drop, temperature, aeration, and reactor scale. *Industrial and Engineering Chemistry Research* 57 (28): 9292–9304.

11 Rajoriya, S., Bargole, S., and Saharan, V.K. (2017). Degradation of reactive blue 13 using hydrodynamic cavitation: effect of geometrical parameters and different oxidizing additives. *Ultrasonics Sonochemistry* 37: 192–202.

12 Brennen, C.E. (2011). *Hydrodynamics of Pumps*. Cambridge University Press.

13 Lobanoff, V.S. and Ross, R.R. (2013). *Centrifugal Pumps: Design and Application*. Elsevier.

14 Volk, M. (2013). *Pump Characteristics and Applications*. CRC Press.

15 Thaker, A.H. and Ranade, V.V. Drop breakage in a single pass through vortex based cavitation device: experiments and modelling. *AIChE Journal* e17512.

16 Krishtop, I., German, V., Gusak, A. et al. (2014). *Numerical Approach for Simulation of Fluid Flow in Torque Flow Pumps*. Trans Tech Publications.

17 Zupanc, M., Kosjek, T., Petkovšek, M. et al. (2013). Removal of pharmaceuticals from wastewater by biological processes, hydrodynamic cavitation and UV treatment. *Ultrasonics Sonochemistry* 20 (4): 1104–1112.

18 De-Nasri, S.J., Nagarajan, S., Robertson, P.K.J., and Ranade, V.V. (2021). Quantification of hydroxyl radicals in photocatalysis and acoustic cavitation: utility of coumarin as a chemical probe. *Chemical Engineering Journal* 420: 127560.

19 De-Nasri, S., Nagarajan, S., PKJ, R., and Ranade, V.V. (ed.) (2018). *Cavitation and Immobilised Photo-Catalysis for Effluent Treatment: A Comparative Study of Individual and Combined Operations*. AIChE.

20 Westbroek, P. and Temmerman, E. (2001). In line measurement of chemical oxygen demand by means of multipulse amperometry at a rotating Pt ring—Pt/PbO2 disc electrode. *Analytica Chimica Acta* 437 (1): 95–105.

21 Li, J., Luo, G., He, L. et al. (2018). Analytical approaches for determining chemical oxygen demand in water bodies: a review. *Critical Reviews in Analytical Chemistry* 48 (1): 47–65.

22 Raman, V. and Abbas, A. (2008). Experimental investigations on ultrasound mediated particle breakage. *Ultrasonics Sonochemistry* 15 (1): 55–64.

23 John, J.J., Kuhn, S., Braeken, L., and Van Gerven, T. (2018). Effect of fluid properties on ultrasound assisted liquid-liquid extraction in a microchannel. *Ultrasonics Sonochemistry* 42: 68–75.

24 Wang, N., Tan, F., Zhao, Y. et al. (2016). Optofluidic UV-Vis spectrophotometer for online monitoring of photocatalytic reactions. *Scientific Reports* 6 (1): 1–8.

25 Stadler, E., Eibel, A., Fast, D. et al. (2018). A versatile method for the determination of photochemical quantum yields via online UV-Vis spectroscopy. *Photochemical and Photobiological Sciences* 17 (5): 660–669.

26 Suryawanshi, N.B., Bhandari, V.M., Sorokhaibam, L.G., and Ranade, V.V. (2017). Developing techno-economically sustainable methodologies for deep desulfurization using hydrodynamic cavitation. *Fuel* 210: 482–490.

27 Amin, L.P., Gogate, P.R., Burgess, A.E., and Bremner, D.H. (2010). Optimization of a hydrodynamic cavitation reactor using salicylic acid dosimetry. *Chemical Engineering Journal* 156 (1): 165–169.

28 Wang, X. and Zhang, Y. (2009). Degradation of alachlor in aqueous solution by using hydrodynamic cavitation. *Journal of Hazardous Materials* 161 (1): 202–207.

29 Hilares, R.T., Kamoei, D.V., Ahmed, M.A. et al. (2018). A new approach for bioethanol production from sugarcane bagasse using hydrodynamic cavitation assisted-pretreatment and column reactors. *Ultrasonics Sonochemistry* 43: 219–226.

30 Soyama, H. and Hoshino, J. (2016). Enhancing the aggressive intensity of hydrodynamic cavitation through a Venturi tube by increasing the pressure in the region where the bubbles collapse. *AIP Advances* 6 (4): 045113.

31 Thoroddsen, S.T., Etoh, T.G., and Takehara, K. (2008). High-speed imaging of drops and bubbles. *Annual Review of Fluid Mechanics* 40: 257–285.

32 Goldstein, R. (2017). *Fluid Mechanics Measurements*. Routledge.

33 Adrian, L., Adrian, R.J., and Westerweel, J. (2011). *Particle Image Velocimetry*. Cambridge University Press.

34 Iida, Y., Yasui, K., Tuziuti, T., and Sivakumar, M. (2005). Sonochemistry and its dosimetry. *Microchemical Journal* 80 (2): 159–164.

35 Sutkar, V.S. and Gogate, P.R. (2009). Design aspects of sonochemical reactors: techniques for understanding cavitational activity distribution and effect of operating parameters. *Chemical Engineering Journal* 155 (1–2): 26–36.

36 Pawar, S.K., Mahulkar, A.V., Pandit, A.B. et al. (2017). Sonochemical effect induced by hydrodynamic cavitation: comparison of venturi/orifice flow geometries. *AIChE Journal* 63 (10): 4705–4716.

37 Morison, K.R. and Hutchinson, C.A. (2009). Limitations of the Weissler reaction as a model reaction for measuring the efficiency of hydrodynamic cavitation. *Ultrasonics Sonochemistry* 16 (1): 176–183.

38 Arakeri, V.H. (1979). Cavitation inception. *Proceedings of the Indian Academy of Sciences Section C: Engineering Sciences* 2 (2): 149–177.

39 Rood, E.P. (1991). Mechanisms of cavitation inception. *Journal of Fluids Engineering* 113 (2): 163–175.

40 Sakamoto, N. and Kamiirisa, H. (2018). Prediction of near field propeller cavitation noise by viscous CFD with semi-empirical approach and its validation in model and full scale. *Ocean Engineering* 168: 41–59.

41 Mancuso, G. (2018). Experimental and numerical investigation on performance of a swirling jet reactor. *Ultrasonics Sonochemistry* 49: 241–248.

42 Peng, X., Wang, B., Li, H. et al. (2017). Generation of abnormal acoustic noise: singing of a cavitating tip vortex. *Physical Review Fluids* 2 (5): 053602.

43 Ceccio, S.L. and Mäkiharju, S.A. (2017). Experimental methods for the study of hydrodynamic cavitation. In: *Cavitation Instabilities and Rotordynamic Effects in Turbopumps and Hydroturbines* (ed. L. d'Agostino and M.V. Salvetti), 35–64. Springer.

44 De, M.K. and Hammitt, F.G. (1982). New method for monitoring and correlating cavitation noise to erosion capability. *Journal of Fluids Engineering* 104 (4): 434–441.

45 He, Y. and Liu, Y. (2011). Experimental research into time–frequency characteristics of cavitation noise using wavelet scalogram. *Applied Acoustics* 72 (10): 721–731.

46 Sarvothaman, V.P. (2020). Hydrodynamic cavitation for effluent treatment: using vortex-based cavitation devices. Doctoral dissertation. Queen's University Belfast.

47 Brennen, C.E. (2014). *Cavitation and Bubble Dynamics*. Cambridge University Press.

48 Quan, K.M., Avvaru, B., and Pandit, A.B. (2011). Measurement and interpretation of cavitation noise in a hybrid hydrodynamic cavitating device. *AIChE Journal* 57 (4): 861–871.

49 Avvaru, B. and Pandit, A.B. (2009). Oscillating bubble concentration and its size distribution using acoustic emission spectra. *Ultrasonics Sonochemistry* 16 (1): 105–115.

50 De Giorgi, M.G., Ficarella, A., and Tarantino, M. (2013). Evaluating cavitation regimes in an internal orifice at different temperatures using frequency analysis and visualization. *International Journal of Heat and Fluid Flow* 39: 160–172.

51 Tabrizi, A.B. and Wu, B. (ed.) (2018). *Noise Investigation of a Cavitating Orifice: Use of cfd Simulation and ffowcs Williams–Hawkings (fw-h) Formulation to Gain Insight into Acoustics Condition*. IOP Publishing.

52 Park, J. and Seong, W. (2017). Experimental study on the effect of number of bubble occurrences on tip vortex cavitation noise scaling law. *Journal of Fluids Engineering* 139 (6).

53 Potočnik, P., Olmos, B., Vodopivec, L. et al. (2021). Condition classification of heating systems valves based on acoustic features and machine learning. *Applied Acoustics* 174: 107736.

54 Martin, C.S., Medlarz, H., Wiggert, D.C., and Brennen, C. (1981). Cavitation inception in spool valves. *Journal of Fluids Engineering* 103 (4): 564–575.

55 Ranade, V.V., Prasad Sarvothaman, V., Simpson, A., and Nagarajan, S. (2021). Scale-up of vortex based hydrodynamic cavitation devices: a case of degradation of di-chloro aniline in water. *Ultrasonics Sonochemistry* 70: 105295.

5

Modeling of Hydrodynamic Cavitation-Based Processes

5.1 Introduction

The physicochemical effects of cavitation are very complex and system-specific (see Figure 5.1). Several examples of different types of applications are discussed in Chapter 1 and in the later chapters of this book. These applications are based on utilizing the generated intense shear, localized hot spots, and hydroxyl radicals by hydrodynamic cavitation (HC) for realizing the desired physicochemical transformations. Most of these transformations involve either physical transformations like droplet/particle breakage or chemical transformations like thermal cracking, pyrolysis, and oxidation of organic pollutants or hydrolysis. In some applications, both physical and chemical transformations occur simultaneously and the interaction of these controls the overall performance (for example, disinfection of water or pre-treatment of lignocellulosic biomass).

These physicochemical transformations realized by cavitation span a very wide range of spatiotemporal scales: from molecular length and time scales to micron scales of cavity, up to reactor/cavitation device-scale processes. Considering the wide range of spatiotemporal scales and the complex underlying physics involving multiphase flows with phase change and chemical reactions, rigorous models fully based on first principles are not yet available. The chemists using cavitation for intensifying various reactions mainly focus on observed performance toward intended application and often use empirical and simplified models to describe the overall process (see for example, Pollet and Ashokkumar [1]). Mathematicians and physicists focus on micro-scale processes occurring in cavitation and develop detailed computational models of those micro-scale processes (see for example, Rasthofer et al. [2]). These models, though rigorous in investigating micro-scale processes under idealized conditions, are not yet directly useful to understand cavitation reactor-scale processes involving significant chemical reactions. Engineers, in general, and chemical engineers, in particular, have attempted to develop models spanning multiple scales and have tried to relate flow models with physicochemical effects (for example, Pandit et al. [3], Gogate and Pandit [4], Ma et al. [5], Sarvothaman et al. [6]). An overall approach and the relevant computational models needed for industrially relevant HC applications are still not readily available. An attempt is made in this chapter to summarize different modeling approaches for simulating relevant HC applications.

Hydrodynamic Cavitation: Devices, Design, and Applications, First Edition.
Vivek V. Ranade, Vinay M. Bhandari, Sanjay Nagarajan, Varaha P. Sarvothaman and Alister T. Simpson.

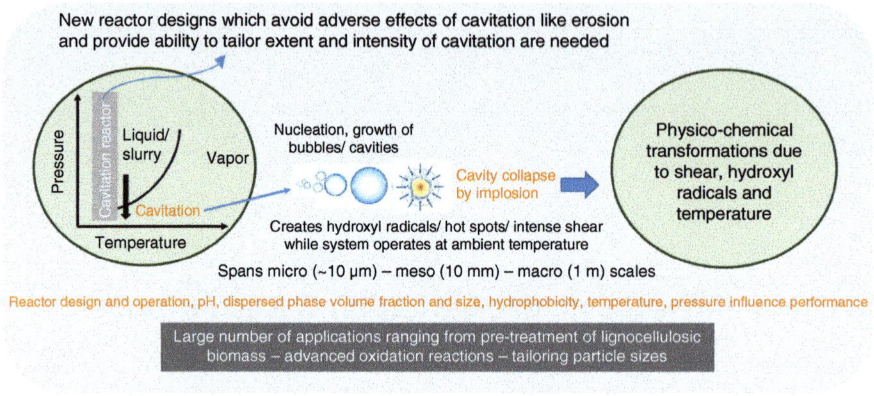

Figure 5.1 Applications of HC – presence of a wide range of scales.

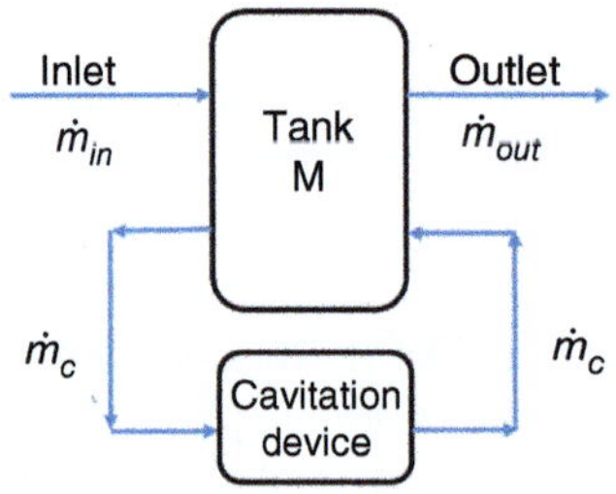

Figure 5.2 Schematic of a typical HC-based application process.

A typical HC-based application process may be represented schematically as Figure 5.2. A batch process may be represented by a complete recycle without any net inlet or outlet streams ($\dot{m}_{in} = \dot{m}_{out} = 0$).

In all such application processes, various fluidic devices (discussed in earlier chapters) are used to generate cavitation. Various physicochemical transformation processes occurring in and around collapsing cavities may be broadly allocated to three regions, as shown schematically in Figure 5.3.

Several studies have been carried out to investigate different aspects of such complex micro-scale processes occurring in and around a collapsing cavity. For example, high temperature (~5000 K), high pressures (~1000 atm), and formation of hydroxyl radicals are reported by Suslick and coworkers [7–9]. The generated radicals react with other gaseous species within the cavity (see for example, Théron et al. [10]) or diffuse into the surrounding liquid and react either in the interfacial zone or in the bulk zone (for example, Flint and Suslick [9]; Ashokkumar et al. [11]). If the liquid contains volatile species, these volatile species will be present in the gaseous cavity. In such a case, either pyrolysis or thermal cracking may occur within the cavity (for example, Ashokkumar et al. [11]). It can thus be seen that various chemical transformations in and around the cavity will be influenced by the characteristics of the chemical species in the considered system (volatility, hydrophobicity, molecular weight, reactivity toward radicals, presence of radical scavengers, propensity to cracking, and so on) as well as variety of other macroscopic factors such as the

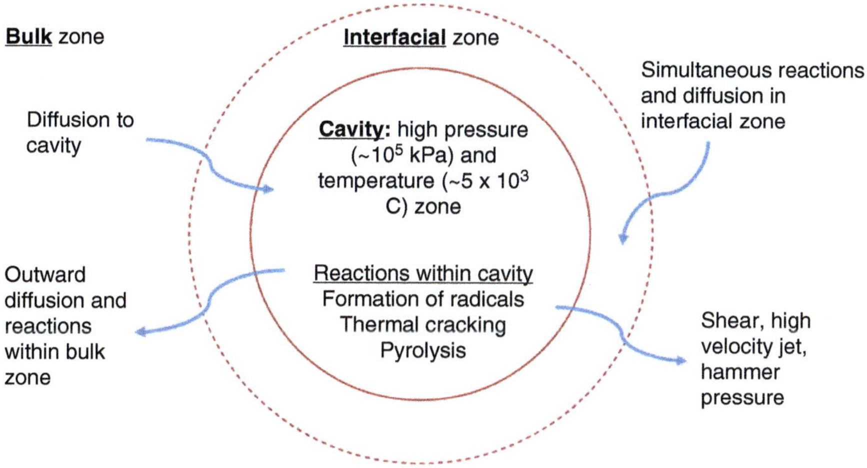

Figure 5.3 Schematic of micro-scale processes occurring in and around a collapsing cavity.

design of the cavitation device and the operating conditions (pH, temperature, pressure drop, and so on).

Most of the studies cited as examples here were published more than two decades ago. In last two decades, several attempts have been made to develop better understanding of micro-scale as well as meso- and macro-scale processes occurring in HC-based applications. However, despite these significant research efforts, comprehensive models for making quantitative predictions of HC and its application are still elusive. Computational models addressing either micro-scale or device-scale processes are reasonably well developed. However, quantitative understanding of the interaction of processes occurring on these widely different scales is not yet adequately established. More often than not, many studies published, even in recent years, still resort to empirical or semi-empirical models. We have therefore included a discussion on various approaches, from empirical to physics-based multi-scale modeling of HC-based transformation processes in this chapter. The prevailing modeling approaches may be broadly grouped into two categories: data-driven or empirical models and physics-based models, as shown schematically in Figure 5.4. There are several variants within each category depicted in Figure 5.4. However, an attempt is made here to capture at least the key aspects of these modeling approaches.

This chapter is organized to discuss modeling of HC-based processes, following the structure shown in Figure 5.2. In the next section, three different types of empirical modeling approaches are discussed. Section 5.3 presents physics-based models, mainly comprising single-cavity models and multi-scale models. An illustrative application or two for each modeling approach is included as well. The presented approaches, model equations, and application examples will be useful to develop effective computational models for design and optimization of a wide range of HC-based processes.

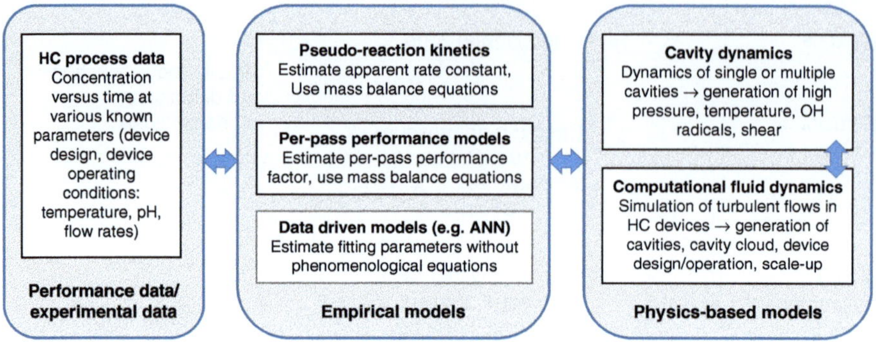

Figure 5.4 Various approaches used for modeling HC systems.

5.2 Empirical Models

Models attempting to describe experimental or performance data using fitted parameters are discussed in this section.

5.2.1 Pseudo-reaction Kinetics Model

In this class of models, the physicochemical transformation of entity S to product P – that may be a reduction in particle size or a decrease in concentration – is modeled as a first-order process with a first-order rate constant k:

$$S \xrightarrow{k} P \tag{R5.1}$$

Several researchers have used this simple model to describe a variety of transformation processes based on ultrasonic and HC. Some of the key studies based on this approach are listed in Table 5.1.

The overall model equations can be formulated for any representative application shown in Figure 5.1 as:

$$\frac{dM}{dt} = \dot{m}_{in} - \dot{m}_{out} \tag{5.1}$$

$$\frac{d(MC)}{dt} = \dot{m}_{in}C_{in} - \dot{m}_{out}C - M\,k\,C \tag{5.2}$$

where k is the effective first-order rate constant, M is mass of liquid in the holding tank, m is mass flow rate, and C is the desired property of interest. The subscripts *in* and *out* denote inlet and outlet, respectively.

For a batch system, the net flow rates are zero and the mass of liquid in the tank, M, is constant. The time variation of outlet concentration can therefore be written as:

$$C = C_o e^{-kt} \tag{5.3}$$

Batch experiments were often interpreted with this model, and an effective first-order rate constant may be obtained by fitting the experimental data of the desired transformation with respect to time using Eq. (5.3). Once the effective rate

Table 5.1 Studies in HC using pseudo-first order kinetics.

Device used	Pollutant (or microorganism) used	Pressure drop range (kPa)	Optimal pressure (kPa)	Rate constant (min^{-1})	Availability of flow rate data	Reference
Slit venturi	Diclofenac sodium	200–400	300	0.0021	Yes	[12]
Slit venturi	Dimethoate	400	400	0.0025	Yes	[13]
Venturi	Ibuprofen	200–550	450	0.0928	Yes	[14]
Circular venturi	Imidacloprid	500–2000	1750	0.0271	Yes	[15]
Circular venturi	Methomyl	200–700	300	0.0028	Yes	[16]
Circular venturi	Orange-4	300–800	500	0.0049	Yes	[17]
Venturi	*Escherichia coli*	100–250	100	0.0180	No	[18]
Venturi	Bisphenol-A	300–700	500	0.0010	Yes	[19]
Venturi	Rhodamine B	100–500	400	0.0043	No	[20]

constant is obtained, the steady-state performance of a continuous system may be obtained as:

$$\dot{m}_{out} = \dot{m}_{in} \text{ and } C = \frac{C_o}{1 + k\tau} \quad \text{where } \tau = \frac{M}{\dot{m}_{in}} \tag{5.4}$$

This approach has been extensively used for interpreting laboratory-scale experimental data on pollutant degradation using ultrasonic or HC. For example, in a recent study of degradation of Rhodamine-B, Ye et al. have reported the first-order rate constant for acoustic and HC [20]. Various other studies in the past decade have interpreted their data by this approach [16, 17, 21, 22].

While this approach is widely used, it has specific disadvantages for interpreting HC-based process performance. This was illustrated by Ranade et al. [23] who carried out experiments on degradation of dichloroaniline (DCA) using the same HC device with two different volumes of holding tank. The observed DCA concentration profiles as a function of time are shown in Figure 5.5.

It can be seen from Figure 5.5 that the concentration profiles with respect to time for two different liquid volumes in the holding tank differ from each other – leading to two different values of effective degradation rate constant ($k_{app} = 1.06 \times 10^{-3}$/min for 0.005 m^3 and $k_{app} = 2.41 \times 10^{-4}$/min for 0.024 m^3). However, it should be noted that the performance of the HC device operated at the same flow rate is not expected to be a function of the liquid volume in the holding tank.

The effective rate constant therefore is not an appropriate measure to quantify performance of a HC device. This needs to be borne in mind while comparing the effective degradation rates reported by different researchers. The per-pass performance factor approach, discussed in the next sub-section, avoids this issue and therefore is recommended over the effective rate constant approach, particularly for HC applications.

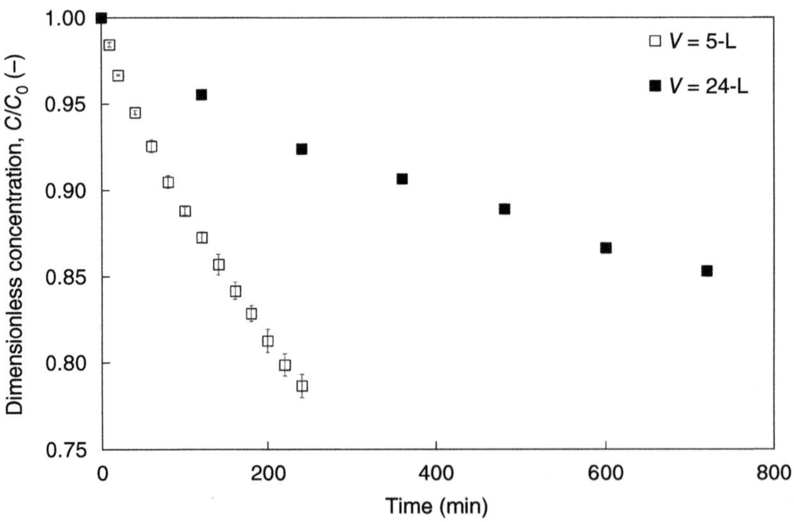

Figure 5.5 Influence of liquid volume on degradation performance, HC device – D3, ΔP: 280 kPa. Liquid volume in the holding tank, V: open square – 0.005 m^3, filled square – 0.024 m^3 (Source: From [23]/with permission of Elsevier).

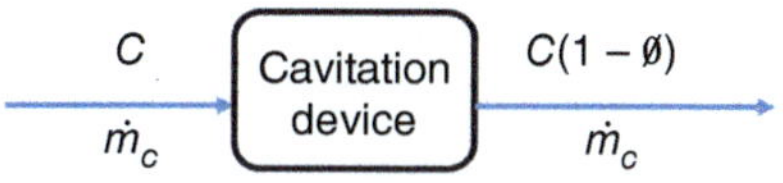

Figure 5.6 Per-pass transformation in a cavitation device.

5.2.2 Per-pass Performance Model

In this approach, an empirical parameter Ø is used to represent the per-pass performance of the HC device. The overall transformation process occurring within a single pass through HC device is therefore represented as shown in Figure 5.6.

The overall model equations can therefore be written as:

$$\frac{d(MC)}{dt} = \dot{m}_{in}C_{in} - \dot{m}_{out}C - \dot{m}_c\, C\emptyset \tag{5.5}$$

The per-pass performance factor Ø represents the extent of physicochemical transformation influencing the key characteristic of interest (C). For example, C could be drop size for drop breakage applications (like producing emulsions involving physical processes like drop breakage). In applications like wastewater treatment, C could be the concentration of a specific pollutant or the chemical oxygen demand (COD) of the effluent. The value of Ø may depend on a variety of complex micro-, meso-, and macro-scale processes, as discussed earlier, and may be obtained from the experimental data.

The Eq. (5.5) may be transformed in terms of number of passes instead of time, which is more appropriate for modeling the HC-based applications represented by Figure 5.2. The number of passes and time are related as:

$$n = \frac{\dot{m}_c\, t}{M} = \beta t \tag{5.6}$$

where β is a ratio of mass flow rate through cavitation device and mass in the holding tank ($\dot{m}_c/M$, s^{-1}). With this transformation, and assuming $\dot{m}_{in} = \dot{m}_{out}$ and M as constant over the operation time, we may write:

$$\frac{dC}{dn} = \frac{(C_{in} - C)}{\beta\,\tau} - C\emptyset \tag{5.7}$$

where τ is residence time based on net mass flow rate ($M/\dot{m}_{in}$). For a batch system ($\dot{m}_{in} = 0$), the overall transformation process may be represented in terms of the number of passes and the per-pass performance factor as:

$$C = C_0\, e^{-n\emptyset} \tag{5.8}$$

For a continuous mode of operation at a steady state, the performance equation becomes:

$$C = \frac{C_{in}}{1 + \beta\tau\emptyset} \tag{5.9}$$

Flow rate through the cavitation device can be estimated from the desired extent of transformation (C/C_{in}) and the system-specific value of the per-pass performance factor (for the given device-operating conditions) as:

$$\beta\tau = \frac{\dot{m}_c}{\dot{m}_{in}} = \frac{1}{\emptyset}\left(\frac{C_{in}}{C} - 1\right) \tag{5.10}$$

Thus, if we desire 50% conversion ($C/C_{in} = 0.5$) and the per-pass performance factor is 0.01, we need the mass flow rate through the cavitation device to be 100 times more than the net flow rate.

The experimental data obtained with the batch system and Eq. (5.8) can be used to estimate the per-pass performance factor. The intrinsic assumption in deriving Eqs. (5.8)–(5.10) is that the per-pass performance factor is constant throughout the overall process. This may not be true particularly when modeling degradation of organic pollutants (which is one of the widely used applications of HC). In such cases, when the considered organic pollutant has a complex molecular structure, the reactivity of the intermediate products formed may be different from the original pollutant. This may cause a variation of the effective per-pass performance factor with the number of passes. This will be especially relevant when the radical production rate is relatively low and the observed change in the pollutant concentration is significant. To account for such variation in the per-pass performance factor, Ranade et al. [23] suggested the following formulation by considering a non-constant per-pass performance factor. In the example discussed by Ranade et al. [23], the overall pollutant degradation was assumed to be predominantly caused by hydroxyl radicals and the overall degradation process occurring in a HC device was represented as:

$$\text{Cavity collapse} \rightarrow \text{OH}^{\cdot}\text{and other radicals (R}^{\cdot}) \tag{R5.2}$$

$$\text{R}^{\cdot} + \text{pollutant} \rightarrow \text{Intermediates} \tag{R5.3}$$

$$\text{R}^{\cdot} + \text{Intermediates} \rightarrow CO_2 + H_2O \tag{R5.4}$$

Because of the possibility of differing reactivity of intermediate products, the per-pass performance (pollutant degradation in this particular case) is assumed to be a function of the number of passes through a cavitation device as [23]:

$$\emptyset = \emptyset_0 f(n) = \frac{\emptyset_0}{(1 + \gamma\, \emptyset_0\, n)^y} \tag{5.11}$$

where $\emptyset_0$ is the initial per-pass degradation factor and $f(n)$ is a correction factor to account for the varying $\emptyset$ over the number of passes (as the pollutant degrades to mineralization via multiple steps). The value of parameter y as zero will reduce this function to unity, and the model will reduce to the simple, constant per-pass performance factor identical to Eqs. (5.8)–(5.10). If the intermediates formed are more reactive than the original pollutant, the γ in Eq. (5.11) will have a positive sign, leading to a reduction in the per-pass performance factor with the subsequent number of passes. If the intermediates formed are less reactive than the original pollutant, the γ in Eq. (5.11) may have a negative sign, leading to an increase in the per-pass degradation factor with the subsequent number of passes. The presence of $\emptyset_0$ in the denominator indicates that the larger the value of the per-pass degradation factor, the higher the impact of different reactivity of intermediates. The empirical parameter y controls the extent of influence of the different reactivity of intermediates.

By using Eqs. (5.11) and (5.5), one can derive performance equations for various special cases:

For $y = 1$:

For the batch system:

$$C_p = C_{p0}\,(1 - n\emptyset_0) \text{ if } \gamma = -1 \tag{5.12}$$

$$C_p = \frac{C_{p0}}{(1 + n\emptyset_0)} \text{ if } \gamma = +1 \tag{5.13}$$

where C_p is a concentration of pollutant p which produces intermediates with different reactivity.

For continuous system:

$$C_p = \frac{C_{pin}}{1 + \frac{\beta\tau\emptyset_0}{(1-n\emptyset_0)}} \text{ if } \gamma = -1 \tag{5.14}$$

$$C_p = \frac{C_{pin}}{1 + \frac{\beta\tau\emptyset_0}{(1+n\emptyset_0)}} \text{ if } \gamma = +1 \tag{5.15}$$

Flow rate through the cavitation device can be estimated from the desired extent of transformation (C_p/C_{pin}) and the system-specific value of the per-pass performance factor (for the given device-operating conditions) as:

$$\beta\tau = \frac{\dot{m}_c}{\dot{m}_{in}} = \frac{(1 - n\emptyset_0)}{\emptyset_0}\left(\frac{C_{pin}}{C_p} - 1\right) \text{ if } \gamma = -1 \tag{5.16}$$

$$\beta\tau = \frac{\dot{m}_c}{\dot{m}_{in}} = \frac{(1 + n\emptyset_0)}{\emptyset_0}\left(\frac{C_{pin}}{C_p} - 1\right) \text{ if } \gamma = +1 \tag{5.17}$$

For $y \neq 1$:

For the batch system:

$$\ln\left(\frac{C_p}{C_{p0}}\right) = -\emptyset_0\left[\frac{(1 + \gamma\,\emptyset_0\,n)^{y-1} - 1}{\emptyset_0\,(y - 1)\,(1 + \gamma\,\emptyset_0\,n)^{y-1}}\right] \tag{5.18}$$

For continuous systems:

$$C_p = \frac{C_{pin}}{1 + \frac{\beta\tau\emptyset_0}{(1-n\emptyset_0)^y}} \text{ if } \gamma = -1 \tag{5.19}$$

$$C_p = \frac{C_{pin}}{1 + \frac{\beta\tau\emptyset_0}{(1+n\emptyset_0)^y}} \text{ if } \gamma = +1 \tag{5.20}$$

$$\beta\tau = \frac{\dot{m}_c}{\dot{m}_{in}} = \frac{(1 - n\emptyset_0)^y}{\emptyset_0}\left(\frac{C_{pin}}{C_p} - 1\right) \text{ if } \gamma = -1 \tag{5.21}$$

$$\beta\tau = \frac{\dot{m}_c}{\dot{m}_{in}} = \frac{(1 + n\emptyset_0)^y}{\emptyset_0}\left(\frac{C_{pin}}{C_p} - 1\right) \text{ if } \gamma = +1 \tag{5.22}$$

In order to illustrate the possibility of non-zero y, the results presented by Ranade et al. [23] are included here. They report the degradation of DCA using HC operated

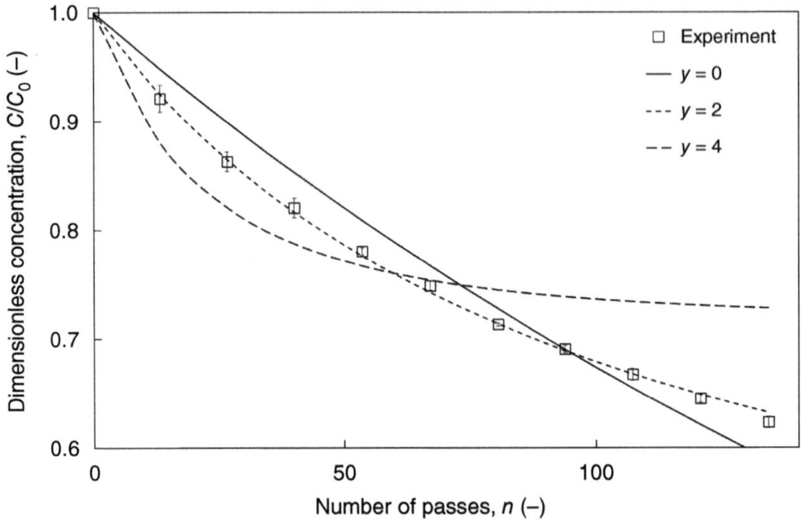

Figure 5.7 Modeling of concentration data with varying per-pass performance factors (Source: Experimental data: From [23]/with permission of Elsevier; Curves: Prediction with Eq. (5.18)).

in a batch mode. The observed degradation performance was described by $y = 2$ (see Figure 5.7).

The approach may be extended to cases where other mechanisms like pyrolysis or intense shear are predominant, following the methodology presented here. Unlike the pseudo-kinetic models (discussed in Section 5.2.1), the per-pass performance models allow one to compare various HC devices on a uniform basis. The possibility of estimating the per-pass performance for non-isothermal operation is developed and can be referred to in the work of Sarvothaman et al. [24]. The approach requires experimental data to obtain a quantitative estimation of the per-pass performance factor. However, it opens up a possibility of developing physics-based models for estimating per-pass performance factors. It also allows a possibility of developing correlations between the per-pass performance factor and key operating parameters, opening up an opportunity for process optimization.

5.2.3 Data-Driven Models

Considering that the models discussed in Sections 5.2.1 and 5.2.2 are empirical and can be used only for describing the experimental data rather than predicting it, other purely data-driven approaches can also be used to describe the performance of HC-based processes. While several data-driven modeling formalisms are available, artificial neural networks (ANNs) appear to be suitable for modeling HC-based processes (see for example, the recent application by Ranade et al. [25]).

ANNs are capable of describing arbitrary functions [26, 27] and are often useful for cases in which it is difficult to establish rigorous phenomenological models such as HC. ANN models may involve complex, recurrent, and multi-layer architectures requiring very large data sets relating input and output variables [28]. However, in many applications of HC, the experiments are complex, time consuming, and expensive, which naturally imposes constraints on the amount of data which could be collected. The limited availability of experimental data often leads to a common error of overfitting, where the ANN model represents the training data set so well that it captures relationships that do not exist. Such overfitting often hinders the objective of interpolation or extrapolation, which are the main goals of developing a model. It is therefore useful to develop as simple an ANN architecture as possible. Typically, ANN models with a single hidden layer which can approximate any function that contains a continuous mapping from one finite space to another [29] provide a good starting point. The maximum number of neurons in a hidden layer is determined by considering the number of weights to be decided in the process of training the network. The number of weights (N_{weight}) is related to the number of neurons (H) in the hidden layer for input size N_{in} and output size N_{out} as:

$$N_{weight} = N_{in} \times H + N_{out} \times H \tag{5.23}$$

The number of records available for training (N_{train}) decides the number of equations (N_{eq}) used to solve while finding the weights. N_{train} should be at least some multiples of the number of weights to be decided, N_{weight}. Otherwise, it causes overfitting, which means better fitting to training data but less generalization for unseen data. A general thumb rule is to take N_{train} as at least ten times the value of N_{weight}. Considering this thumb rule, Eq. (5.23), and assuming that 70% of the available data is used for training, the number of neurons to be used in the single hidden layer, H, can be calculated as:

$$H \leq \frac{0.7 \times N_{data}}{(N_{in} + N_{out}) \times 10} \tag{5.24}$$

Here, N_{data} is the number of experimental data points available.

Using this general guidance, several network models with a variable number of neurons can be created and tested for their performance. The usual practice is to develop different versions of ANN models comprising a different number of neurons in the hidden layer. The wellness of fit is usually evaluated using the coefficient of determination, R^2, and mean square error, MSE, of test data as:

$$R^2 = 1 - \frac{\sum_{i=1}^{N_{train}} \left(X_m^i - X_o^i\right)^2}{\sum_{i=1}^{N_{train}} \left(X_o^i - \overline{X}_0\right)^2} \tag{5.25}$$

$$MSE = \frac{\sum_{i=1}^{N_{test}} \left(X_m^i - X_o^i\right)^2}{N_{test}} \tag{5.26}$$

where X is a modeled variable, the superscript i denotes the ith record and the subscripts m and o denote modeled and observed data, respectively. N_{train} is the number of records used for training and N_{test} is the number of records used for testing. It can be seen that the value of R^2 becomes one when the modeled values exactly match the observed values. The appropriate number of neurons in the hidden layer are selected by ensuring that the coefficient of determination (R^2) is higher than 0.99

Table 5.2 Key parameters of the ANN models.

Network Property	BIOGAS model	WATER model
Input layer size	2	2
Output layer size	1	1
Hidden layer size	10	4
Hidden layer transfer function	"tansig"	"tansig"
Output layer transfer function	"purelin"	"purelin"
Data pre/post-processing	"mapminmax" [0 1]	"mapminmax" [0 1]
Training function	"trainlm"	"trainlm"
Performance function	"mse"	"mse"
Data division (ratio)	"dividerand" (70:15:15)	"dividerand" (70:15:15)
Train parameter goal	0.01	0.0

Source: From [25]/with permission of Elsevier.

and error in estimating the test data is minimum. Appropriate data conditioning and cleaning techniques are recommended before using the experimental data for training ANN models. It is often useful to utilize known information based on prior experience/intuitive understanding or from known phenomenology (for example, initial and boundary conditions, if known). Such known information may be used to artificially augment the data for reinforcing training of ANN models.

Application of this methodology was recently illustrated by Ranade et al. [25] for two very different applications of HC: pre-treatment of waste biomass for enhancing biogas yield and degradation of organic pollutants in water. ANN models with a single hidden layer were developed and trained using MATLAB tools. The available experimental data was divided into three parts as 70 : 15 : 15, to train, validate, and test for both the applications. The key parameters used for establishing the ANN models are listed in Table 5.2.

The developed ANN models were shown to be capable of interpolating the experimental data very well (see for example, Figure 5.8a and b reproduced from Ranade et al. [25]). The models were also able to extrapolate the data to some extent.

While the models discussed in this section have been shown to be able to describe experimental data, these are essentially empirical models and cannot really provide new insights beyond available experimental data. In contrast to these, models based on underlying physics may enhance our understanding and may be helpful for identifying optimization strategies, though they may not describe the experimental data as accurately as empirical models. Such physics-based models are discussed in the following section.

5.3 Physics-Based Models

The performance of any HC-based process critically depends on the number of generated cavities as well as the location and intensity of their collapse. The

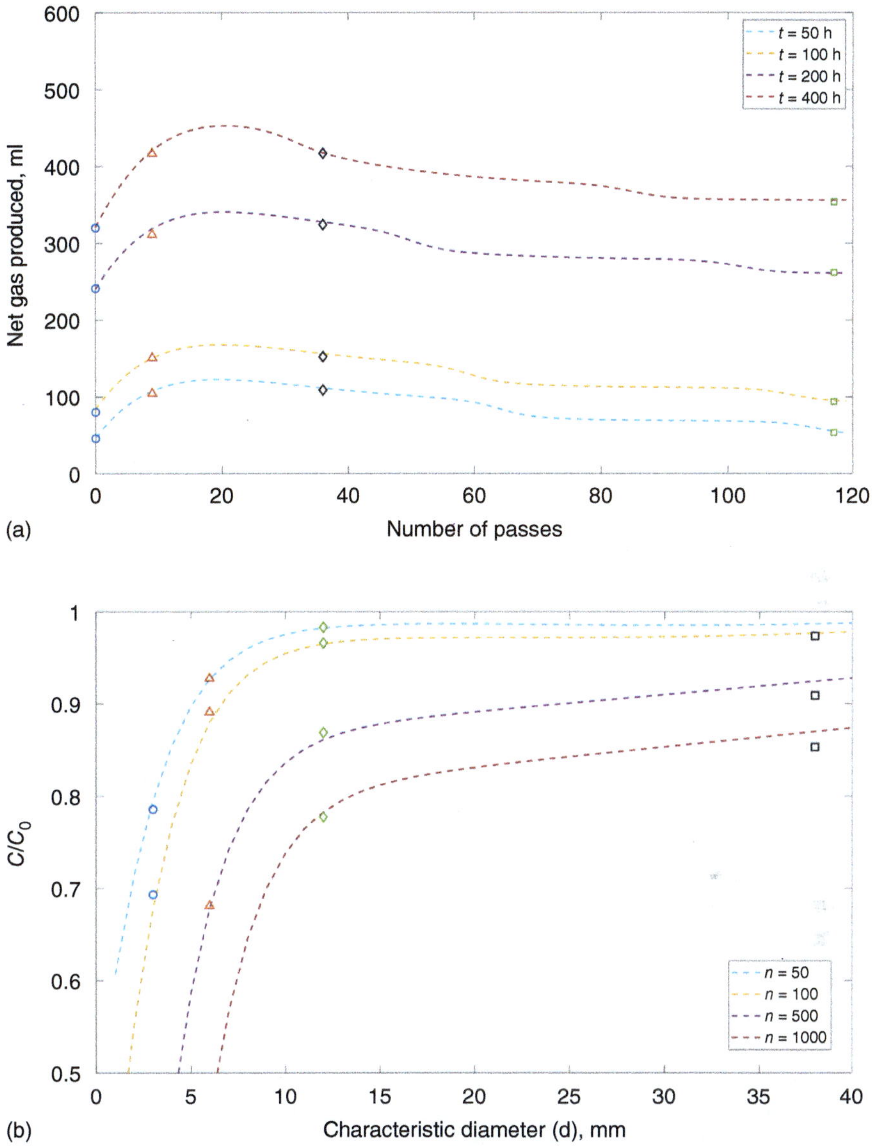

Figure 5.8 Application of ANN model for interpolation of results. (a) Influence of number of passes at four different time instants. (b) Influence of characteristic diameter of HC reactor at four values of number of passes (Source: From [25]/with permission of Elsevier).

process naturally spans over a wide range of spatiotemporal scales: from a single cavity (~10^{-6} m, ~10^{-4} s), to a cluster of cavities (10^{-3} m scale, ~10^{-3} s), to the cavitation reactor scale (~10^{0} m, ~10^{-1} s). The overall performance of HC-based processes requires thorough understanding of the device-scale flow processes, micro-scale processes of generation, growth and collapse of cavities, and interaction of device-scale flows with generated cavities (meso-scale processes). This overall picture is shown schematically in Figure 5.9.

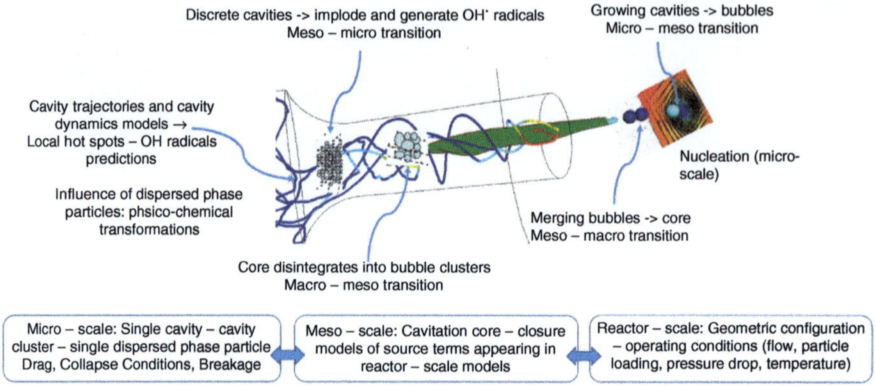

Figure 5.9 Multiple scales and multi-layer modeling of HC-based processes.

Modeling of device-scale fluid dynamics of various HC devices is discussed in Chapters 2 and 3. In this section, the detailed modeling of micro-scale processes is discussed. We also discuss how the understanding of micro-scale processes can be combined with the device-scale flow models to develop multi-scale and/or multi-layer models for simulating overall performance of HC-based processes. As discussed in Chapters 2 and 3, device-scale flow in HC devices can be modeled using multiphase computational fluid dynamics (CFD) models [30–32]. Micro-scale processes can be modeled based on the dynamics of a single or a group of cavities. Cavity dynamics models solve mass, momentum, and energy balance equations to estimate the radius, temperature, and pressure profiles within a single cavity in a varying pressure field [33–37]. The CFD and cavity dynamics models are mostly treated as separate models and on some occasions have been coupled [6, 33, 38] to describe cavitation device performance. The successes and limitations of these modeling approaches and their application will be discussed in the following two sub-sections.

5.3.1 Cavity Dynamics Models

The main objective of the cavity dynamics models is to estimate the physical and chemical effects caused by collapse of a cavity. Once these effects are estimated, the results may then be used to estimate the overall performance of HC-based application. Lord Rayleigh first proposed the single cavity/bubble dynamics equation [39, 40], known as the Rayleigh–Plesset (R-P) equation. This equation essentially represented the liquid-side momentum balance and allowed tracking the variation of bubble radius as a response to the fluctuating pressure field. The R-P equation was subsequently modified to include the effects of surface tension and viscosity (Gilmore's equation [41]). Keller–Miksis (KM) further modified it to account for large pressure oscillations [42]. The KM equation is typically used for simulating single-bubble dynamics [43–46]. All these equations consider a single spherically symmetric bubble. More sophisticated CFD-based models (using volume-of-fluid methods) which are capable of accounting for compressibility and asymmetry have been proposed recently [47]. However, these models usually are not effective in estimating the chemical effects of cavitation and are not discussed in this section.

The single-bubble dynamics model can be coupled with either the detailed chemical reaction kinetics [43, 48–50] or the Gibbs free-energy minimization method [6, 14, 38] to simulate generation of radicals and other chemical effects. The physical effects of cavitation are attributed to shockwaves [51, 52], shear [53], and jet formation [51, 54]. Supponen et al. [52] proposed a semi-empirical model to estimate energy dissipated as shockwaves. Peng et al. [54] used the single-bubble dynamics model to estimate breakage of rocks due to cavitating jets. However, all these approaches using the single-bubble model did not account for the complex physics of asymmetric collapse [24]. Recently, Pandit et al. [55] have presented a comprehensive bubble dynamics model to estimate the chemical and physical effects of cavitation. Their model is briefly discussed in the following section. The reader may refer to the original publication for more details.

5.3.1.1 Model Equations Governing Single-Bubble Dynamics

The Keller–Miksis (KM) equation may be used to simulate bubble dynamics:

$$\emptyset_1 R\frac{d^2R}{dt^2} + \frac{3}{2}\emptyset_2\left(\frac{dR}{dt}\right)^2 = \frac{\emptyset_3}{\rho_L}\left(P^B - P_L^{\infty}(t)\right) + \left(\frac{R}{\rho_L c_L}\right)\frac{dP^B}{dt} - \left(\frac{4\nu_L}{R}\right)\frac{dR}{dt} - \frac{2\sigma_L}{\rho_L R} \tag{5.27}$$

This second-order differential equation may be reduced to two first-order simultaneous ordinary differential equations as:

$$\frac{dR}{dt} = s \tag{5.28}$$

$$\frac{ds}{dt} = \left(\frac{\emptyset_3\left(P^B - P_L^{\infty}(t)\right)}{\emptyset_1 \rho_L R}\right) + \left(\frac{1}{\emptyset_1 \rho_L c_L}\right)\frac{dP^B}{dt} - \left(\frac{4\nu_L s}{\emptyset_1 R^2}\right) - \left(\frac{2\sigma_L}{\emptyset_1 \rho_L R^2}\right) - \left(\frac{3\emptyset_2 s^2}{2\emptyset_1 R}\right) \tag{5.29}$$

$$\emptyset_1 = 1 - \frac{s}{c_L} \quad \emptyset_2 = 1 - \frac{s}{3c_L} \quad \emptyset_3 = 1 + \frac{s}{c_L} \tag{5.30}$$

A gas-phase equation of state (EoS) is needed to solve the preceding equations. Previous studies have considered the Sauve–Redlich–Kwong EoS [48], the van der Waal's (VdW) EoS [43, 44] or some variations thereof [45, 46], and the ideal gas (IG) EoS [44]. It was previously reported that the choice of the EoS did not affect the basic physics of cavitation but only affected the final quantitative values [44, 48]. Here, the VdW EoS is used:

$$P^B = \frac{N_t^B k\, T^B}{\frac{4\pi}{3}R^3 - \frac{N_t^B}{N_A}b} - a\frac{\left(\frac{N_t^B}{N_A}\right)^2}{\left(\frac{4\pi}{3}R^3\right)^2} \tag{5.31}$$

$$a = \sum_i a_i {x_i^B}^2;\; b = \sum_i x_i^B b_i \tag{5.32}$$

The mass transfer of water molecules from the surrounding liquid into the cavitation bubble is usually modeled using a diffusion-limited mass-transfer model. The bubble's internal contents may be assumed to be perfectly mixed (homogeneous).

$$\frac{dN_i^B}{dt} = -4\,\pi\,R^2\,D_i\left(\frac{\partial C_i}{\partial r}\right)_{r=R} = 4\,\pi\,R^2\,D_i\left(\frac{C_i^i - C_i^B}{l_{diff}}\right) \tag{5.33}$$

$$l_{diff} = \min\left(\sqrt{\frac{RD_i}{|S|}},\ \frac{R}{\pi}\right) \tag{5.34}$$

The heat transfer processes during bubble dynamics may be modeled as [43]:

$$\frac{d}{dt}\left(\sum_{i=1}^{N_c} C_{v,i}N_i^B T^B\right) = \left[\left(-4\pi R^2 \lambda\left(\frac{\partial T}{\partial r}\right)_{r=R}\right) - \left(4\pi R^2 P^B \frac{dR}{dt}\right) + \left(\sum_{i=1}^{N_c} h_i \frac{dN_i^B}{dt}\right)\right] \tag{5.35}$$

$$\frac{dT^B}{dt} = \frac{1}{C^*_{v,mix}}\left[\left(-4\,\pi\,R^2\,\lambda\left(\frac{\partial T}{\partial r}\right)_{r=R}\right) - \left(4\,\pi\,R^2 P^B \frac{dR}{dt}\right) + \left(\sum_{i=1}^{N_c}(h_i - T^B C_{v,i})\frac{dN_i^B}{dt}\right)\right] \tag{5.36}$$

$$C^*_{v,mix} = \sum_{i=1}^{N_c}(C_{v,i} - \Psi_i)N_i^B \tag{5.37}$$

$$\Psi_i = k\sum_{j=1}^{N_\theta}\delta_{i,j}\frac{\left\{(\exp(\delta_{i,j}))^2\left(2\delta_{i,j} - \delta_{i,j}^2\right) - \left(\delta_{i,j}^2 + 2\delta_{i,j}\right)\exp(\delta_{i,j})\right\}}{(\exp(\delta_{i,j}) - 1)^3};\ \delta_{i,j} = \frac{\theta_{i,j}}{T^B} \tag{5.38}$$

$$\left(\frac{\partial T}{\partial r}\right)_{r=R} = \frac{T^\infty - T^B}{l_{th}} \tag{5.39}$$

$$l_{th} = \min\left(\sqrt{\frac{R\kappa}{S}},\ \frac{R}{\pi}\right) \tag{5.40}$$

The solution of these basic mass, momentum, and energy balance equations requires estimation of various physicochemical properties. The relevant equations for these are included in Appendix 5.A. The model equations may be solved numerically to track evolution of the bubble radius and the composition, pressure, and temperature of the bubble. This information may then be used to estimate chemical and physical effects of cavitation, briefly discussed in the following section.

5.3.1.2 Estimation of Generation of Hydroxyl Radicals

The extreme pressure and temperature conditions generated at collapse result in several reactions within the collapsing bubble which lead to the production of hydroxyl radicals from the water molecules. Previous studies have used either detailed chemical kinetics models [43, 48–50] or the minimization of Gibbs free energy (MGFE) method [6, 38, 45, 46] to estimate the rate of generation of radical species at bubble collapse as a function of various operating parameters. Both methods have uncertainties associated with them: the reliability of either kinetic parameters or thermodynamic parameters for the range of extreme conditions generated at bubble collapse. Considering these uncertainties, the MGFE method may be used to obtain a meaningful upper bound on the rate of hydroxyl radical production. The composition at equilibrium of all the possible species may be calculated by the minimizing the total Gibbs Free Energy based on the pressure, temperature, and the bubble's composition, calculated using the model equations discussed earlier. Various commercial or open source tools are available for implementing the MGFE method once the pressure, temperature, and composition are known (for example, ASPEN, FactSage Equilib tool).

5.3.1.3 Estimation of Hammer Pressure/Jet Velocity Due to Collapse

Bubble collapse in unbounded liquid releases intense energy, generated in the form of a spherical shock wave [14]. If the bubble collapses nearer to a wall or a surface, the collapse becomes asymmetric and a strong jet is formed. This high-velocity jet and the resulting intense shear results in various physical effects of cavitation such as erosion and particle breakage. Assuming absence of any heat or mass transfer during the collapse phase, the initial bubble energy may be taken as equal to the energy dissipated due to bubble collapse. The bubble internal energy may be estimated as:

$$E^B = \sum_{i=1}^{N_c} C_{v,i} N_i^B T^B \tag{5.41}$$

In the absence of viscous dissipation, sound loss, or light emission, the collapse energy is generally dissipated as a combination of a shockwave and a jet hammer. Hence, the corresponding energies may be related to the bubble internal energy at collapse as:

$$E_{C,0}^B = E_S + E_J \tag{5.42}$$

The shockwave pressure at the collapse location is very high (~10 000 bar) but falls very rapidly as the shockwave travels away from the bubble [56]. Hence, a bubble collapsing in an unbounded liquid does not contribute to the cavitation erosion as much as the jet hammer generated by a bubble collapsing near a wall. It was also shown by Supponen et al. [52] that as the bubble approaches the wall, the shockwave energy decreases to zero. Hence, it can be concluded that the collapse energy would entirely be dissipated as the jet hammer energy for a bubble close to the wall. In other words, the kinetic energy of the liquid which, in an unbounded liquid, would typically go into the compression of the bubble, due to non-spherical collapse dissipates itself as the jet hammer energy. The jet hammer energy in general may be written as the kinetic energy of the small liquid column piercing the bubble, as shown in Eq. (5.43):

$$E_J = \frac{1}{2}\rho\pi R_J^2 R_{max} U_J^2 = z_2 \rho\pi R_{max}^3 U_J^2 \tag{5.43}$$

It was assumed in this case that the velocity of the jet consists of a cylindrical liquid column which extends for a length corresponding to the maximum bubble radius (R_{max}). Further, it was assumed that the jet radius (R_J) was also proportional to the maximum bubble radius (R_{max}). The parameter z_2 used in Eq. (5.43) was estimated to be 0.0183. Hence, the maximum jet velocity may be estimated by assuming that the collapse energy is entirely converted into the kinetic energy for the jet as:

$$U_{J,max} = \sqrt{\frac{E^B_{C,0}}{z_2 \rho_L R^3_{max}}} \tag{5.44}$$

$$P_{J,max} = \sqrt{\frac{E^B_{C,0} \rho_L c^2_L}{4 z_2 R^3_{max,0}}} \tag{5.45}$$

The impact pressure generated by the liquid jet may typically be approximated by using the jet hammer formula [52] as:

$$P_J = f_J \rho_L c_L U_J \tag{5.46}$$

The factor f_J appearing in Eq. (5.46) may vary in the range from 0 to 1. For instance, for biological/organic materials, the factor would be closer to zero due to lower densities. For applications pertaining to cellular biology or droplet breakage, the factor would be closer to that for water (0.5) as the compositions are like water. For impact with metal surfaces, the factor would be closer to one. For typical HC applications, $f_J = 0.5$ will be reasonable.

5.3.1.4 Illustrative Results from Cavity Dynamics Models

The model equations discussed here may be solved numerically using appropriate initial conditions. The details of the numerical solution may be found in Pandit et al. [55]. The fluctuating pressure field for ultrasonic cavitation may be represented as:

$$P^{\infty}_L(t) = P^{\infty}\left(1 - P'_A \sin(\omega t)\right) \tag{5.47}$$

$$P'_A = \frac{P_A}{P^{\infty}}; \omega = 2\pi f \tag{5.48}$$

For simulating HC, turbulent pressure fluctuations may be approximated by similar sinusoidal representation; or, if information on turbulent pressure fluctuations experienced by cavities/bubbles is available (for example, via Lagrangian simulation of cavity trajectories), it may be used to specify a fluctuating pressure field experienced by the bubble. The cavity dynamics model may be used to generate relevant performance maps. As an example, the contour plots of the logarithm of generated OH radicals with respect to the amplitude of pressure fluctuations (P'_A) as the primary variable and the ambient temperature, frequency of fluctuations, and the initial bubble radius each as the secondary variables are shown in Figure 5.10. For all cases considered, it was found that the OH radical generation increased with increasing P'_A, and an OH generation of the order 10^{10} ·OH per collapse event may be obtained over most of the operating space. It was found that increasing ambient temperatures enhanced the OH radical generation at intermediate values of P'_A. Decreasing the driving frequency increased the OH generation. However, the effect was most pronounced only at higher P'_A. Increasing the initial bubble radius was seen to increase the generation of OH radicals but only at lower values of P'_A, and was seen to enable the earlier onset of cavitation with respect to P'_A. For all plots, it may be observed that the Blake threshold pressure (see Pandit et al. [55] for more

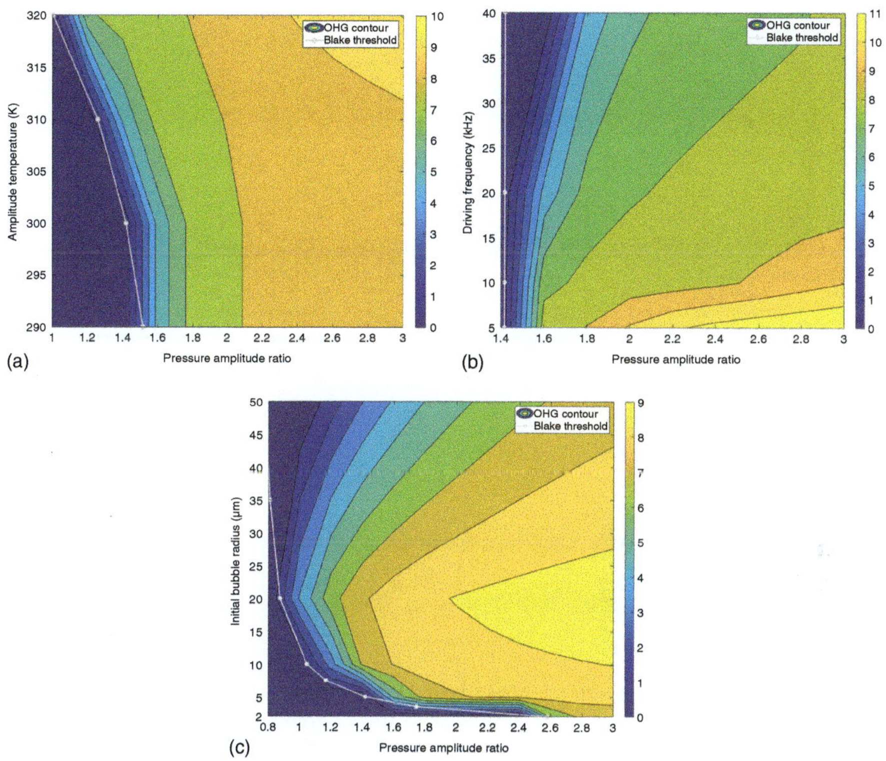

Figure 5.10 Contour plots (plotted with Blake threshold) of the logarithm of ·OH generation for HC as a function of the pressure amplitude ratio and (a) ambient temperature, (b) driving frequency, and (c) initial bubble radius (Source: From [55]/with permission of Elsevier).

details) may be used as a criterion for predicting the onset of "useful cavitation" for the given process conditions.

The cavity dynamics model can also be used to estimate jet hammer pressure and jet velocity, which may be used to estimate physical effects like reduction in particle size or droplet breakage. For example, the influence of key parameters on jet pressure is shown in Figure 5.11.

The generation of hydroxyl radicals or jet hammer pressure estimated using cavity dynamics models can be used for estimating macroscopic performance such as degradation of organic pollutants or size reduction. Attempts have been made to develop empirical correlations based on the results of detailed cavity dynamic models. For example, Gogate and Pandit [35] have presented an empirical correlation for cavity collapse pressure as a function of orifice configuration and inlet pressure. Tao et al. [57] have reported a correlation of generated hydroxyl radicals. Considering strong nonlinearities and complex underlying physics, models based on ANN may be more suitable for developing less compute-intensive correlations (see for example, Ranade et al. [25]). With the ready availability of faster computing resources, the direct solution of cavity dynamic models can also be readily used for design exercises.

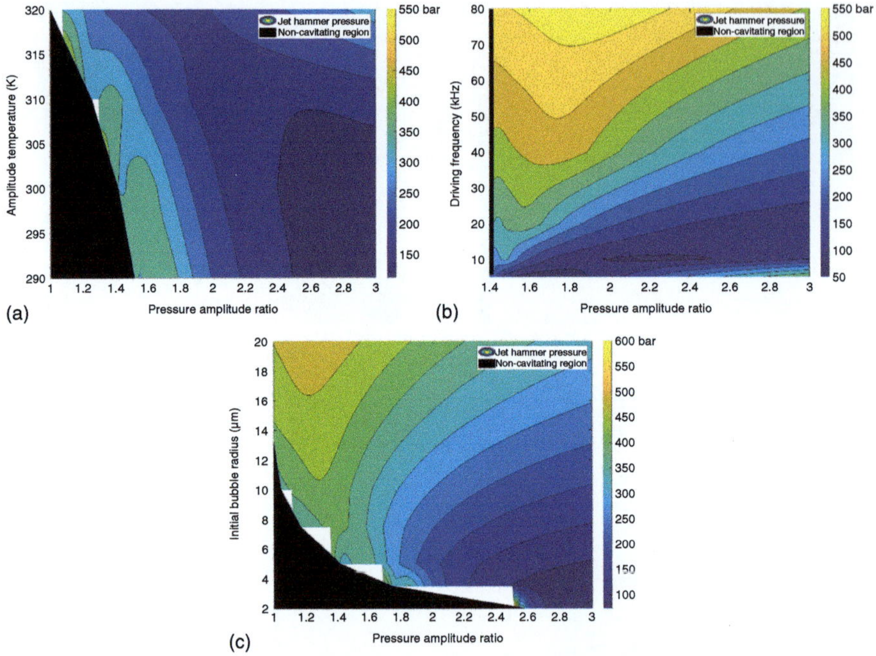

Figure 5.11 Contour plots (plotted with non-cavitating region) of the jet hammer pressure for HC as a function of the pressure amplitude ratio and (a) ambient temperature, (b) driving frequency, and (c) initial bubble radius (Source: From [55]/with permission of Elsevier).

The insights obtained through the cavity dynamics models provide overall guidelines on process improvement and optimization. The models can be used to select the appropriate operating temperature and tailor the pressure fluctuations experienced by cavities by manipulating flow in the cavitation device or the initial cavity size by manipulating dissolved gases and other mechanisms controlling nucleation. The composition of dissolved gases may also be tailored to manipulate the physicochemical effects of cavity collapse (see for example, Rooze et al. [58]). The potential application of cavity dynamics models for developing quantitative models for simulating performance of HC-based processes is discussed in the next sub-section.

5.3.2 Multi-scale/Multi-layer Models for Simulating Performance of Cavitation Processes

The micro-scale flow processes of cavity collapse need to be combined with the meso-scale and macro-scale processes for developing a model for simulating overall performance of cavitation processes. Any model which aims to simulate overall performance of a HC device or a process based on it has to take into account the following:

- Cavity collapse dynamics that leads to the high temperatures and pressures in the bubble to produce intense shear, reactive intermediates, as well as pyrolysis products (discussed in Section 5.3.1),

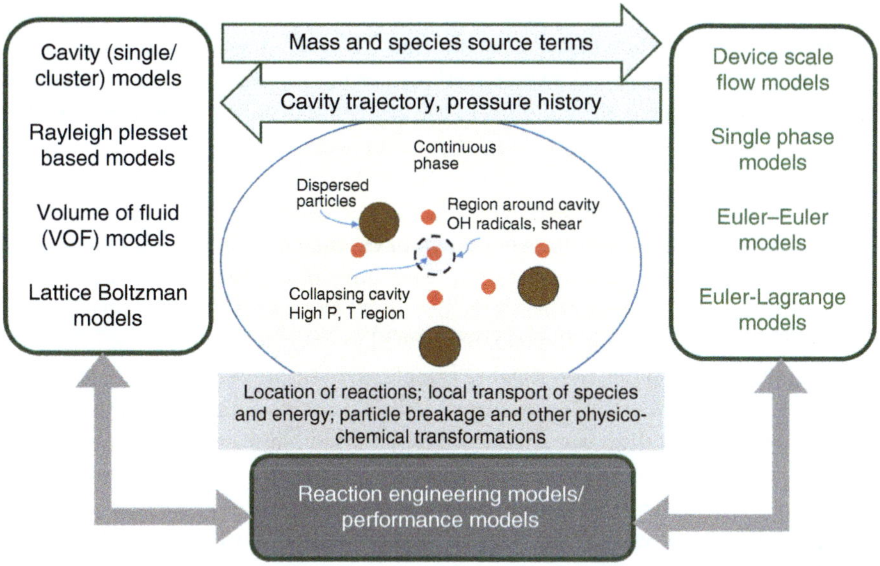

Figure 5.12 Overall approach to develop performance models for design and optimization of cavitation reactors.

- micro-scale and meso-scale models for simulating the impact of products of cavity collapse (shear, high temperature and pressure zone, radicals, pyrolysis, etc.) on the surrounding medium and transformation processes, and
- device-scale flow models for simulating the rate of generation of cavities, trajectories of cavities, location of cavity collapse, and overall performance (the device-scale flow models are discussed in Chapters 2 and 3 for different types of cavitation devices).

Various components and the interactions among them for such overall performance models are shown schematically in Figure 5.12.

The primary building block of multi-scale performance model of HC reactor involves developing a relationship between the micro-scale cavity-scale processes and likely physicochemical transformations. As mentioned previously, HC essentially acts as an energy concentrator. The collapsing cavities generate localized hot spots, high pressure regions, radicals, and intense shear, as discussed in Section 5.3.1. At the end of the collapse, the cavities' contents (a variety of generated radicals and other species) and energy are released into the surrounding liquid phase, leading to several physicochemical transformations. For developing any performance model, it is therefore essential to relate the conditions of collapsing cavity and the transformations occurring because of those. There is also interphase mass transfer of species from liquid phase to vaporous cavities. Localized hot spots generate radicals and may also lead to pyrolysis reactions within cavities. Significant efforts have been made to understand the influence of collapsing cavities on transformation processes occurring in the medium surrounding them.

Most of the early modeling studies focused on the region surrounding the cavities and ignored the influence of device-scale flow or made drastic simplifications of device-scale flow. One of the key assumptions used in these studies was that at the end of the cavity collapse, the contents of the cavity are instantaneously mixed with the surrounding medium. Once the assumption of instantaneous release is accepted, then the amounts of products released by the collapsing cavities can be calculated by following the models discussed in Section 5.3.1. If the number of cavities collapsing per unit time per unit volume are known or estimated, then the total amount of reactive content being released per unit time per unit volume of the liquid can be calculated. The mass balances for these species and those present in the liquid can then be easily written by following the usual chemical reaction engineering approach.

Most of the early studies investigated transformation processes using ultrasonic cavitation for a batch reactor. For example, iodine formation from potassium iodide due to ultrasonic cavitation was investigated by Prasad-Naidu et al. [59], and Rajan et al. [60] investigated interesting modifications of iodine formation reaction in the presence of carbon tetrachloride. These early efforts on developing reaction engineering models for ultrasonic reactors have been discussed in the book by Pandit [3] and are therefore not included here. It was soon realized, particularly for ultrasonic cavitation, that the cavity collapse and the generated physicochemical effects are significantly influenced by the presence of nearby cavities. Kanthale et al. [61] extended cavity dynamics models for a cluster of cavities and reported empirical correlations of collapse pressure and other relevant parameters. Gogate and Pandit [4] have reviewed these efforts and their potential use in designing cavitation reactors.

It is essential to link such models of a single cavity or of a cluster of cavities with the device-scale flow for simulating performance of HC. The generation and transport of cavities within HC devices/reactors have been modeled with different levels of computational flow models. The flow modeling studies may be broadly divided into two groups: the first group focusses on detailed flow characteristics using sophisticated CFD models. For example, Chahine and coworkers [5, 62] have presented a multi-scale framework which merges the Lagrangian approach for tracking nuclei and microbubbles with the macro-scale Eulerian approach for resolving larger cavities, using a bridging scheme between the disparate scales. These approaches have been shown to be able to successfully reproduce cavity formation. These studies however are far too disconnected with the performance of the cavitation reactor. On the other hand, the second group focuses on simulating performance of cavitation reactors using rather simplified CFD models. For example, Capocelli et al. [38] used a simplified one-dimensional CFD model for flow in cavitating venturi for qualitatively capturing the trend observed in performance of the cavitation reactor. Pawar et al. [33] have used single-phase flow simulations at a single operating condition for four HC devices.

The most commonly used ways of coupling cavity dynamics models and device-scale flow models may be grouped as:

- Use single-phase CFD models to simulate liquid flow through cavitation devices and use the simulated flow field to simulate cavity trajectories (for example, Capocelli et al. [38] and Pawar et al. [33])

- Use multiphase CFD models with cavitation sub-models for simulating generation and distribution of cavities within the considered domain (for example, Simpson and Ranade [32, 63])

In the first group, cavity trajectories are simulated by assuming that cavities will follow the flow field of liquid. The mean pressure experienced by cavities is superimposed by random pressure fluctuations, the magnitude of which is determined by the local values of turbulent kinetic energy. The fluctuating pressure field experienced by trajectories can then be used for simulating cavity dynamics and the resulting generation of hydroxyl radicals and shear. It is important to consider differences between ultrasonic cavitation and HC while solving cavity dynamics models (see Section 5.3.1). It is essential to carefully analyze different timescales of cavity collapse, pressure fluctuations, and variation in mean pressure. The results of the cavity dynamics model are quite sensitive to the initial size of the cavity. There have been attempts to include distribution of initial cavity sizes instead of considering a single-size cavity. For example, Capocelli et al. [38] used a size distribution of initial cavity size, R_0, in the range of 20–250 μm, based on the work of Liu and Brennen [64]. Attempts have been made to qualitatively relate the results of cavity dynamics models with the performance of the cavitation reactor. Sharma et al. [65] have developed an empirical correlation for temperature, pressure, and radical generation at collapse as a function of initial radius of cavity, inlet pressure to the HC device, and diameter of an orifice hole. Unfortunately, the applicability of such correlations is quite restricted and different HC devices are seldom directly comparable [66]. Another potential method of coupling cavity dynamics models with device-scale flow was used by Sarvothaman et al. [6], where they approximated the fluctuations in the bulk pressure experienced by the cavity as:

$$P^B = f\,P_v + \rho\,k_R\,\sin(\omega_R\,t) \tag{5.49}$$

where P^B is the pressure experienced by a cavity, P_v is vapor pressure of liquid, f is an empirical parameter, and k_R and ω_R are average turbulent kinetic energy and turbulent frequency in the cavity collapse region R. The rationale behind this approximation was that the cavity collapse time scales are usually quite small, and considering typical mean velocities in cavitation devices, the distance travelled by cavities till their collapse is rather small. It is therefore reasonable to approximate values of turbulent kinetic energy and turbulent frequency averaged over the small cavity collapse region R. It can be seen that the amplitude of pressure fluctuations is proportional to the liquid density and turbulent kinetic energy, and the frequency of pressure fluctuations is directly related to turbulent frequency. The detailed flow field and the region of cavity collapse can then be obtained using the multiphase flow models (as described in Chapters 2 and 3). These device-scale flow results can provide the number of cavities generated in a HC reactor. When coupled with cavity dynamics models via Eq. (5.49), it is possible to estimate the overall products (hydroxyl radicals/shear) generated within the cavitation reactor.

In order to develop a model for simulating overall performance of the HC device/process, it is essential to develop a per-pass performance model for the overall products (hydroxyl radicals/shear) generated within the cavitation reactor. Sarvothaman et al. [6] have presented such a model for estimating per-pass pollutant degradation performance. The approach is, however, quite general and can be extended for other desired transformation processes. Once the per-pass pollutant degradation factor ϕ is estimated, it can then be used with the models discussed in Section 5.2.2. This approach is briefly discussed in the following section.

Consider a special case of batch operation of a HC-based pollutant degradation process, shown schematically in Figure 5.2. The per-pass pollutant degradation factor ϕ can be related to the generation of hydroxyl radicals in the cavitation device (G, kmol/s) and their subsequent reactions with the pollutant. Considering that radicals have very small lifetime compared to the residence time in the cavitation device (and downstream piping up to the holding tank), it may be safely assumed that all radicals are consumed within the reactor. Radicals are highly reactive and therefore, in addition to reacting with organic pollutants present in liquid, radicals may react with any other scavengers present in the liquid (including water itself). Therefore, only a fraction of generated radicals (say, δ) will be used for degrading the pollutant. This fraction may be written as:

$$\delta = \frac{k_2 C}{(k_2 C + k_S C_S)} \tag{5.50}$$

where k_2 and k_S are rate constants of reactions between pollutant (C) and other scavengers (C_S) with hydroxyl radicals, respectively. The per-pass degradation factor can then be related to the generation rate as:

$$Q\,C\emptyset = G\,\delta \quad \text{where } Q = \frac{\dot{m}_c}{\rho} \tag{5.51}$$

The per-pass degradation coefficient can then be written as:

$$\emptyset = \frac{\left(\frac{G}{Q\,C_S}\right)}{\left(\frac{k_S}{k_2} + \frac{C}{C_S}\right)} \tag{5.52}$$

The per-pass degradation factor will depend on the generation rate of hydroxyl radicals (G), flow rate through cavitation device (Q), chemical nature of pollutant (reactivity with hydroxyl radicals, k_2), concentration of the pollutant (C), concentration of scavengers (C_S), and the relative rate reactivity of pollutant and scavengers (k_2/k_S) with hydroxyl radicals. It can be seen that for low concentrations of pollutant ($C \ll C_S$), the value of the per-pass degradation factor will be independent of pollutant concentration. It will essentially depend on the effective generation rate of hydroxyl radicals, G, and the ratio of k_S to k_2. As the reactivity of pollutant decreases with respect to reactivity of the other existing scavengers, the per-pass degradation factor decreases.

The radical generation rate G depends on the flow characteristics of the cavitation device which controls the number density of cavitation events and the intensity of cavity collapse (which controls the generation of hydroxyl radicals per collapsing cavity). The generation rate of hydroxyl radicals may be related to the number of cavities collapsing per second in the cavitation device (n, s^{-1}) and hydroxyl radicals generated per collapse, m_{OH} (kmol) as:

$$G = n\, m_{OH} \text{ kmol/s} \tag{5.53}$$

It should be noted that hydroxyl radicals generated per collapse depend on a variety of factors such as the initial radius of cavity and the bulk pressure fluctuations experienced by the cavity (as it is moving within the cavitation device). All the cavities generated will not be of the same initial radius, and all the generated cavities (even with those with same initial radius) will not experience the same bulk pressure fluctuations because of inherently turbulent flow. It may be therefore necessary to introduce a proportionality factor in the previous equation. The number of cavities generated per second may be indirectly related to the flow rate through the device, gas volume fraction, and the representative volume of a cavity as:

$$n \propto \frac{Q\, \varepsilon_{GR}}{V_{in}} \text{ s}^{-1} \tag{5.54}$$

where V_{in} is the volume of the cavity at inception. The term ε_{GR} denotes the effective volume fraction of cavities. It is important to carefully select a region to estimate the appropriate value of ε_{GR}.

Computational fluid dynamics (CFD) models may allow detailed simulation of three-dimensional, transient flow field of HC devices. Some examples and illustrative results obtained with such CFD models are presented and discussed in Chapters 2 and 3. A detailed discussion of modeling of turbulent, multiphase flows with phase change is out of the scope of this chapter. However, the basic approach, some guidance on selecting appropriate models, and model equations used for typical CFD simulations of HC devices are presented in Appendix 5.A. Such CFD models with appropriate numerical solution strategies may be used for obtaining the key flow parameters required for estimating performance of HC devices. For the sake of illustration, a case of the vortex-based cavitation devices considered by Sarvothaman et al. [6] is briefly described here. The flow in the vortex-based cavitation device was simulated by following the model of Simpson and Ranade [67]. A sample of results in the form of contours of gas volume fraction and trajectories is shown in Figure 5.13.

It can be seen that formation of the gas core is captured correctly by the CFD models. The cavities are generated in this low pressure region – at the core of the vortex. These cavities then travel away from the core following the turbulent fluctuations of the flow. In order to understand the region over which cavities travel and collapse, it will be useful to examine relevant time and space scales. The results from cavity dynamics models typically indicate cavity lifetimes of the order of 2×10^{-4} s, which translates to cavities travelling distances of the order of ~1–2 mm from inception to collapse. To illustrate these relatively short travel distances, sample results of simulated trajectories released from an iso-surface of constant vapor volume fraction (0.5) for the time of 2×10^{-4} s are shown in Figure 5.10b. The trajectories are colored by the ratio of absolute pressure and vapor pressure, indicating that the maximum pressure recovery attained by the cavities prior to final collapse is typically less than three times of vapor pressure over these travel distances.

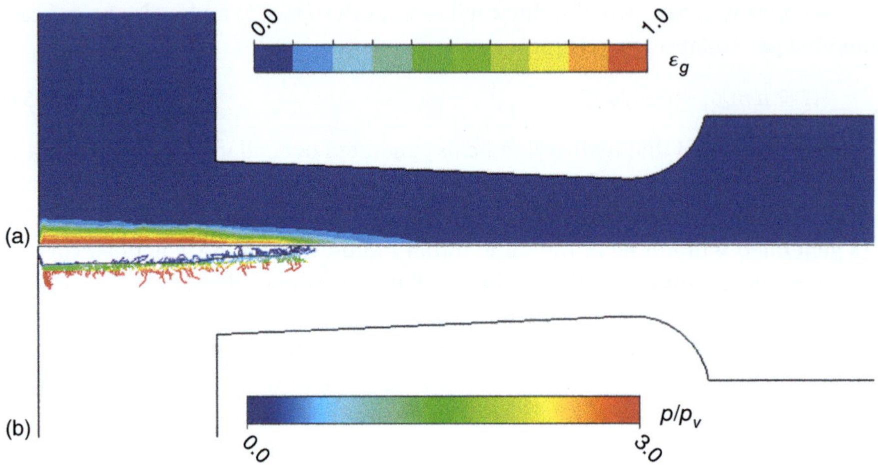

Figure 5.13 A sample of simulated results using the CFD model; 6 mm cavitation device, throat velocity = 3 m/s, pressure drop = 250 kPa. (a) Gas volume fraction. (b) Trajectories (Source: From [6]/with permission of Elsevier).

It should be noted that the source and sink terms used in the cavitation model considered in the CFD model are based on approximated cavity dynamics equations. Therefore, the simulated distribution of gas volume fractions using the CFD model can also be used to quantify the relevant region of generation of cavities and collapsing of cavities. Considering that the CFD model used in simulating the results shown in Figure 5.13 was based on the Eulerian framework and did not explicitly account for coalescence of cavities to form a gas core, some approximation is needed to identify the gas core. Earlier studies on multiphase flows with the Eulerian models have used a cut-off of gas volume fractions ranging from 0.5 to 0.66 for identifying gas–liquid interface [68]. Following those, Sarvothaman et al. [6] assumed a volume fraction of 0.5 to identify a boundary of gas core and dispersion. Cavities were assumed to be generated on the surface of this identified gas core. The generated cavities travel in the regions of high pressure through the turbulent flow field and eventually collapse. As cavities collapse, the volume fraction of gas starts reducing. Beyond a certain collapsing zone, the gas volume fraction will be virtually zero. The outer boundary of the cavity collapse region was estimated by taking a cut-off value of gas volume fraction as 0.1. Of course, there will be some cavities which may travel beyond this region. However, most of the cavities may be assumed to collapse in this region. The boundaries of the gas core and the cavity-collapse region based on this assumed cut-off in gas volume fraction for the case considered in Figure 5.13 are shown in Figure 5.14. Once we identify the region of collapsing cavities, one may use the simulated CFD results to estimate the relevant quantities of interest. The volume fraction of cavities, ε_{GR}, required for estimating number of cavitation units as well as effective sound speed in the collapsing region, can then be calculated from the CFD simulations for a range of operating flow rates and for different device scales.

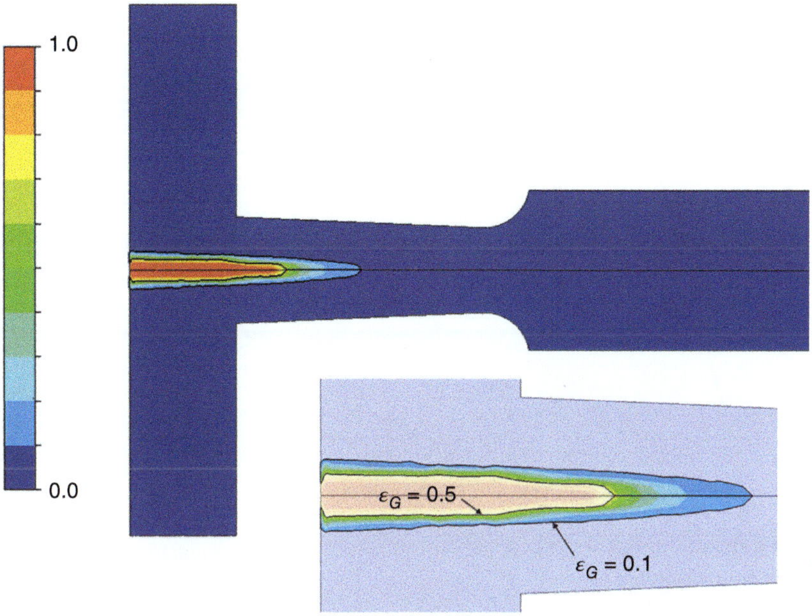

Figure 5.14 Regions of gas core ($\varepsilon_G > 0.5$) and cavity collapse ($0.5 > \varepsilon_G > 0.1$); 6 mm cavitation device, throat velocity = 3 m/s, pressure drop = 250 kPa (Source: From [6]/with permission of Elsevier).

Once the relevant gas volume fraction ε_{GR} is estimated, with some mathematical rearrangement, the per-pass degradation factor may be obtained as:

$$\emptyset \propto \frac{\varepsilon_{GR}\left(\frac{C_{\mathrm{OH}}}{C_S}\right)}{\left(\frac{k_S}{k_2} + \frac{C}{C_S}\right)} \tag{5.55}$$

where C_{OH} is concentration of the hydroxyl radical at the cavity collapse ($\propto \frac{m_{\mathrm{OH}}}{V_{in}}$). If concentration of pollutant is much smaller than C_S and $k_S \sim k_2$, we can simplify this equation as:

$$\emptyset = \alpha\, \varepsilon_{GR}\left(\frac{C_{\mathrm{OH}}}{C_S}\right) \tag{5.56}$$

where α is the proportionality constant. The concentration of scavengers may be assumed as constant and may be lumped with the proportionality constant. For the sake of dimensional consistency, C_S may be assumed to be 1 kmol/m^3. The Eq. (5.56) provides a direct link between the key flow characteristics of the cavitation device/reactor and the overall performance of the cavitation reactor. The values of gas volume fractions appearing in Eq. (5.56) may be obtained from CFD models of the cavitation device. C_{OH} may be estimated by solving cavity dynamics models and Eq. (5.49).

It should be noted that the mathematical modeling framework described here still requires information about initial cavity size for estimating values of per-pass degradation factor. The value of initial cavity size (or size distribution) is rather difficult to

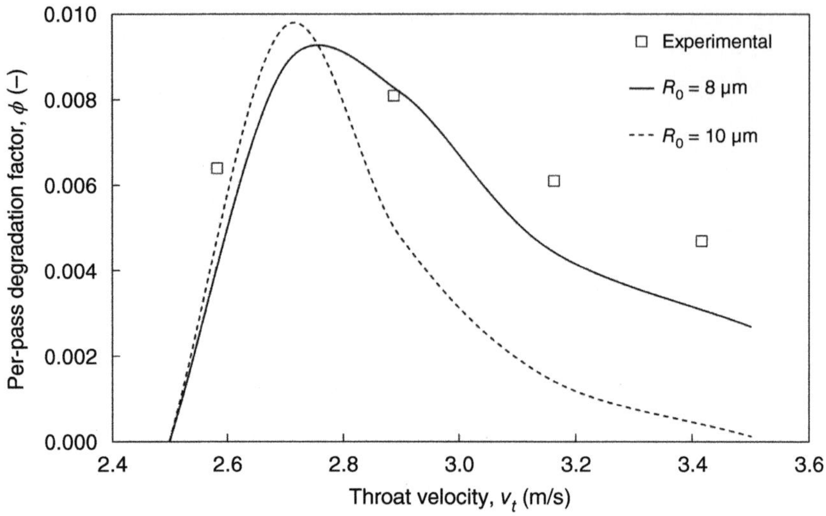

Figure 5.15 Influence of throat velocity on performance of cavitation reactor (Source: From [6]/with permission of Elsevier).

estimate. It may therefore be practical to treat the initial cavity size as an adjustable parameter. With such an approach, Sarvothaman et al. [6] were able to simulate the influence of pressure drop across a vortex-based cavitation device on the degradation performance of the device. Their results are shown in Figure 5.15 for illustrating the potential of this approach.

The approach described here is quite powerful and has the potential to make high-fidelity computational models for simulating and optimizing HC devices and processes. Significant further work on improving our ability to simulate cavitating flows and particularly on developing "performance" models relating observable performance with cavity collapse conditions is needed. Some of these needs are briefly discussed in Section 5.5.

5.4 Modeling of Heterogeneous Systems Treated with HC

There are several applications of HC devices and processes for heterogeneous systems containing dispersed solid particles or liquid droplets; for example, biomass pre-treatment [69–72], extraction of bioactives from algae [73], and sulfur removal [74, 75]. The approach discussed in Section 5.3 may be extended for such heterogeneous systems. For example, recently, Thaker and Ranade [76] have used detailed computational flow models for simulating drop breakage in vortex-based cavitation devices.

Thaker and Ranade [76] extended the models discussed in Appendix 5.1 for dispersed liquid–liquid systems. The three-phase flow (water–oil–vapor) was modeled using the mixture model approach as:

$$\frac{\partial \rho_m}{\partial t} + \nabla \cdot \rho_m \vec{v}_m = 0 \tag{5.57}$$

where $\vec{v}_m$ is the mass-averaged velocity and defined as

$$\vec{v}_m = \frac{\sum_{k=1}^{q} \alpha_k \rho_k \vec{v}_k}{\rho_m} \tag{5.58}$$

where mixture density ρ_m is defined as $\vec{v}_m = \sum_{k=1}^{q} \alpha_k \rho_k$, where α_k is the volume fraction of phase k, and q is the total number of phases.

$$\frac{\partial \rho_m \vec{v}_m}{\partial t} + \nabla \cdot \rho_m \vec{v}_m \vec{v}_m = -\nabla P + \nabla \cdot \left[\mu_m \left(\nabla \vec{v}_m + \vec{v}_m^{\,T}\right)\right] + \rho_m g + F - \nabla \cdot \left[\sum_{k=1}^{q} \alpha_k \rho_k \vec{v}_{dr,k} \vec{v}_{dr,k}\right] \tag{5.59}$$

where F and μ_m are body force and viscosity of the mixture, respectively.

$$\mu_m = \sum_{k=1}^{q} \alpha_k \mu_k \tag{5.60}$$

The drift velocity $\vec{v}_{dr}$ is related to slip velocity $\vec{v}_{slip}$ ($\vec{v}_{slip} = \vec{v}_k - \vec{v}_A$), as:

$$\vec{v}_{dr,k} = \vec{v}_{slip,k} - \sum_{k=1}^{q} \frac{\alpha_k \rho_k}{\rho_m} \vec{v}_{kA} \tag{5.61}$$

where $\vec{v}_{kA}$ is velocity of phase k relative to continuous phase A.

$$\vec{v}_{slip,k} = \frac{24}{Re} \frac{(\rho_k - \rho_m) d_k^2}{18 \mu_A C_D} \left(g - \frac{Dv_m}{Dt}\right) \tag{5.62}$$

The drag coefficient C_D is defined as the following [111]:

$$C_D = \begin{cases} \frac{24}{Re}(1 + 0.15 Re^{0.687}) & Re \leqslant 1000 \\ 0.44 & Re > 1000 \end{cases} \quad \text{where } Re = \frac{\rho_m \vec{v}_{slip,k} d_k}{\mu_A} \tag{5.63}$$

The turbulence models and cavitation models discussed in Appendix 5.1 may be used with such multiphase simulations. Those equations are not repeated here. The particle or droplet breakage can be simulated using population balance approaches. The evolution of particles/drop population may be tracked using the discrete method (method of class) [77]. This method represents continuous *particle size distribution (PSD)* in terms of discrete size groups. The population balance equations may be represented in terms of the volume fraction of drop size i:

$$\frac{\partial \rho_m \alpha_{k,i}}{\partial t} + \nabla \cdot (\rho_m \alpha_{k,i} \vec{v}_m) = \rho_O V_i (B_i - D_i) \tag{5.64}$$

where $\alpha_{k,i}$ is the oil volume fraction of a representative group of drop size i of phase k. B_i and D_i are the rates of birth and loss of particles of size i due to the breakage per unit volume, respectively. For the sake of simplicity, the population balance equation represented by (5.64) does not contain any coalescence or agglomeration terms. In any case, in cavitating flows generating intense shear, coalescence may not be significant.

$$\alpha_{k,i} = N_i V_i \tag{5.65}$$

where N_i is total number of particles of size i per unit volume and is calculated as:

$$N_i(t) = \int_{V_i}^{V_{i+1}} n(V, t) dV \tag{5.66}$$

where V_i is volume of representative group for drop size i.

The volume fraction of all the groups (α_k) is defined as:

$$\alpha_O = \sum_{i=1}^{M} \alpha_{O,i} \tag{5.67}$$

The birth and loss terms were discretized as:

$$B_i = \sum_{j=i+1}^{M} p_j g(V_j) N_j \beta(V_i/V_j) \qquad \text{and } D_i = g(V_i) N_i \tag{5.68}$$

where $g(V_j)$ is the breakage frequency that represents the number of parent drops with volume V_j breaking per unit time. p_j is the number of daughter droplets generated by breakage of parent drop with volume V_j. $\beta(V_i/V_j)$ is the daughter particle distribution function that defines the probability of generating daughter drop of volume V_i from the breakage of parent drop with volume V_j. Various breakage models and daughter particle distribution functions are available, which may be selected based on the dispersed system under consideration. For example, Thaker and Ranade [76] used the breakage model of Laakkonen et al. [78] which uses the following expression for breakage frequency $g(V_j)$. The expression of $g(V')$ considered from Laakkonen et al. [78] is:

$$g(V_j) = C_3 \varepsilon^{1/3} \mathrm{erfc}\left(\sqrt{C_4 \frac{\sigma}{\rho_A \varepsilon^{2/3} d^{5/3}} + C_5 \frac{\mu_k}{\sqrt{\rho_A \rho_k} \varepsilon^{1/3} d^{4/3}}}\right) \tag{5.69}$$

where d is particle diameter and μ_k is viscosity of the dispersed phase. C_3, C_4, and C_5 are model parameters. Using this model, Thaker and Ranade [76] were able to simulate drop breakage in liquid–liquid emulsions occurring in vortex-based cavitation devices. A sample of their results in the form of simulated droplet size distributions after one pass through cavitation devices at different operating pressure drops across cavitation devices is shown in Figure 5.16 to illustrate the application of the model discussed here.

The model was also able to simulate influence of dispersed phase volume fraction on Sauter mean diameter and energy efficiency of drop breakage quite adequately (see Figure 5.17). The original paper may be referred to for additional details.

The modeling approach illustrated here for droplet breakage can be extended to other applications such as cavi-crystallization or biomass pre-treatment where solid particle breakage occurs, with or without accompanying chemical transformations.

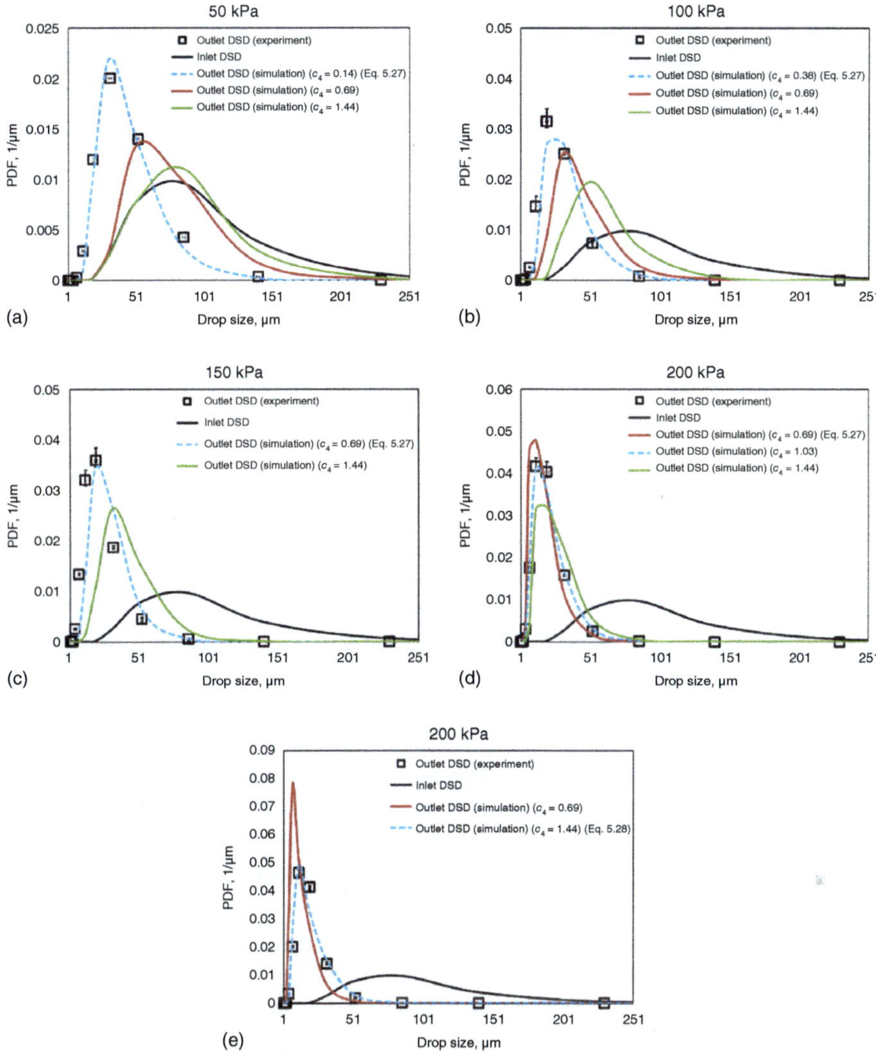

Figure 5.16 Measured and predicted drop size distribution for different pressures (α_k= 0.05). (a) Pressure drop = 50 kPa, (b) Pressure drop = 100 kPa, (c) Pressure drop = 150 kPa, (d) Pressure drop = 200 kPa, (e) Pressure drop = 250 kPa (Source: From [76]/with permission of John Wiley & Sons).

5.5 Summary and Outlook

The complex processes occurring in HC still pose stiff challenges for developing predictive computational models. In this chapter, the key prevailing approaches used for modeling applications of HC are discussed after broadly classifying them into empirical and physics-based approaches. Fully data-driven or semi-empirical approaches are discussed in Section 5.2. The per-pass performance approach was found to be more appropriate for evaluating hydrodynamic-cavitation based processes than the

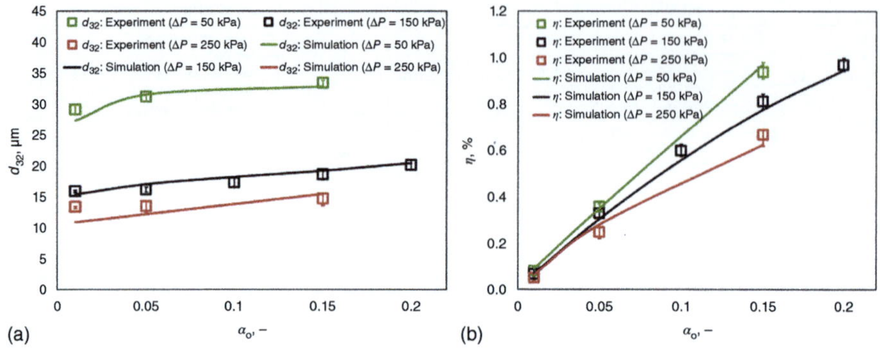

Figure 5.17 Comparison of measured and simulated (a) Sauter mean diameter (d_{32}) and (b) drop breakage efficiency at different oil volume fractions (Source: From [76]/with permission of John Wiley & Sons).

pseudo-kinetics approach. ANN and other data-driven modeling formalisms may also be used to simulate performance of HC-based processes. However, adequate data (quality as well as quantity) are essential for applying such approaches. None of these empirical or semi-empirical approaches are useful for scale-up and developing new HC devices/reactors. Physics-based models, though not yet adequate to make a priori predictions, can provide useful insights and can be used for overall optimization of devices/reactors and process performance. We hope that the presented approaches, models, and examples included in this chapter will be useful for selecting the appropriate approach for the design and optimization of a wide range of HC-based processes.

To make further progress toward development of predictive models, it is essential to develop a thorough understanding of the inception of cavities, their trajectories, and their eventual violent collapse, apart from their interaction with the surrounding, leading to myriad physicochemical transformations. In this chapter, we discussed cavity dynamics models based on the classical Rayleigh–Plesset (R-P) cavity dynamics model [79]. The model presented in this chapter [55], however, was restricted to dynamics of a cavity in an infinite medium (symmetric expansion and collapse). It is important to extend the cavity dynamics models for simulating asymmetric breakup. For simulating asymmetric collapse, it is essential to use full-fledged CFD models rather than simplified one-dimensional equations used for simulating symmetric cavity dynamics. A finite volume method (VOF) may be used for this purpose. Recently, Dular and coworkers [80] have attempted to simulate asymmetric collapse of a cavity near a liquid droplet. These simulations were still based on the assumption of axis-symmetric configuration. Fully asymmetric, three-dimensional transient flow models are necessary for quantitative predictions of physicochemical transformations caused by asymmetric cavity collapse. New approaches for simulating micro-scale processes in cavitation are being developed. One of the promising options appears to be a pseudo-phase Lattice Boltzmann (PPLB) approach, which naturally captures phase separation and interface formation/collapse. This has a potential to realize a breakthrough in understanding for cavity growth and collapse processes on a microscale. The PPLB method [81] uses appropriate equation of state and spontaneously generates phase segregation (liquid–vapor interfaces) without requiring front-tracking methods. The phase segregation, generation, and motion of phase interfaces is based on interactions modeled at the scale in between the molecular level and the continuum approach exploited in common CFD. Through iterative improvements in recent years, PPLB methods have reached a stage where they can be used to replicate the complex bubble growth and collapse in 2D simulations [82, 83], and the approach has been extended to 3D studies of individual collapsing cavities [84]. The method also shows promise in predicting collapsing cavity behavior in proximity to flat solid surfaces [85]. Recently, full 3D finite volume-based CFD simulations using the volume of fluid (VOF) approach are being carried out to understand the influence of nearby surfaces (either rigid or flexible) on cavity collapse. For example, Trummler et al. [86] have simulated the near-wall collapse of a cavity with full 3D domain. Such finite volume methods or PPLB-based approaches with full 3D simulations will hopefully provide better insights and quantitative information on asymmetric cavity collapse and the resulting physicochemical transformations.

In addition to accounting for asymmetric collapse, it is also essential to quantify the influence of neighboring cavities. During the HC process, cavities rarely exist in isolation and almost invariably are present in the form of a cluster of cavities. Bubbles or cavities generated through micro-scale processes interact with each other, which leads to the formation of mesoscale structures, their coalescence, and breakup. Bulk behavior emerges from mesoscale structures (inherently transient in space and time) and significantly influences performance of processes/reactors. Recently, Rasthofer et al. [87] simulated a collapse of a cloud of 12 500 cavities. Several attempts are being made to develop new cavitation models. For example, recently, Shi et al. [88] presented a new cavitation model based on inter-cavity interactions within a cloud of cavities. Such models and approaches will pave the way toward truly multi-scale models to encompass micro-, meso-, and macro-scale processes occurring in HC and relate those to overall process performance. It is hoped that such multi-scale models combined with the state-of-the-art machine learning approaches will lead to the development of digital twins of HC devices and processes. The models and discussion presented in this chapter will be useful during the journey on this ambitious path.

Abbreviations

ANN	Artificial neural network
BIOGAS	ANN model for describing influence of HC-based pre-treatment on biomethane generation
COD	Chemical oxygen demand
DCA	Dichloroaniline
EoS	Equation of state
KM	Keller–Miksis

MGFE	Minimization of Gibbs free energy
PPLB	Pseudo-phase Lattice Boltzmann
PSD	Particle size distribution
R-P	Rayleigh Plesset
VdW	van der Waal's
WATER	ANN model for describing HC-based degradation of DCA in water

Symbols

a	Pressure correction factor for van der Waal's equation (Pa-m^6/kmol^2)
b	Volume correction factor for van der Waal's equation (m^3/kmol)
B_i	Rate of birth of size i due to the breakage per unit volume
$C_1 - C_2$	constants for cavitation model (–)
$C_3 - C_5$	constants for breakage frequency model (–)
C_0	Initial value of desired property
C_D	Drag coefficient (–)
C_{in}	Value of desired property of interest in the inlet stream
C_{OH}	Concentration of hydroxyl radical at collapse (kmol/m^3)
C_{out}	Value of desired property of interest in the outlet stream
C_{p0}	Initial concentration of pollutant p which produces intermediates with different reactivity
C_{pin}	Inlet concentration of pollutant p which produces intermediates with different reactivity
C_p	Concentration of pollutant p which produces intermediates with different reactivity
C_p	Heat capacity at constant pressure (J/K-no of molecules)
C_s	Concentration of scavenger (kmol/m^3)
C_v	Heat capacity at constant volume (J/K-no of molecules)
c	Velocity of sound (m/s)
C	Desired property of interest
C	Concentration (no of molecules/m^3)
d_{max}	maximum stable diameter (μm)
d_{32}	Sauter mean diameter (μm)
$d_{@\overline{\varepsilon}=1}$	Sauter mean diameters at $\overline{\varepsilon} = 1$ m^2/s^3 (μm)
$d_{@E=1}$	Sauter mean diameters at E= 1 J/kg (μm)
D_i	Rate of loss of size i due to the breakage per unit volume
D	Diffusion coefficient (m^2/s)
E	Energy (J)
f	Degrees of freedom (–)
f	Driving frequency (Hz)
$f(n)$	Correction factor to account for varying per-pass degradation factor (–)
g	Gravitational acceleration (m^2/s)
G	Generation rate of hydroxyl radicals (kmol/s)

h	Molecular enthalpy of ith species at the interface (J/no of molecules)
H	Number of neurons in the hidden layer of ANN model
k_2	Second-order rate constant (M^{-1} s^{-1})
k_{app}	Apparent first-order constant (s^{-1})
k_R	Turbulence kinetic energy (m^2/s^2)
k_S	Second-order rate constant of reaction between hydroxyl radicals and other reactants than the pollutant of interest (M^{-1} s^{-1})
k	First-order rate constant (s^{-1})
k	Boltzmann's constant (J/K-no of molecules)
l_{diff}	Diffusion length for mass transfer (m)
l_{th}	Diffusion length for heat transfer (m)
$\dot{m}_c$	Mass flow rate through the cavitation device operated in batch mode (kg/s)
$\dot{m}_{in}$	Mass flow rate for inlet stream in a continuous HC process (kg/s)
$\dot{m}_{out}$	Mass flow rate for outlet stream in a continuous HC process (kg/s)
m_{OH}	Hydroxyl radicals generated per collapse (kmol/s)
M	Mass of liquid in holding tank (kg)
MSE	Mean square error
n	Number of passes through cavitation device (–)
n	Number of cavities collapsing per second in the cavitation device (s^{-1})
N_{data}	Number of experimental data points (records) available
N_A	Avogadro's constant (6.023×10^{26}) (no of molecules/kmol)
N_i	total number of drops per unit volume ($1/m^3$)
N	Number of molecules
N_{data}	Number of experimental data points available
N_{eq}	Number of equations required to be solved to decide the weights for ANN model
N_{in}	Number of neurons in hidden layer for input size
N_{out}	Number of neurons in hidden layer for output size
N_{train}	Number of records available to train the ANN model
N_{test}	Number of records available to test the ANN model
N_{weight}	Number of weights to be decided for the ANN model
ΔP	Pressure drop across cavitation device (kPa)
P	Entity represented as the product
P	Pressure (Pa)
P	Pressure (kPa), in the context of Section 5.4
Q	Recirculating flow through the cavitation device (m^3/s)
r	Radial coordinate (–)
R	Regression coefficient
R	Radius of bubble (m)
R^2	Coefficient of determination
Re	Reynolds number (–)
s	Velocity of the bubble wall (m/s)
S	Entity represented as the reactant

t	Time (s)
T	Temperature (K)
U	Velocity (m/s)
v	Velocity (m/s)
V_{in}	Volume of cavity at the inception (m^3)
V	Volume of liquid in holding tank (m^3)
X	Modeled variable
x	Mole fraction (–)
y	Parameter indicating the extent of deviation upon intermediate formation (–)
z	Constant for jet velocity calculations (–)

Greek Symbols

α	Proportionality constant (–)
α	Volume fraction (–)
β	Ratio of mass flow rate through the cavitation device and mass in the holding tank (s^{-1})
δ	Variable for the calculation of heat capacity (–)
δ	Fraction of generated radicals used for degrading pollutant (–)
ε_{GR}	Gas volume fraction (–)
γ	Parameter indicating positive or negative deviation for varying per-pass degradation constant (–)
λ	Thermal conductivity (W/m-K)
$\emptyset_1$	Keller–Miksis Correction Factor 1 (–)
$\emptyset_2$	Keller–Miksis Correction Factor 2 (–)
$\emptyset_3$	Keller–Miksis Correction Factor 3 (–)
$\emptyset_0$	Initial per-pass degradation factor (–)
$\emptyset$	Per-pass performance constant (–)
ν	Kinematic viscosity (m^2/s)
ω_R	Turbulent frequency (kHz)
ω	Angular frequency – acoustic cavitation (rad/s)
ρ	Density (kg/m^3)
σ	Surface tension (N/m)
Ψ	Correction term for contribution for $C^*_{v,mix}$(J/K)
θ	Characteristic vibrational temperature (K)
τ	Residence time based on ratio of mass of liquid and inlet mass flow rate (s)
μ	Viscosity (kg/m-s)

Subscripts

0	Initial condition
A	Amplitude of oscillations
A	Aqueous phase

c	Pertaining to cavity collapse conditions
i, j	Indices to represent species
J	Cavitation jet
k	Total number of phases
L	Pertaining to the bulk medium-liquid
m	Model (Section 5.2.3)
m	Mixture (Section 5.4)
max	Maximum
mix	Pertaining to a mixture
o	Observed
O	Organic phase
q	Total number of phases
S	Shockwave
t	Total number or throat
$test$	Data set used for testing
$train$	Data set used for training
v	Vapor

Superscripts

$-$	Average over all the points in the data set
∞	Pertaining to far-field bulk medium conditions-liquid
B	Pertaining to bubble internal contents
i	ith data point (Section 5.2.3)
i	Pertaining to the liquid–bubble interface (Section 5.3.1)
$*$	Corrected term
$'$	Reduced or normalized term

5.A CFD Modeling of Cavitating Flows

Various approaches for modeling of multiphase flows are discussed in the book by Ranade [68]. Usually, complex, turbulent cavitating flows are modeled using the mixture model. In this model, the liquid–gas/vapor cavity mixture is treated as a single fluid, comprised of a homogeneous mixture of two phases. The continuity equation for the mixture flow is written as:

$$\frac{\partial}{\partial t}(\rho_m) + \nabla\boldsymbol{\cdot}(\rho_m \vec{u}_m) = 0 \tag{A5.1}$$

where ρ_m is the mixture density, and $\vec{u}_m$ is the mass-averaged mixture velocity. The corresponding momentum equation for the mixture flow, assuming that both phases share the same velocity field, is written as:

$$\frac{\partial}{\partial t}(\rho_m \vec{u}_m) + \nabla\boldsymbol{\cdot}(\rho_m \vec{u}_m \vec{u}_m) = -\nabla p + \nabla\boldsymbol{\cdot}\left[\mu_m\left(\nabla\vec{u}_m + \nabla\vec{u}_m^T\right)\right] + \rho_m\vec{g} + \vec{F} \tag{A5.2}$$

where $\vec{u}_m$ is the mixture velocity vector, μ_m is the mixture viscosity, $\rho_m\vec{g}$ is the gravitational body force, and the term $\vec{F}$ accounts for additional external body forces applied to the fluid volume (i.e. that may arise from interaction with dispersed phases). In the Reynolds-averaged (RANS) approaches, the velocity terms in Eqs. (A5.1) and (A5.2) are replaced by the sum of their mean ($\overline{u}$) and instantaneous (u') components, $u = \overline{u} + u'$, and an ensemble average taken. This averaging process results in additional Reynolds stresses, $-\rho\overline{u_i u_j}$. In order to close the momentum equations, the Reynolds stress terms require additional mathematical models. One approach is to use a Reynolds Stress Model (RSM), which involves solving separate transport equations for each of the additional Reynolds stresses (six in total for three-dimensional flows). More typical in RANS approaches is to employ the Boussinesq hypothesis, which approximately relates the Reynolds stresses to the mean velocity gradients in the flow as follows:

$$-\rho\overline{u_i u_j} = \mu_t\left(\frac{\partial u_i}{\partial j} + \frac{\partial u_j}{\partial i}\right) - \frac{2}{3}\rho k\delta_{ij} \tag{A5.3}$$

Here the subscripts i and j represent two mutually perpendicular directions, k is the turbulent kinetic energy, and δ_{ij} is the Kronecker delta, which is introduced to make the formula applicable to the normal stresses where $i = j$. Most importantly, this expression also introduces the concept of the turbulent viscosity, μ_t. This is not a physical property, but rather a scalar which has a value proportional to the local turbulence properties. This quantity is typically modeled through additional transport equations for the turbulent kinetic energy (k) and turbulent dissipation rate (ε), or the specific dissipation rate (ω).

The choice of turbulence closure model is crucially important, particularly so in cavitating regimes. In typical HC devices, the flow features include separation and reattachment, large density gradients, the formation of jets and shear layers and adverse pressure gradients. Cavitating flows are also inherently transient [89–91], exhibiting unsteady cavitation cloud shedding downstream of the throat. It has been observed that (for example, see Reboud et al. [90]) the standard two-equation models, in particular the k-ε model, tend to overestimate the turbulent viscosity in the cavitating region, leading to unrealistic predictions of a stable cavity zone. An arbitrary pre-multiplier to the turbulent viscosity term, μ_t, has been proposed for improving capture of unsteady effects. Arbitrary modifiers of this type have yielded success for other researchers in this field, with a number of examples showing successful capture of cavitation cloud shedding with two-equation RANS approaches [91–93]. Similar to these axial flow-based devices, high-swirl ratio cavitating vortices present a different set of modeling challenges, and are also known to exhibit periodic unsteady behavior [94, 95]. Arndt et al. have highlighted some of the complexities in cavitating vortices which feature unsteady behavior over a broad frequency spectra, accompanied by complex cavity collapse characteristics [96, 97]. Pandare et al. [98] highlighted the presence of a periodic, precessing vortex, exhibiting frequencies of the order of 60 Hz. Their study indicated that the *k-ε model* under-predicts the swirl velocity magnitudes, while the *k-ω SST and RSM* models showed close agreement, successfully capturing transient oscillations. Based on the studies of Pandare and Ranade [98], and Simpson and Ranade [6], the Shear Stress Transport (*SST) k-ω* model of Menter [99] may be recommended. This model has demonstrated superior predictions to other two-equation approaches in situations involving flow separation and adverse pressure gradients (see for example Bardina et al. [100]). The rationale for selecting this turbulence model is further discussed later in Section 3.2, while discussing the results from different turbulence models. The turbulent viscosity, μ_t, in the *SST k-ω* model is defined as:

$$\mu_t = \frac{\rho k}{\omega} \tag{A5.4}$$

The transport equations for k and ω are then written as follows:

$$\frac{\partial}{\partial t}(\rho k) + \frac{\partial}{\partial x_i}(\rho k u_i) = \frac{\partial}{\partial x_j}\left(\Gamma_k \frac{\partial k}{\partial x_j}\right) + G_k - Y_k + S_k \tag{A5.5}$$

and:

$$\frac{\partial}{\partial t}(\rho \omega) + \frac{\partial}{\partial x_i}(\rho \omega u_i) = \frac{\partial}{\partial x_j}\left(\Gamma_\omega \frac{\partial \omega}{\partial x_j}\right) + G_\omega - Y_\omega + S_\omega \tag{A5.6}$$

G represents the generation, Γ_k and Γ_ω are effective diffusivity of k and ω, respectively, and Y is the dissipation due to turbulence [101]. S_k and S_w are additional user-defined source terms.

Most of the linear/axial flow-based HC devices exhibit some level of flow separation downstream of the restriction. Swirling/rotating flow-based devices may exhibit a stagnation zone in the vortex core. Additional measures are therefore usually needed to mitigate against over-prediction of the turbulence production term, G_k. The limiting formulations for the production term, G_k, proposed by Menter [102], may be used:

$$G_k = \min\left[G_k, C_{lim}\rho k\right] \tag{A5.7}$$

where the limiting coefficient C_{lim} is set as 0.9 in the original publication. An additional limiting function, proposed by Kato [103], may also be applied. This is based on the observation that high turbulence kinetic energy production in stagnation regions is generally caused by the calculation of excessive shear strain rate, γ. As such, this is corrected by the local vorticity rate, Ω, which tends to lower values in stagnation regions:

$$G_k = \mu_t \gamma \Omega \tag{A5.8}$$

In addition to turbulence modeling, it is essential to model phase change from liquid to vapor and vice versa while simulating cavitating flows. Various mass-transfer models have been proposed to describe the cavitation process, with the most commonly used approaches based on reduced forms of the Rayleigh–Plesset equation. Examples include the models proposed by Schnerr and Sauer [104], Zwart et al. [105], and Singhal et al. [106]. The simulated results are not found to be sensitive to a specific choice of cavitation model. Considering various applications and validations for a wide range of flow cases, the model of Singhal et al. may be used, at least as a starting model (see examples in [106, 107]). It has an advantage that the bubble number per unit volume, n, need not be prescribed as input. In this model, the vapor volume fraction is computed locally from a transport equation for the vapor mass fraction, f, [14], which introduces an additional pair of mass source and sink terms for the evaporation (R_e) and condensation (R_c) of the vapor:

$$\frac{\partial}{\partial t}(\rho_m f) + \nabla\cdot(\rho_m \vec{v}_m f) = \nabla\cdot(\Gamma \nabla f) + R_e - R_c \tag{A5.9}$$

where

$$\frac{1}{\rho_m} = \frac{f}{\rho_v} + \frac{1-f}{\rho_l} \tag{A5.10}$$

Here, ρ_m, ρ_v, and ρ_l refer to the densities of the mixture, vapor, and liquid, respectively. The vapor volume fraction, α, can then be calculated as follows:

$$\alpha = f\frac{\rho_m}{\rho_v} \tag{A5.11}$$

In the Singhal approach [106], the evaporation and condensation terms are derived from a reduced form of the Rayleigh–Plesset equation, commonly shortened to the R-P equation [79]:

$$\rho\left[R_B\frac{d^2R_B}{dt^2} + \frac{3}{2}\left(\frac{dR_B}{dt}\right)^2\right] = p_B - p - \frac{2\sigma}{R_B} - \frac{4\mu}{R_B}\frac{dR_B}{dt} \tag{A5.12}$$

In this expression, R_B *is the bubble radius*, p_B refers to bubble surface pressure, and p is the local liquid phase pressure. Full derivation of the R-P equation is presented in Brennen [79], and numerous examples of solutions of the R-P equation exist in literature, for example Alehossein and Qin [108]. Commonly, when modeling cavitation mass transfer mechanisms in CFD codes, the R-P equation for bubble growth is used to approximate void propagation. To derive mass transfer terms compatible with the Eulerian–Eulerian multiphase approach, the surface tension, viscous damping, and higher-order acceleration terms in Eq. (A5.12) are neglected to produce a mass transfer rate term of the following form:

$$\frac{dR_B}{dt} \cong (-1)^n\sqrt{\frac{2}{3}\frac{(p_B - p)}{\rho_l}} \tag{A5.13}$$

where $n = 1$ during bubble expansion/evaporation, and $n = 2$ during the condensation phase. Using this approach, Singhal et al. [106] derived a simplified vapor transport equation:

$$\frac{\partial}{\partial t}(\rho_m f) + \nabla\cdot(\rho_m \vec{v}_m f) = (n4\pi)^{1/3} + (3\alpha)^{1/3}\frac{\rho_v\rho_l}{\rho}\left[\frac{2}{3}\left(\frac{p_B - p}{\rho_l}\right)\right]^{1/2} \tag{A5.14}$$

Equation (5.19) is referred to as the *Reduced Bubble Dynamics Formulation*. All terms in this expression, except n, the bubble number density, are either known constants or dependent variables. To avoid having to specify a bubble number density, the phase change expression is rewritten in terms of bubble radius:

$$R_e = \frac{3\alpha}{R_B}\frac{\rho_v\rho_l}{\rho_m}\left[\frac{2}{3}\left(\frac{p_B - p}{\rho_l}\right)\right]^{1/2} \tag{A5.15}$$

The typical bubble size, R_B, is taken to be equal to the limiting (maximum possible) bubble size using a correlation commonly used in the nuclear industry:

$$R_B = \frac{0.061 We\sigma}{2\rho_l u_{rel}^2} \tag{A5.16}$$

where We is the Weber number and σ is the surface tension; the Weber number is given by the following expression:

$$We = \frac{\rho_l u_l R_B}{\sigma} \tag{A5.17}$$

In which ρ_l and u_l are the liquid density and velocity, respectively. In the Singhal model, the square of the relative velocity term is approximated as a linear characteristic velocity, $u_{rel}^2 = u_{ch} = \sqrt{k}$. The relative velocities and turbulent velocity fluctuations are of the same order (1–10%), and as such $\sqrt{k}$ is considered to be a suitable approximation for u_{ch}. This produces the following final pair of phase change rate terms:

$$R_e = C_1 \frac{\max(1.0, \sqrt{k})}{\sigma} \rho_v \rho_l \left[\frac{2}{3}\left(\frac{p - p_v}{\rho_l}\right)\right]^{1/2} (1 - f_v - f_g) \tag{A5.18}$$

$$R_c = C_2 \frac{\max(1.0, \sqrt{k})}{\sigma} \rho_v \rho_l \left[\frac{2}{3}\left(\frac{p - p_v}{\rho_l}\right)\right]^{1/2} f_v \tag{A5.19}$$

Here C_1 and C_2 are empirical constants, for which Singhal et al. [106] recommend values of 0.2 and 0.01, respectively, following their extensive studies on a range of sharp-edged orifice and hydrofoil flows. Additionally, in order to take account of the effect of the magnitude of local turbulent pressure fluctuations (p'_{turb}), a modified term for the local saturated vapor pressure p'_v is calculated as a function of the local turbulent kinetic energy, k, and the liquid saturated vapor pressure, p_v, as given by the following relationships:

$$p'_{turb} = 0.39 \rho_m k \tag{A5.20}$$

$$p'_v = p_v + p'_{turb}/2 \tag{A5.21}$$

where ρ_m is the mixture density, k is the turbulent kinetic energy, and the constant value of 0.39 is taken from [106].

These model equations may then be solved using well established solvers (see Ranade [68] for more details). Once the flow field is obtained, it may be used to simulate trajectories of individual cavities using the Lagrangian approach [68]. Since the multiphase flow field is computed using the models described previously, usually one-way coupling is assumed while simulating cavity trajectories. The cavity trajectories are driven by the primary flow gradients and turbulence quantities. The particle trajectories are computed by integrating the force balance for a discrete particle of a series of discrete time steps; the force balance is given as:

$$\frac{d\vec{u}_p}{dx} = \frac{\vec{u} - \vec{u}_p}{\tau_r} + \frac{\vec{g}(\rho_p - \rho)}{\rho_p} + \vec{F} \tag{A5.22}$$

This equates the particle inertia with the forces acting on the particle. The first term on the right-hand side of the equation is the drag force per unit mass of the particle, and the second term is the force due to gravity. The final term, $\vec{F}$, is an additional acceleration term, through which additional force terms can be incorporated into the overall balance to account for phenomena such as virtual mass, pressure gradient forces, and particle rotation. The influence of continuous phase turbulence on the tracked particles can be accounted for by separating the velocity, u, into the sum of the mean and instantaneous components:

$$u = \overline{u} + u' \tag{A5.23}$$

Different models are available to approximate instantaneous velocity from the simulated mean velocity field and turbulence quantities (for example, see Ranade [68]). Generally, the simple, the discrete random walk model, or the "eddy lifetime" model is used. In this approach, each discrete particle is considered to interact with a succession of discrete turbulent eddies which modify their instantaneous velocities [101]. This involves introducing two modeled terms: first, the random fluctuating component of velocity is calculated as a function of the local turbulent kinetic energy value:

$$u' = \zeta\sqrt{2k/3} \tag{A5.24}$$

where ζ is a normally distributed random number. Second, the concept of a particle eddy lifetime, T_L, is introduced to define the time intervals over which this random fluctuating component is updated. This "eddy lifetime" is approximated as a function of the local turbulence frequency:

$$T_L \approx 0.15\frac{k}{\varepsilon} \tag{A5.25}$$

Additional limits can be placed on the maximum time step size; in this study, a minimum of five time steps is also imposed across any given computational cell.

The book by Ranade [68], and several other excellent resources are available on discussions of numerical solution of multiphase CFD models and best practices associated with it (for example, Marchisio and Fox [109] and Yeoh and Tu [110]). The discussion on numerics and other aspects of CFD simulations is therefore not included here.

References

1 Pollet, B.G. and Ashokkumar, M. (ed.) (2019). Fundamental and applied aspects of ultrasonics and sonochemistry. In: *Introduction to Ultrasound, Sonochemistry and Sonoelectrochemistry*, 1–19. Springer.

2 Rasthofer, U., Wermelinger, F., Hadijdoukas, P., and Koumoutsakos, P. (2017). Large scale simulation of cloud cavitation collapse. *Procedia Computer Science* 108: 1763–1772.

3 Shah, Y.T., Pandit, A.B., and Moholkar, V.S. (1999). *Cavitation Reaction Engineering*. Springer Science & Business Media.

4 Gogate, P.R. and Pandit, A.B. (2005). A review and assessment of hydrodynamic cavitation as a technology for the future. *Ultrasonics Sonochemistry* 12 (1–2): 21–27.

5 Ma, J., Hsiao, C.-T., and Chahine, G.L. (2017). A physics based multiscale modeling of cavitating flows. *Computers & Fluids* 145: 68–84.

6 Sarvothaman, V.P., Simpson, A.T., and Ranade, V.V. (2019). Modelling of vortex based hydrodynamic cavitation reactors. *Chemical Engineering Journal* 377: 119639.

7 Suslick, K.S., Hammerton, D.A., and Cline, R.E. (1986). Sonochemical hot spot. *Journal of the American Chemical Society* 108 (18): 5641–5642.

8 Suslick, K.S. (1989). The chemical effects of ultrasound. *Scientific American* 260 (2): 80–87.

9 Flint, E.B. and Suslick, K.S. (1991). The temperature of cavitation. *Science* 253 (5026): 1397–1399.

10 Théron, P., Pichat, P., Guillard, C. et al. (1999). Degradation of phenyltrifluoromethylketone in water by separate or simultaneous use of TiO 2 photocatalysis and 30 or 515 kHz ultrasound. *Physical Chemistry Chemical Physics* 1 (19): 4663–4668.

11 Ashokkumar, M., Mulvaney, P., and Grieser, F. (1999). The effect of pH on multibubble sonoluminescence from aqueous solutions containing simple organic weak acids and bases. *Journal of the American Chemical Society* 121 (32): 7355–7359.

12 Bagal, M.V. and Gogate, P.R. (2014). Degradation of diclofenac sodium using combined processes based on hydrodynamic cavitation and heterogeneous photocatalysis. *Ultrasonics Sonochemistry* 21 (3): 1035–1043.

13 Thanekar, P., Lakshmi, N.J., Shah, M. et al. (2020). Degradation of dimethoate using combined approaches based on hydrodynamic cavitation and advanced oxidation processes. *Process Safety and Environmental Protection* 143: 222–230.

14 Musmarra, D., Prisciandaro, M., Capocelli, M. et al. (2016). Degradation of ibuprofen by hydrodynamic cavitation: reaction pathways and effect of operational parameters. *Ultrasonics Sonochemistry* 29: 76–83.

15 Raut-Jadhav, S., Saharan, V.K., Pinjari, D. et al. (2013). Synergetic effect of combination of AOP's (hydrodynamic cavitation and H2O2) on the degradation of neonicotinoid class of insecticide. *Journal of Hazardous Materials* 261: 139–147.

16 Raut-Jadhav, S., Saini, D., Sonawane, S., and Pandit, A. (2016). Effect of process intensifying parameters on the hydrodynamic cavitation based degradation of commercial pesticide (methomyl) in the aqueous solution. *Ultrasonics Sonochemistry* 28: 283–293.

17 Gore, M.M., Saharan, V.K., Pinjari, D.V. et al. (2014). Degradation of reactive orange 4 dye using hydrodynamic cavitation based hybrid techniques. *Ultrasonics Sonochemistry* 21 (3): 1075–1082.

18 Arrojo, S., Benito, Y., and Tarifa, A.M. (2008). A parametrical study of disinfection with hydrodynamic cavitation. *Ultrasonics Sonochemistry* 15 (5): 903–908.

19 Choi, J., Cui, M., Lee, Y. et al. (2018). Hydrodynamic cavitation and activated persulfate oxidation for degradation of bisphenol A: Kinetics and mechanism. *Chemical Engineering Journal* 338: 323–332.

20 Ye, Y.-F., Zhu, Y., Lu, N. et al. (2021). Treatment of rhodamine B with cavitation technology: comparison of hydrodynamic cavitation with ultrasonic cavitation. *RSC Advances* 11 (9): 5096–5106.

21 Yi, C., Lu, Q., Wang, Y. et al. (2018). Degradation of organic wastewater by hydrodynamic cavitation combined with acoustic cavitation. *Ultrasonics Sonochemistry* 43: 156–165.

22 Braeutigam, P., Franke, M., Schneider, R.J. et al. (2012). Degradation of carbamazepine in environmentally relevant concentrations in water by Hydrodynamic-Acoustic-Cavitation (HAC). *Water Research* 46 (7): 2469–2477.

23 Ranade, V.V., Prasad Sarvothaman, V., Simpson, A., and Nagarajan, S. (2021). Scale-up of vortex based hydrodynamic cavitation devices: a case of degradation of di-chloro aniline in water. *Ultrasonics Sonochemistry* 70: 105295.

24 Sarvothaman, V.P., Nagarajan, S., and Ranade, V.V. (2018). Treatment of solvent contaminated water using vortex-based cavitation: influence of operating pressure drop, temperature, aeration, and reactor scale. *Industrial & Engineering Chemistry Research* 57 (28): 9292–9304.

25 Ranade, N.V., Nagarajan, S., Sarvothaman, V., and Ranade, V.V. (2021). ANN based modelling of hydrodynamic cavitation processes: biomass pre-treatment and wastewater treatment. *Ultrasonics Sonochemistry* 72: 105428.

26 Himmelblau, D.M. (2000). Applications of artificial neural networks in chemical engineering. *Korean Journal of Chemical Engineering* 17 (4): 373–392.

27 Panerati, J., Schnellmann, M.A., Patience, C. et al. (2019). Experimental methods in chemical engineering: artificial neural networks–ANNs. *The Canadian Journal of Chemical Engineering* 97 (9): 2372–2382.

28 Venkatasubramanian, V. (2019). The promise of artificial intelligence in chemical engineering: Is it here, finally? *AIChE Journal* 65 (2): 466–478.

29 Heaton, J. (2008). *Introduction to Neural Networks with Java*. Heaton Research, Inc.

30 Abbasi, E., Saadat, S., Jashni, A.K., and Shafaei, M.H. (2020). A novel method for optimization of slit Venturi dimensions through CFD simulation and RSM design. *Ultrasonics Sonochemistry* 67: 105088.

31 Dastane, G.G., Thakkar, H., Shah, R. et al. (2019). Single and multiphase CFD simulations for designing cavitating venturi. *Chemical Engineering Research and Design* 149: 1–12.

32 Simpson, A. and Ranade, V.V. (2019). 110th anniversary: comparison of cavitation devices based on linear and swirling flows: hydrodynamic characteristics. *Industrial & Engineering Chemistry Research* 58 (31): 14488–14509.

33 Pawar, S.K., Mahulkar, A.V., Pandit, A.B. et al. (2017). Sonochemical effect induced by hydrodynamic cavitation: comparison of venturi/orifice flow geometries. *AIChE Journal* 63 (10): 4705–4716.

34 Moholkar, V.S. and Pandit, A.B. (2001). Numerical investigations in the behaviour of one-dimensional bubbly flow in hydrodynamic cavitation. *Chemical Engineering Science* 56 (4): 1411–1418.

35 Gogate, P.R. and Pandit, A.B. (2000). Engineering design methods for cavitation reactors II: hydrodynamic cavitation. *AIChE Journal* 46 (8): 1641–1649.

36 Apfel, R.E. (1981). Acoustic cavitation. In: *Methods in Experimental Physics*, vol. 19, 355–411. Elsevier.

37 Flynn, H.G. (1964). Physics of acoustic cavitation in liquids. *Physical Acoustics* 1 (Part B): 57–172.

38 Capocelli, M., Musmarra, D., Prisciandaro, M., and Lancia, A. (2014). Chemical effect of hydrodynamic cavitation: simulation and experimental comparison. *AIChE Journal* 60 (7): 2566–2572.

39 Rayleigh, L.V.I.I.I. (1917). On the pressure developed in a liquid during the collapse of a spherical cavity. *The London, Edinburgh, and Dublin Philosophical Magazine and Journal of Science* 34 (200): 94–98.

40 Plesset, M.S. and Prosperetti, A. (1977). Bubble dynamics and cavitation. *Annual Review of Fluid Mechanics* 9 (1): 145–185.

41 Gilmore, F.R. (1952). The growth or collapse of a spherical bubble in a viscous compressible liquid. PhD thesis. California Institute of Technology, Pasadena.

42 Keller, J.B. and Miksis, M. (1980). Bubble oscillations of large amplitude. *The Journal of the Acoustical Society of America* 68 (2): 628–633.

43 Toegel, R., Gompf, B., Pecha, R., and Lohse, D. (2000). Does water vapor prevent upscaling sonoluminescence? *Physical Review Letters* 85 (15): 3165.

44 Kerboua, K. and Hamdaoui, O. (2017). Computational study of state equation effect on single acoustic cavitation bubble's phenomenon. *Ultrasonics Sonochemistry* 38: 174–188.

45 Chakma, S. and Moholkar, V.S. (2013). Numerical simulation and investigation of system parameters of sonochemical process. *Chinese Journal of Engineering* 2013: 362682.

46 Gireesan, S. and Pandit, A.B. (2017). Modeling the effect of carbon-dioxide gas on cavitation. *Ultrasonics Sonochemistry* 34: 721–728.

47 Koch, M., Lechner, C., Reuter, F. et al. (2016). Numerical modeling of laser generated cavitation bubbles with the finite volume and volume of fluid method, using OpenFOAM. *Computers & Fluids* 126: 71–90.

48 Storey, B.D. and Szeri, A.J. (2000). Water vapour, sonoluminescence and sonochemistry. *Proceedings of the Royal Society of London, Series A: Mathematical, Physical and Engineering Sciences* 456 (1999): 1685–1709.

49 Kalmár, C., Klapcsik, K., and Hegedűs, F. (2020). Relationship between the radial dynamics and the chemical production of a harmonically driven spherical bubble. *Ultrasonics Sonochemistry* 64: 104989.

50 Yasui, K., Tuziuti, T., Lee, J. et al. (2008). The range of ambient radius for an active bubble in sonoluminescence and sonochemical reactions. *The Journal of Chemical Physics* 128 (18): 184705.

51 Philipp, A. and Lauterborn, W. (1998). Cavitation erosion by single laser-produced bubbles. *Journal of Fluid Mechanics* 361: 75–116.

52 Supponen, O., Obreschkow, D., Kobel, P. et al. (2017). Shock waves from non-spherical cavitation bubbles. *Physical Review Fluids* 2 (9): 093601.

53 Sawant, S.S., Anil, A.C., Krishnamurthy, V. et al. (2008). Effect of hydrodynamic cavitation on zooplankton: A tool for disinfection. *Biochemical Engineering Journal* 42 (3): 320–328.

54 Peng, K., Tian, S., Li, G. et al. (2018). Bubble dynamics characteristics and influencing factors on the cavitation collapse intensity for self-resonating cavitating jets. *Petroleum Exploration and Development* 45 (2): 343–350.

55 Pandit, A.V., Sarvothaman, V.P., and Ranade, V.V. (2021). Estimation of chemical and physical effects of cavitation by analysis of cavitating single bubble dynamics. *Ultrasonics Sonochemistry* 77: 105677.

56 Pecha, R. and Gompf, B. (2000). Microimplosions: cavitation collapse and shock wave emission on a nanosecond time scale. *Physical Review Letters* 84 (6): 1328.

57 Tao, Y., Cai, J., Huai, X., and Liu, B. (2018). Global average hydroxyl radical yield throughout the lifetime of cavitation bubbles. *Chemical Engineering & Technology* 41 (5): 1035–1042.

58 Rooze, J., Rebrov, E.V., Schouten, J.C., and Keurentjes, J.T.F. (2013). Dissolved gas and ultrasonic cavitation–a review. *Ultrasonics Sonochemistry* 20 (1): 1–11.

59 Naidu, D.V.P., Rajan, R., Kumar, R. et al. (1994). Modelling of a batch sonochemical reactor. *Chemical Engineering Science* 49 (6): 877–888.

60 Rajan, R., Kumar, R., and Gandhi, K.S. (1998). Modelling of sonochemical oxidation of the water-KI-CCl4 system. *Chemical Engineering Science* 53 (2): 255–271.

61 Kanthale, P.M., Gogate, P.R., Pandit, A.B., and Wilhelm, A.M. (2003). Cavity cluster approach for quantification of cavitational intensity in sonochemical reactors. *Ultrasonics Sonochemistry* 10 (4–5): 181–189.

62 IIsiao, C.-T., Ma, J., and Chahine, G.L. (2017). Multiscale tow-phase flow modeling of sheet and cloud cavitation. *International Journal of Multiphase Flow* 90: 102–117.

63 Simpson, A. and Ranade, V.V. (2018). Modelling of hydrodynamic cavitation with orifice: Influence of different orifice designs. *Chemical Engineering Research and Design* 136: 698–711.

64 Liu, Z. and Brennen, C.E. (1995). Models of cavitation event rates. *International Symposium on Cavitation, CAV '95*, 321–328 (May 1995), Deauville, France.

65 Sharma, A., Gogate, P.R., Mahulkar, A., and Pandit, A.B. (2008). Modeling of hydrodynamic cavitation reactors based on orifice plates considering hydrodynamics and chemical reactions occurring in bubble. *Chemical Engineering Journal* 143 (1–3): 201–209.

66 Šarc, A., Stepišnik-Perdih, T., Petkovšek, M., and Dular, M. (2017). The issue of cavitation number value in studies of water treatment by hydrodynamic cavitation. *Ultrasonics Sonochemistry* 34: 51–59.

67 Simpson, A. and Ranade, V.V. (2019). Flow characteristics of vortex based cavitation devices: computational investigation on influence of operating parameters and scale. *AIChE Journal* 65: e16675.

68 Ranade, V.V. (2002). *Computational Flow Modeling for Chemical Reactor Engineering*. Academic Press.

69 Hilares, R.T., Dionízio, R.M., Muñoz, S.S. et al. (2020). Hydrodynamic cavitation-assisted continuous pre-treatment of sugarcane bagasse for ethanol production: effects of geometric parameters of the cavitation device. *Ultrasonics Sonochemistry* 63: 104931.

70 Bimestre, T.A., Júnior, J.A.M., Botura, C.A. et al. (2020). Theoretical modeling and experimental validation of hydrodynamic cavitation reactor with a Venturi tube for sugarcane bagasse pretreatment. *Bioresource Technology* 311: 123540.

71 Nagarajan, S. and Ranade, V.V. (2019). Pretreatment of lignocellulosic biomass using vortex-based devices for cavitation: influence on biomethane potential. *Industrial & Engineering Chemistry Research* 58 (35): 15975–15988.

72 Hilares, R.T., de Almeida, G.F., Ahmed, M.A. et al. (2017). Hydrodynamic cavitation as an efficient pretreatment method for lignocellulosic biomass: a parametric study. *Bioresource Technology* 235: 301–308.

73 Wu, Z., Ferreira, D.F., Crudo, D. et al. (2019). Plant and biomass extraction and valorisation under hydrodynamic cavitation. *Processes* 7 (12): 965.

74 Baradaran, S. and Sadeghi, M.T. (2019). Intensification of diesel oxidative desulfurization via hydrodynamic cavitation. *Ultrasonics Sonochemistry* 58: 104698.

75 Suryawanshi, N.B., Bhandari, V.M., Sorokhaibam, L.G., and Ranade, V.V. (2017). Developing techno-economically sustainable methodologies for deep desulfurization using hydrodynamic cavitation. *Fuel* 210: 482–490.

76 Thaker, A.H. and Ranade, V.V. Drop breakage in a single pass through vortex based cavitation device: experiments and modelling. *AIChE Journal* e17512.

77 Kumar, S. and Ramkrishna, D. (1996). On the solution of population balance equations by discretization – I. A fixed pivot technique. *Chemical Engineering Science* 51 (8): 1311–1332.

78 Laakkonen, M., Alopaeus, V., and Aittamaa, J. (2006). Validation of bubble breakage, coalescence and mass transfer models for gas–liquid dispersion in agitated vessel. *Chemical Engineering Science* 61 (1): 218–228.

79 Brennen, C.E. (2014). *Cavitation and Bubble Dynamics*. Cambridge University Press.

80 Orthaber, U., Zevnik, J., and Dular, M. (2020). Cavitation bubble collapse in a vicinity of a liquid-liquid interface – basic research into emulsification process. *Ultrasonics Sonochemistry* 68: 105224.

81 Chen, L., Kang, Q., Mu, Y. et al. (2014). A critical review of the pseudopotential multiphase lattice Boltzmann model: methods and applications. *International Journal of Heat and Mass Transfer* 76: 210–236.

82 Yang, J. (2015). Simulation on cavitation bubble collapsing with lattice Boltzmann method. *Journal of Applied Mathematics and Physics* 3 (8): 947.

83 Chen, X. (2010). Simulation of 2D cavitation bubble growth under shear flow by lattice Boltzmann model. *Communications in Computational Physics* 7 (1): 212.

84 Su, Y., Tang, X., Wang, F. et al. (2017). Three-dimensional cavitation bubble simulations based on lattice Boltzmann model coupled with Carnahan-Starling equation of state. *Communications in Computational Physics* 22 (2): 473–493.

85 Shan, M., Zhu, Y., Yao, C. et al. (2017). Modeling for collapsing cavitation bubble near rough solid wall by mulit-relaxation-time pseudopotential lattice Boltzmann model. *Journal of Applied Mathematics and Physics* 5 (6): 1243–1256.

86 Trummler, T., Schmidt, S.J., and Adams, N.A. (2021). Effect of stand-off distance and spatial resolution on the pressure impact of near-wall vapor bubble collapses. *International Journal of Multiphase Flow* 141: 103618.

87 Rasthofer, U., Wermelinger, F., Karnakov, P. et al. (2019). Computational study of the collapse of a cloud with 12 500 gas bubbles in a liquid. *Physical Review Fluids* 4 (6): 063602.

88 Shi, Y., Luo, K., Chen, X. et al. (2021). A new cavitation model considering inter-bubble action. *International Journal of Naval Architecture and Ocean Engineering* 13: 566–574.

89 Sato, K., Taguchi, Y., and Hayashi, S. (2013). High speed observation of periodic cavity behavior in a convergent-divergent nozzle for cavitating water jet. *Journal of Flow Control, Measurement & Visualization* 2013: 102–107.

90 Reboud, J.-L., Stutz, B., and Coutier, O. (1998). Two phase flow structure of cavitation: experiment and modeling of unsteady effects. *3rd International Symposium on Cavitation CAV1998*, vol. 26. Grenoble, France.

91 Charrière, B. and Goncalves, E. (2017). Numerical investigation of periodic cavitation shedding in a Venturi. *International Journal of Heat and Fluid Flow* 64: 41–54.

92 Charrière, B., Decaix, J., and Goncalvès, E. (2015). A comparative study of cavitation models in a Venturi flow. *European Journal of Mechanics – B/Fluids* 49: 287–297.

93 Li, D., Grekula, M., and Lindell, P. (2009). A modified SST k-ω turbulence model to predict the steady and unsteady sheet cavitation on 2D and 3D hydrofoils. *CAV2009 - 7th International Symposium on Cavitation* (16–20 August 2009), Ann Arbor, MI. https://hdl.handle.net/2027.42/84288.

94 Hreiz, R., Gentric, C., and Midoux, N. (2011). Numerical investigation of swirling flow in cylindrical cyclones. *Chemical Engineering Research and Design* 89 (12): 2521–2539.

95 Yin, J., Li, J., Ma, Y. et al. (2015). Study on the air core formation of a gas–liquid separator. *Journal of Fluids Engineering* 137 (9): 1–9.

96 Arndt, R.E.A., Arakeri, V.H., and Higuchi, H. (1991). Some observations of tip-vortex cavitation. *Journal of Fluid Mechanics* 229: 269–289.

97 Arndt, R., Pennings, P., Bosschers, J., and Van Terwisga, T. (2015). The singing vortex. *Interface Focus* 5 (5): 20150025.

98 Pandare, A. and Ranade, V.V. (2015). Flow in vortex diodes. *Chemical Engineering Research and Design* 102: 274–285.

99 Menter, F.R. (1994). Two-equation eddy-viscosity turbulence models for engineering applications. *AIAA Journal* 32 (8): 1598–1605.

100 Bardina, J., Huang, P., Coakley, T. et al. (1997). Turbulence modeling validation. *28th Fluid Dynamics Conference, American Institute of Aeronautics and Astronautics* (6.1997-2121). https://doi.org/10.2514/6.1997-2121.

101 Fluent, A. (2016). *17.0 ANSYS Fluent Theory Guide*. Canonsburg, PA: ANSYS Inc.

102 Menter, F.R., Kuntz, M., and Langtry, R. (2003). Ten years of industrial experience with the SST turbulence model. *Turbulence, Heat and Mass Transfer* 4 (1): 625–632.

103 Kato, M. (1993). The modelling of turbulent flow around stationary and vibrating square cylinders. *Turbulent Shear Flow* 1: 10–14.

104 Schnerr, G.H. and Sauer, J. (ed.) (2001). *Physical and Numerical Modeling of Unsteady Cavitation Dynamics*. New Orleans: ICMF.

105 Zwart, P.J., Gerber, A.G., and Belamri, T. (2004). A two-phase flow model for predicting cavitation dynamics. *Fifth International Conference on Multiphase Flow* (30 May 2004), Yokohama, Japan (paper number 152).

106 Singhal, A.K., Athavale, M.M., Li, H., and Jiang, Y. (2002). Mathematical basis and validation of the full cavitation model. *Journal of Fluids Engineering* 124 (3): 617–624.

107 Ebrahimi, B., He, G., Tang, Y. et al. (2017). Characterization of high-pressure cavitating flow through a thick orifice plate in a pipe of constant cross section. *International Journal of Thermal Sciences* 114: 229–240.

108 Alehossein, H. and Qin, Z. (2007). Numerical analysis of Rayleigh–Plesset equation for cavitating water jets. *International Journal for Numerical Methods in Engineering* 72 (7): 780–807.

109 Marchisio, D.L. and Fox, R.O. (2007). *Multiphase Reacting Flows: Modelling and Simulation*. Springer.

110 Yeoh, G.H. and Tu, J. (2019). *Computational Techniques for Multiphase Flows*. Butterworth-Heinemann.

111 Schiller, L. and Naumann, A. (1933). Drag coefficient for spherical shape. *VDI Zeits.* 13: 318.

Part IV

Applications of Hydrodynamic Cavitation

6

Disinfection of Water

6.1 Introduction

Disinfection of water is essential to remove pathogenic microorganisms that are responsible for a number of waterborne diseases, as 88% of the diseases in the developing world are caused by unsafe drinking water. Although drinking water typically has a large number of contaminants such as chemicals, color, metal impurities, fluoride, arsenic, and nitrates, the most serious contamination is due to the presence of harmful microorganisms and pathogens for the reasons of associated health risks to humans. The list of harmful microorganisms occurring in drinking water mainly includes pathogenic bacteria (e.g., *E. coli*), viruses (e.g., Hepatitis A virus), protozoa, and helminthes (e.g., *Cryptosporidium parvum*). The World Health Organization (WHO) estimates that more than 3.4 million people die every year from waterborne diseases caused by bacterial contamination of water and that disinfection using suitable methods is essential for providing safe drinking water to the human population [1–3]. The significant waterborne viruses include Rotavirus, Norovirus, Astrovirus, Sapovirus, Adenovirus, Hepatitis A and E, and Enterovirus, many of which are responsible for gastroenteritis. The problems of unsafe drinking water exist worldwide and are severe in regions such as Asia [4]. The world population is increasing, along with growing industrialization. The increase in the population is reflected in environmental pollution by sewage water, apart from other biological pollutants; whereas the industrial growth results in industrial effluents that are released into the water bodies, with a large number of organic and inorganic pollutants that adversely impact life. The overall standards of drinking-water may vary worldwide and no single and universally acceptable approach for water treatment has been found. Thus, as emphasized by the World Health Organization, "The potential health consequences of microbial contamination are such that its control must always be of paramount importance and must never be compromised"[5]. A large number of waterborne diseases have been identified over the years, such as amoebiasis, shigellosis, cholera, typhoid, and hepatitis A or E. This clearly highlights the importance of providing safe drinking

Hydrodynamic Cavitation: Devices, Design, and Applications, First Edition.
Vivek V. Ranade, Vinay M. Bhandari, Sanjay Nagarajan, Varaha P. Sarvothaman and Alister T. Simpson.

water to safeguard millions of lives. According to the prescribed norms of WHO, the fecal and total coliform bacterial count in drinking water should be zero.

Naturally, a large number of processes have been well established over the years with standard operating practices to disinfect water and render it safe for human consumption. In view of the huge variations in terms of the capital investment and operating costs of the different technologies, apart from the limitations/disadvantages of each, the selection as well as the methods of practical implementation of disinfection technology vary from place to place, between developed countries and developing countries, rural and urban areas, and so on. A number of guidelines are also provided periodically by various organizations such as the World Health Organization and local governing bodies (WHO; EPA). A large number of newer processes/technologies are also being constantly developed to reduce the cost of treatment and to effectively eliminate harmful microorganisms, with measures for their easy implementation at different locations so that safe drinking water can be made available to everyone.

Hydrodynamic cavitation is one technology that has the potential to destroy harmful microorganisms, reduce the cost of disinfection when compared to existing chemical processes, and eliminate the formation of harmful byproducts associated with many of the chemical methods of disinfection that are most popularly being used today. Hydrodynamic cavitation can be used as a standalone technology without use of any chemicals, and can either be used with process intensification or in integration with existing water disinfection technologies (for reducing the overall cost and increasing efficiency), or in hybrid form with appropriate additives for increased competitiveness. Hydrodynamic cavitation, *per se*, is not a new technology since its application for disinfection of water has been reported for over two decades. A large number of cavitating reactors were discussed and their efficacy for elimination of different types of microorganisms was evaluated. Many theoretical/mathematical models were evaluated to enable prediction of disinfection behavior, many using computational fluid dynamics (CFD) studies for estimating the flow behavior. Unfortunately, most of these studies were not successful to the extent required to enable commercial implementation. Thus, it is imperative to not just review the progress made in this regard so far, but also to highlight the aspects that can enable practical application of hydrodynamic cavitation in real life, either as an alternative to the existing practices or in a combination with existing processes. A hybrid process of hydrodynamic cavitation using natural oils having antimicrobial properties certainly shows clear promise in this regard [6].

The importance of disinfection and of hydrodynamic cavitation for disinfection has attracted wide attention in recent years again, after the initial reports that were not so encouraging, and the subject has been reviewed recently in different forms by various researchers [7–10]. Since quantitative microbial risk assessment for all types of harmful pathogens can be a Himalayan task, the conventional approach is that researchers report studies on the most common reference pathogens from each class (e.g., one each from bacteria, virus, and protozoa) so that the risk assessment is made in this regard and the data/results can be extended over other important pathogens within each class – thereby adopting a generalized approach. The important and most relevant questions that need to be addressed are therefore:

1. The characteristics of the water to be treated (source, contamination, pollutants).
2. The characteristics of harmful pathogens (nature, concentration, etc.).
3. The desired quality of the water (most stringent for drinking water).
4. The nature of the treatment technology (use of chemicals, additives, efficiency, etc.).
5. Harmful effects of technology (disinfection by-products, chemicals, etc.).
6. Place and ease of implementation of technology (rural/urban, household, etc.).
7. Cost of disinfection (single or hybrid method, scalability, operation, maintenance).

The present chapter is not intended to provide an exhaustive review, but to provide **a fresh, in-depth analysis** of the disinfection work reported so far – essential from the point of view of practical application – and in this regard, the past, present, and future aspects have been highlighted. An exercise of this type would facilitate not just the implementation of the hydrodynamic cavitation technology for disinfection, but will also facilitate newer developments in the form of reactors, process modifications, process integrations, and techno-economic feasibility evaluations vis-à-vis other existing as well as emerging water disinfection technologies. In fact, considering the most recent COVID-19 pandemic worldwide, it should also be possible to use a hydrodynamic cavitation process, in simple or in hybrid form, for elimination of harmful viruses from water sources, although no such work has been reported as of November 2020.

6.2 Conventional Methods of Disinfection

The common and newer methods of disinfection are listed subsequently to facilitate further discussion on the utility of hydrodynamic cavitation in disinfection. Figure 6.1 gives a more qualitative understanding of these processes.

Chemical

- Chlorination using chlorine
- Chlorine dioxide, chloramines
- Hypochlorite-based, for example, sodium hypochlorite
- Ozone
- Hydrogen peroxide
- Per acetic acid
- Electrochemical

Physical

- Filtration (e.g., sand filtration)
- UV radiation
- Membrane separation
- Thermal/solar

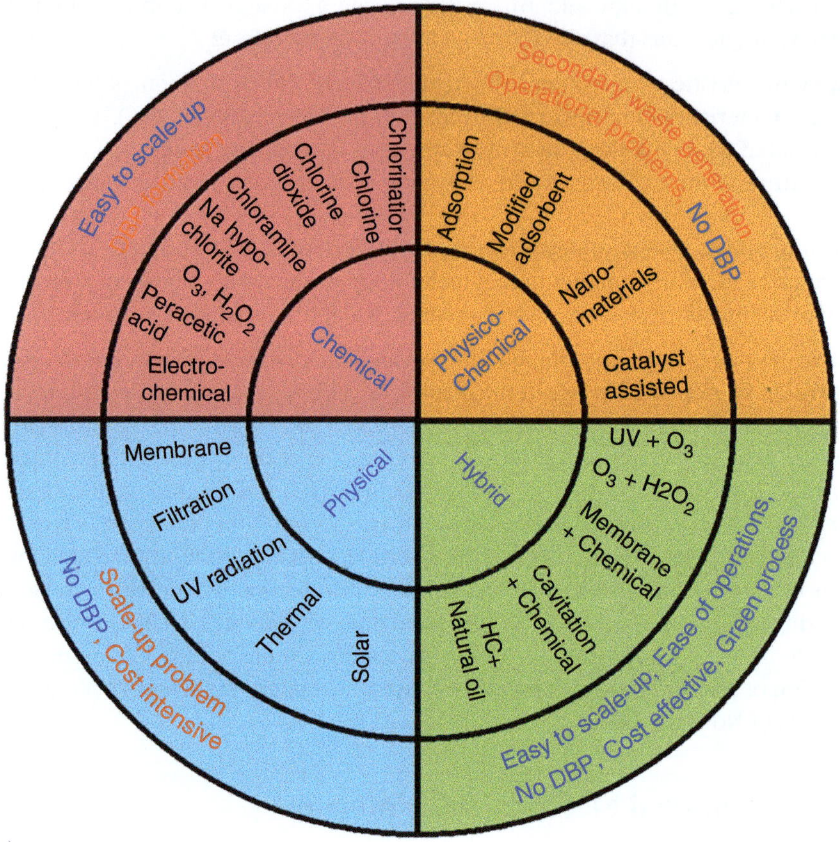

Figure 6.1 Different methods of disinfection (DBP: disinfection by-products).

Physicochemical

- Adsorption (e.g., chemical/metal modified adsorbents and nanomaterials)
- Cavitation
- Catalyst-assisted disinfection

Hybrid

- UV + Chemical methods (e.g., UV and ozonation)
- Membrane + Chemical
- Chemical + Chemical (e.g., ozone and hydrogen peroxide)
- Cavitation + Chemical (e.g., cavitation and hydrogen peroxide, ozone, etc.)
- Hybrid cavitation process using natural oils

A large number of conventional methods have been used to disinfect water. The description of the conventional methods such as chlorination, and of their action mechanism, etc. is beyond the scope of this discussion and has been well reported in many of the reviews on disinfection of water. To summarize the salient features and to create a background for evaluation of hydrodynamic cavitation technology in proper perspective, the following essential details are to be considered.

Most of the conventional chemical methods can eliminate pathogenic bacteria up to a certain extent, but can also produce harmful by-products [11, 12]. The physical methods include heating, radiation, microwave, filtration, UV irradiation, and plasma. The common drawbacks of physical methods include higher cost and long treatment times. UV irradiation has insufficient light scattering ability and is ineffective toward bacterial photoreactivation repair mechanisms [13]. UV can be cost effective; however, it requires a high dose for the complete destruction of microorganisms, is ineffective when colloidal and total suspended solids are present, and also needs regular maintenance. Many researchers have reported biosynthesis of metal nanoparticles using medicinally potent plant/leaf extracts and their potential applications [14, 15]. Kiruba et al. [16] carried out green synthesis of silver nanoparticles using West Indian cherry extract to exploit its antibacterial activity toward pathogenic bacteria. Membrane filtration for water disinfection is not very effective, has fouling problems, and requires frequent replacement of membrane, which ultimately increases the cost of treatment [10].

The most common chemical method of water disinfection is chlorination [2, 5, 7, 10]. Chlorination – the use of hypochlorites/chlorine dioxides – largely proceeds through dissolution of chlorine in water followed by the formation of hydrochloric and hypochlorous acid or the high oxidizing powers of these; the disinfection action is attributed to a modification of the structure of microorganisms, the possible bacterial elimination by inactivation of enzymatic systems, or the interruption of protein synthesis. Many common oxidants such as ozone have higher disinfection efficiency compared to chlorination, due to extreme reactivity; they exhibit the ability to effectively destroy the cell wall of the microorganisms. Chlorination can effectively kill many harmful microorganisms by damaging the cell membrane and thereby disrupt the cell respiration and DNA activity, leading to cell death. It is easy to use and operate as compared to many other methods and is also not cost-intensive. The chemical treatment may produce an unpleasant smell due to by-products which also can be mutagenic and carcinogenic in nature [7]. Chemical methods are also disadvantageous due to the mass-transfer limitations and difficulty in removing biocide-resistant bacteria, bacteria residing in biofilm, or any sediment. The chemical methods are believed to reduce the risk, but may not be effective in totality. Consequently, they cannot ensure the production of totally safe drinking water. Thus, a combination of various methods is typically required for satisfactory results. Arrojo et al. [13] reported hydrodynamic cavitation with orifice and venturi as cavitating devices for disinfection of water. Gondal et al. [17] investigated photocatalytic activity for disinfection and found dependence of the rate of disinfection on the concentration of ZnO photocatalyst. Asuncion et al. [18] prepared zinc oxide nanoparticles for disinfection of water. Bondarenko et al. [19] discussed the adverse effects when nanoparticles such as AgNP, CuO, and ZnO are released into water streams – causing serious damage to the crustaceans, algae, and fish – consequently damaging aquatic flora and fauna. Huang et al. [20] developed biogenic Silver Nanoparticles by *Cacumen Platycladi* extract that exhibit antibacterial activity towards *E. coli* and *S. aureus*. These aspects have not received attention, though some of the known properties of silver and copper nanomaterials have been exploited for commercial water purification applications. Piccirillo et al. [21] reported the composition and antibacterial properties of *Ginjacherryare's* stems and leaves to Gram-negative and Gram-positive bacteria. Kirti et al. [22] reported novel biomass-derived nanocomposites with magnetic properties for easy disinfection as well as for easy separation. They found that such materials can be highly cost-effective compared to conventionally reported nanomaterials.

According to Environment Canada's Municipal Water and Wastewater Survey of 2004, the following is the contribution of various disinfection methods in terms of worldwide use [6]:

- Chlorination as a disinfection technique 93.4%,
- Chloramines as a disinfection technique 3.7%
- Chlorine dioxides as a disinfection technique 2.2%
- UV and Ozone as a disinfection technique 5.9%

6.2.1 Major Drawbacks in Continuing the Use of Conventional Methods

The problems associated with the chemical methods of disinfection are many and vary according to the nature of the chemicals used for disinfection. However, a common and serious disadvantage with most of the chemical methods is the formation of disinfection by-products (DBP) that are extremely harmful and can be carcinogenic, such as trihalomethanes, haloacetic acids, bromate, and chlorite. The practice of the most common method – chlorination – for water disinfection involves addition of chlorine in appropriate proportions such that it kills the bacteria and harmful microorganisms. The practice is more than a century old, and is still being used worldwide (WHO). The addition can be in the form of chlorine gas, sodium hypochlorite solution (NaOCl), or solid calcium hypochlorite ($Ca(OCl)_2$). There are a number of disadvantages to this method. First, it requires a dose at an appropriate level, since an overdose of chemical certainly is harmful. Many treatment plants add chlorine well over the required dose. The duration of treatment and chlorine concentration need to be properly evaluated depending on the level of contamination. Secondly, chlorination is not effective for all the harmful microorganisms and some protozoan cysts are typically resistant to chlorination, which therefore requires the employment of additional/different methods of disinfection. The most serious concern is, however, the formation of trihalomethanes and other disinfection by-products which are carcinogenic in nature and increase risk of bladder, colon and rectal cancer. In many places, therefore, chlorination is being discontinued as a method of disinfection, especially in Europe.

It is also important to understand that apart from chlorination, many other chemical methods are also not completely safe. For example, ozone, though it is a powerful oxidant and is known/used for water disinfection, can produce toxic by-products such as aldehydes and other compounds during ozonation which can be potentially harmful [23].

One easy-to-implement alternative to lower the formation of toxic byproducts in chemical disinfection is to reduce the use of harmful chemicals, for example, reduce use of chlorine or ozone by integrating the synergistic effects of various methodologies such as advanced oxidation processes. It is also technically possible to combine the use of different oxidants (some are highly powerful, such as hydroxyl radicals – explained later in the section on process intensification) and physical processes such as ultraviolet radiation. However, such attempts, although reported in the literature, are not very effective in completely eliminating the problems. It is also necessary to obtain satisfactory rates of disinfection, typically achieved through the increased dose of the chemicals or oxidants. Thus, the most crucial aspects are the efficacy of the method in destroying the microorganisms (not just the model microorganisms, but the real-life complex composition) and the rates of disinfection – both critical for taking the process to commercial market.

6.2.2 Emerging Newer Methods of Disinfection

Broadly, as listed earlier, the disinfection methods are chemical (chlorination, ozonation, chemical treatment, etc.) and physical/physicochemical (UV light, membrane, cavitation, high-intensity ultrasound, etc.). The processes can also be classified as conventional (e.g., chlorination) and hybrid/advanced processes (e.g., advanced oxidation). The typical variables are concentration, exposure time, nature and number of microorganisms, etc. Further, the performance and disinfection efficiency are also affected by several interferences by the type of ferrous and manganese ions, nitrites, and sulfides that are typically present in surface water/ground water apart from organic substances/pollutants (oxidation of organic pollutants can be competitive in the processes such as oxidation). The cost of treatment can become a major concern, especially in the developing countries. Increased levels of contaminants, especially emerging contaminants and antimicrobial-resistant bacteria (AMR) pose serious challenges not just for the life but also for the conventional disinfection methodologies. The problem statement of today can be simply stated as "Find methodology that can provide safe and healthy drinking water at low (acceptable) cost with substantial ease of operation/scale-up and without producing harmful disinfection byproducts." The emerging technologies can come from any of the existing conventional methods such as coagulation, adsorption, advanced oxidation, and membrane separation, or from chemical and biological treatment. The emerging technologies require obtaining newer materials and methods to eliminate the current problems or disadvantages associated with these methods. For example, newer nanocomposites using silver or copper can provide a disinfection methodology in the form of portable cartridges for making safe water available in rural areas. Graphene-based filters can also provide an affordable and long-lasting solution in future, if material and engineering modifications can be developed. Modified membranes and membrane applications for water disinfection can also be emerging technologies for practical use. Essential natural oils can form a viable alternative in the biological treatment area since many of these possess antimicrobial properties. The use of natural oils in advanced oxidation and cavitation was recently reported to increase the disinfection efficiency and reduce the cost [6]. Medicinal plants are a source of antimicrobial active compounds that can be exploited by way of extracting the specific components or through natural oils. The mechanism of disinfection in using natural oils is not very clearly understood, and is generally attributed to the presence of substances such as phenolics, alcohols, aldehydes, ketones, and acids such as eugenol, 1,8-eucalyptol, cinnamaldehyde, ricinoleic acid, etc. The active ingredients, which have antimicrobial properties, can destroy the microbial metabolism through various actions such as destroying the permeability of bacterial membranes, leading to loss of electrolytes, or by reaction with the phospholipids of the cell membrane and altering its permeability, and denaturing the cell protein, causing cell death. Integrating biological methods, such as use of natural oils, with existing physical/physicochemical methods can be promising, if the intricacies can be resolved for the various parameters of disinfection, for example, nature of microorganism, concentration, physical and chemical attributes of the process such as pH, temperature, and so on. More research in this regard can provide possible technological alternatives to the existing methods.

6.3 Disinfection of Water by Cavitation

The disinfection of water using cavitation in general, and hydrodynamic cavitation in particular, has been reported for different types of microorganisms including different bacteria and viruses. The microorganisms can be destroyed both by the physical and the chemical effects that are described in detail in the following sections. Though cavitation is described in general and compared using different methods, the main discussion is centered on hydrodynamic cavitation using mechanical devices. The early studies on hydrodynamic cavitation for disinfection were mainly academic, exploring the cavitation reactor and achieving efficacy of the method for removal of various model microorganisms. There are a number of issues that are to be considered for effective implementation of the process, including type of the microorganism, nature of the cavitation device, design of the reactor, process conditions apart from cost consideration, and ease of operation. The process is highly complex and there are no theoretical models available for reliable a priori prediction of the disinfection behavior. The definition of cavitation number is not satisfactory and hence not applicable for all types of devices, the disinfection performance is not always commensurate with the calculated/predicted cavitation numbers, and the interaction of different process parameters is not fully understood. These and many such aspects make the subject of disinfection of water highly difficult when the implementation of cavitation methodology is considered.

6.3.1 Cavitation Process Principle

Cavitation has been considered to be a promising disinfection methodology for over two decades and a large number of research studies have been reported using methods such as acoustic cavitation and hydrodynamic cavitation. Figure 6.2 depicts contemporary use of the cavitation methodology in disinfection, with qualitative

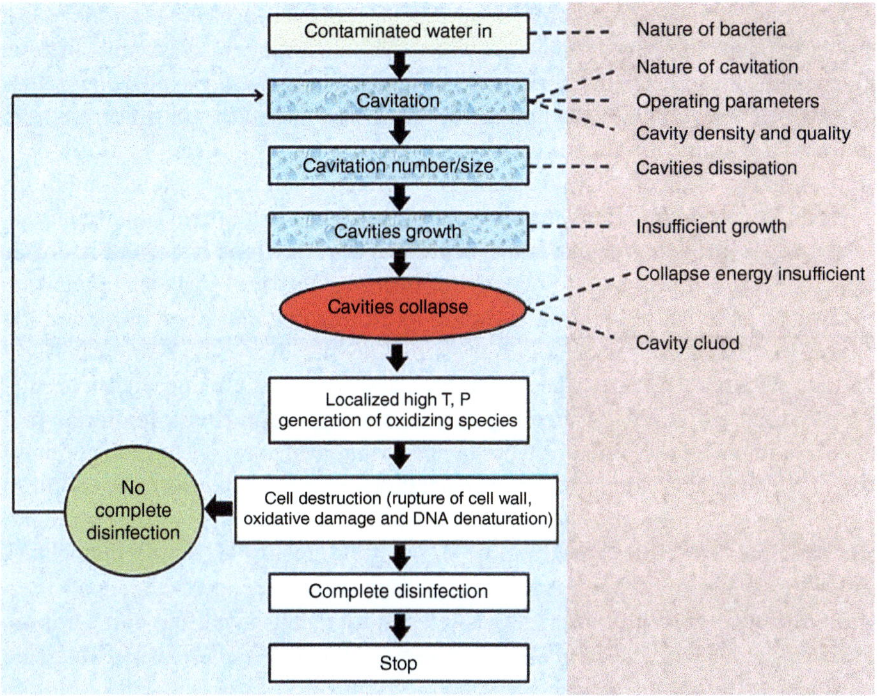

Figure 6.2 Flow chart for the implementation of hydrodynamic cavitation in disinfection.

description. In spite of the huge volume of literature available on water disinfection using cavitation, its practical applications are still far from being satisfactory due to complexity in the design, poor understanding of the fundamental principles, and most importantly, limitations in terms of techno-economic feasibility. The application of cavitation methods in this regard, in general, and hydrodynamic cavitation, in particular is, by and large, purely empirical [24].

Cavitation is both a physical and a chemical method of treatment (physico-chemical) which requires inter-related synergy between the two for its effective implementation. By definition, cavitation is a process in which cavities are allowed to form (liquid flashing into vapor), grow, and then collapse in a specified manner such that the collapse of the cavities results in intense pressure and temperature conditions at the point of collapse, driving water molecules to split homolytically in to H· and OH· radicals. The hydroxyl radicals thus generated, being strong oxidizing agents, participate in the oxidation of the species [25, 26]. The collapse of the cavities is also termed as implosion and the temperatures can be of the order of 10 000 K and pressure of the order of 5000 atm at the point of collapse. However, it needs to be understood that not all cavities are productive and not all implosions are fruitful. Thus, the efficiency of the cavitation process depends on the number of cavities, size of the cavities (extent of growth), and productive collapse frequency. The visible effect of the cavities' implosion is seen in the form of shockwaves, microjets, and high shear stress.

Considering the physicochemical nature of the overall cavitation process, a number of practical pathways have been found for cavitation. These mainly include acoustic cavitation, hydrodynamic cavitation, particle cavitation, and optic cavitation. Acoustic cavitation is the application of ultrasound waves to cause pressure variation in a liquid to generate cavities. Acoustic cavitation can certainly be a practical option as an effective method for sterilization, but has serious limitations for treating large volumes and has poor scalability due to engineering and economic constraints. Hydrodynamic cavitation is produced by any type of cavitating device such as valves, orifice, vortex diode, or venturi that obstructs/disturbs liquid flow (velocity) so as to cause pressure variation in a flowing liquid, thereby generating cavities. In optic cavitation, liquid bursts into cavities due to high-intensity light or a laser, whereas in the particle cavitation, any type of elementary particle such as a proton is employed for creating cavities through rupturing a liquid, as in a bubble chamber. Of the different types of cavitation, from a practical utility point of view, only acoustic and hydrodynamic cavitation can be considered to be useful as of today. Again, for the large-scale operations in water and wastewater treatment, hydrodynamic cavitation appears to be the only promising option. Orifice plates, venturi tubes, and rotor-stator reactors (e.g., high speed homogenizers), apart from hybrid disinfection techniques (process integration using cavitation with aeration, additives, chemical disinfectants, etc.) are some of the most commonly investigated techniques with multiple design variations.

An important application of cavitation is in water disinfection, where the microorganisms are killed due to the physical rupture of cells and also due to the oxidizing environment. The physical and chemical effects can be further augmented by process intensification for increased efficiency and improved process economics. Figure 6.3 illustrates the conventionally understood cavitation mechanism for disinfection.

Apart from disinfection, hydrodynamic cavitation is very useful in wastewater treatment– sewage and industrial effluent treatment – for the removal of organic pollutants – chemicals, dyes, pesticides, phenolics, pharmaceuticals, etc. [26–35], sludge treatment [36], in desulfurization of fuels [37, 38], food processing [39], biodiesel synthesis [40, 41], delignification of wheat straw for paper manufacturing [42], and many such new applications involving the role of hydrodynamic cavitation, mainly as an advanced oxidation process.

Hydrodynamic cavitation is a complex phenomenon. Typically, a special pressure coefficient – the cavitation number C_v, – is used to describe the degree of cavitation. It is one of the most important parameters to describe the intensity of cavitation. The definition is rather arbitrary and has two types of cavitation numbers:

$$C_v = \frac{P_d - P_v}{0.5\,\rho\, v^2} \tag{6.1}$$

The other form of cavitation number based on pressure drop is given as.

$$C_v = \frac{p_d - p_v}{p_u - p_d} \tag{6.2}$$

Figure 6.3 The working principle of the cavitation disinfection process.

Generation of active radicals and oxidising species

Homolytic water cleavage during cavity implosion

$$H_2O \rightleftarrows HO^{\cdot} + H^{\cdot}$$

$$O_2 \rightleftarrows 2O^{\cdot}$$

$$O^{\cdot} + H_2O \longrightarrow 2HO^{\cdot}$$

$$H^{\cdot} + O_2 \longrightarrow HOO^{\cdot}$$

$$2HO^{\cdot} \longrightarrow H_2O_2$$

$$2HOO^{\cdot} \longrightarrow H_2O_2 + O_2$$

Disinfection due to oxidation, Shockwave

Physico-chemical effect - Oxidation

$HO^{\cdot}$ + microorganism $\longrightarrow$ Cell death

Physical effect

Shock wave + microorganism $\longrightarrow$ Cell death

High T & P $\longrightarrow$ Cell death

where, p_d, p_u and p_v are downstream, upstream, and vapor pressures of the fluid, respectively, ρ = density of fluid and v = flow velocity.

An increase in upstream pressure should decrease C_v and increase the number of cavitation events. Under ideal conditions, cavitation occurs when $C_v < 1$. In a model simulation, Gogate and Pandit indicated that cavities generated at $C_v > 2$ do not collapse, thereby providing no desirable chemical effects. The number of cavities generated and the cavitation events are high at a lower cavitation number while the intensities due to cavity collapse are low. For cavitation number more than one, liquid is indicated as not prone to cavitation, and when the number is less than one, the fluid energy is sufficiently applied for the creation of the vapor phase, essential for the physical format of cavitation involving the formation, growth, and collapse of vapor bubbles. Again, conventionally, it has been considered that the lower the cavitation number, the greater is the effect of cavitation, though this has not been conclusively proven in most studies. It is to be noted that cavitation can also take place at $C_v > 1$ due to the presence of dissolved gases or suspended particles [43]. Cavitation number increases with an increase of the orifice hole diameter [44, 45]. Braeutigam et al. [46] showed that the optimal value of C_v decreases with an increase of pipe diameter and is also reciprocally related to the velocity of the fluid.

Thus, the values of cavitation number are typically less than 1. The usual definition of cavitation number, though straightforward in the case of linear flow, is not applicable to the devices such as vortex diode in which vortex/rotational flow is employed. Therefore, no matter what cavitation devices are employed, it should be ensured that they operate in the cavitation regime.

Pressure drop (ΔP) is a critical parameter in hydrodynamic cavitation. For any type of cavitating device, it is essential to determine the point of inception of the cavitation. The flow rate and pressure drop data are essential characterization required for this purpose, and the optimum pressure drop for operation is selected. For cavitation devices using linear flow, the characterization can be simple and can be done using the cavitation number which is based on vapor pressure and throat velocity. Therefore, for orifice or venturi, the cavitation number, typically less than 1, can be used to identify the possible inception of cavitation. For cavitating devices employing rotational flows, the conventional cavitation number defined previously is not relevant, for example, vortex-based cavitation devices, and it is essential to ensure the reactor operation in the cavitation regime. The pressure drop dictates the performance and cost of the cavitation process.

The rates of disinfection can be drastically altered using different reactor geometries. For example, water disinfection using vortex diode (employing vortex flow for cavitation) was found to be significantly higher compared to the conventional device, orifice (employing linear flow for cavitation). It was also found to be higher by an order of magnitude than that for acoustic cavitation. The orifice typically requires significantly higher pressures compared to the vortex diode, implying higher energy requirement.

6.3.2 Present Status

Water disinfection by cavitation – especially using hydrodynamic cavitation – has been discussed for over two decades. However, practical implementation of any of the methods is still not evident mainly due to the complexity involved in predicting the cavitation behavior and the appropriate process parameters for the destruction of different types of microorganisms. The common types of cavitation processes reported include disinfection using orifice, venturi, rotor-stator reactors (e.g., high-speed homogenizers), and very recently, using the vortex diode [6, 47, 48]. *E. coli* was the most common bacteria investigated in most studies.

Save et al. [49, 50] reported the early studies on cavitation for disinfection. More successful studies were later reported by Jyoti and Pandit [23], in which a combination of ozone and cavitation for disinfection of water was studied to prove synergistic processes as a promising method for water disinfection. Both acoustic and hydrodynamic cavitation processes were discussed and compared. Disinfection of the heterotrophic plate count (HPC) bacteria and indicator microorganisms such as total coliforms, fecal coliforms, and fecal streptococci was studied using an ultrasonic horn with a frequency of 22 kHz for acoustic cavitation and a ball valve as a cavitating device for hydrodynamic cavitation (pressures from 1.7 to 5.2 bar). The bacterial concentrations were however very small and only up to 90% disinfection could be obtained within one hour using individual methods, which could be increased to 100% using a combination of ozone with cavitation. It was also observed that acoustic cavitation accelerates the decomposition of ozone. Through their cost analysis, it was suggested that hybrid methods like the use of hydrodynamic cavitation + ozone and acoustic cavitation + ozone can be potentially economical techniques. As a continuation of application of the hydrodynamic cavitation method, the destruction of microbes (zooplankton in sea water) using orifice and a theoretical understanding of the same were also reported in 2008 and it was suggested that cavitation and/or turbulent fluid shear are effective tools for sea water disinfection, with more than 80% elimination of the zooplankton present in the sea water [51].

While Jyoti and Pandit explored various types of devices for disinfection, mostly orifice [23, 52–54], Liu et al. [55] reported multi-hole orifice as a cavitating device for the disinfection of water (*E. coli*), by which close to 99% removal was obtained in one hour. Balasundaram and Harrison [56] used orifice configuration for the study on cell destruction. Similarly, Arrojo et al. [13] studied different devices for disinfection efficiency, such as orifice plates and venturi tubes, varying the numbers of holes and other such design parameters. However, the results were not convincing. Loraine et al. [57] also attempted to compare disinfection efficiency for different reactors. Liu et al. [55] used a very large number of holes – 49 – in orifice to evaluate disinfection performance. In orifice as cavitation reactor, many of the authors also tried to evaluate the different geometries of the opening, such as circular, triangular, and rectangular, apart from different single- and multiple-hole arrangements.

Most of the studies practically had no commonalities, including the nature of cavitating devices, initial concentration of bacteria, nature of bacteria, the operating conditions such as pressures/pressure drop, cavitation numbers, and so on, making the comparison and quantitative analysis practically difficult, and were unable to arrive at exploitable conclusions. The heart of the hydrodynamic cavitation technology is the "Cavitating device or Reactor" – the design and application of which is key to the success of the cavitation technology in disinfection.

6.3.3 Hydrodynamic Cavitation and Cavitation Devices/Reactors

Hydrodynamic cavitation using reactors, with or without moving elements, is a complex technology that is distinctly different from acoustic cavitation in which bubbles grow and collapse in quiescent water (Figure 6.4). The hydrodynamic cavitation process, depending on the type of device, involves complex non-linear interactions between bubbles and turbulence, which have not been understood even today though computational fluid dynamics (CFD) simulations are being used to understand and interpret the behavior, for cavitation effect that is not only related to disinfection but also for a variety of applications including degradation of organics in wastewater treatment. In principle, cavitation can be effected by any cavitating device described subsequently and can even occur due to simple, commonly used flow devices such as pumps or valves. Usually, however, the contribution of pumps or valves in the cavitation process is negligible when well-designed cavitation devices are used.

Cavitation devices/reactors can include those devices having moving parts such as rotating impeller and those devices without moving parts. The cavitation devices

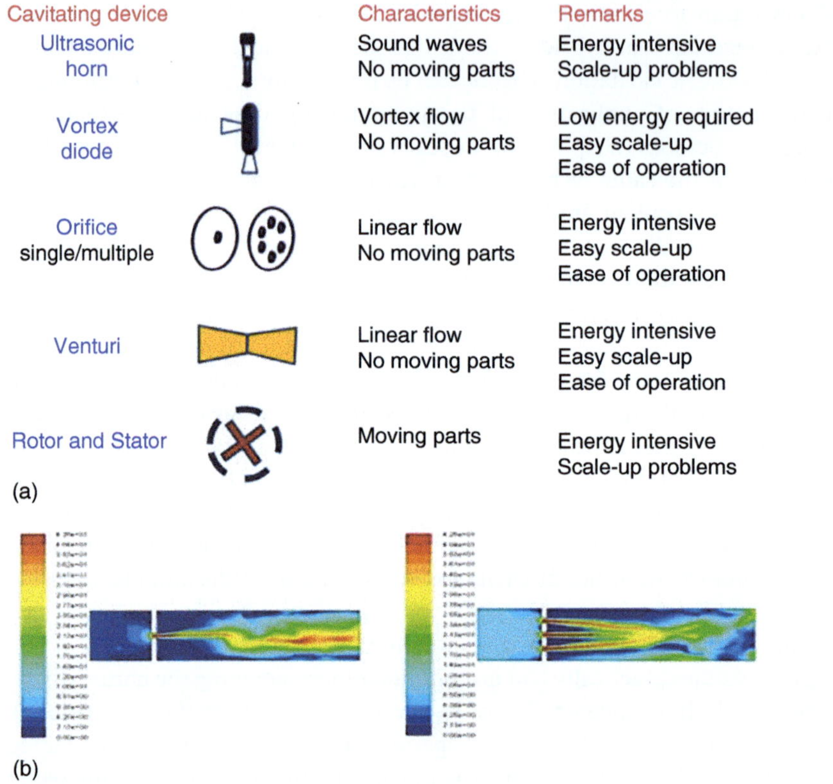

Figure 6.4 Common cavitating devices for disinfection of water. (a) Different types of cavitating devices. (b) Flow visualization of single hole and multiple hole orifice.

with moving parts typically have a complex design compared to those without moving parts, are expensive, and also have a higher operating and maintenance cost. The rotor type of cavitation devices are claimed to have improved efficiencies, though not yet well proven; the cavitation mechanism is not clearly understood. Mezule et al. [58] used a moving-element (rotor) for generating cavitation for the disinfection of water (*E. coli*); the disinfection efficiency was, however, not satisfactory. Therefore, it is felt that cavitating devices without moving parts, at this point of time, would be more beneficial for realizing the cavitation and the disinfection. In view of the large number of designs available and the theoretical tools for newer designs and application data in various applications, it may be possible to select cavitation devices without moving parts as promising devices for realizing the full potential of hydrodynamic cavitation for large-scale applications at a significantly lowered cost. The cavitation devices without moving parts, though they appear simple in the physical form, can be complex due to the flow pattern, which can be more simply classified as linear flow or rotational flow. The conventional devices – orifice or venturi – employ linear flow for effecting cavitation and can have very specific designs based on the number of holes (single or multiple), shape of holes (circular, triangular, or rectangular), different forms of converging and diverging sections, and so on. Some of the successful patterns are:

- Orifice with single hole of different hole diameter.
- Orifice/valves with variations in the open area of constriction.
- Venturi cavitation reactors with different convergent– divergent sections.
- Vortex diode with vortex flow – tangential and inverse flow.

The fluid is forced through single or multiple restrictions to create the desired pressure drop for liquid to flash into vapor, thereby effecting cavitation. Thus, the conventional devices employ constrictions so that the velocity at the throat (narrow-flow area) is large enough to generate cavities that subsequently collapse in a downstream region, resulting in the implosion of the cavities to initiate subsequent actions of hemolytic water molecule cleaving, hydroxyl radical formation, and oxidation. In the other type of rotational flow devices, the pressure drop and the cavitating region are centered on axis of the rotation. In the newer type of cavitating device, the vortex diode, which is a device that does not employ any moving parts, the fluid enters tangentially to generate a highly swirling flow to form a low-pressure region and to generate cavities which subsequently escape the chamber via an outlet port and then collapse as they enter the high-pressure region. The subsequent actions following the implosion of the cavities are same in any type of cavitating device, with or without moving parts, whether linear flow or rotational flow; however, the efficacy of the cavitation process depends on number density and collapse intensity of the generated cavities apart from the cost of process. It is to be noted that the designs – linear flow or rotational flow – are also required to eliminate/minimize the risk of associated damage to the cavitating device.

A large number of cavitation devices have been reported with huge variation in their design and efficacy for different types of applications. Broadly, these can be categorized in terms of linear flow-based devices such as orifice, venture, etc., and rotational flow-based devices such as vortex diode. The typical flow diagram for implementing hydrodynamic cavitation using a vortex diode as a cavitating device is shown in Figure 6.5 along with the schematic of design of vortex diode [26, 38, 48, 59].

The performance of hydrodynamic cavitation and that of disinfection is dictated by the type and design of the reactors for the disinfection efficiency, energy requirement, ease of operation, scalability, and most importantly, the cost – which would ultimately determine its applicability in disinfection. Thus, the design and development of high-performance hydrodynamic cavitating devices/reactors is of paramount importance, which also requires elucidation of the corresponding disinfection mechanisms and effect of process parameters – issues that have been largely but not fully understood even today. Sun et al. [9] has reviewed the different cavitation reactors for disinfection and found the highest reports for orifice reactors (49%), followed by the rotor-stator type (13%), venturi (10%), high-pressure homogenizers (10%), blenders (8%), liquid whistle reactors (3%), and finally the one developed recently by our research group, vortex diode (2%). It is evident that cavitating devices without moving elements are the most commonly researched

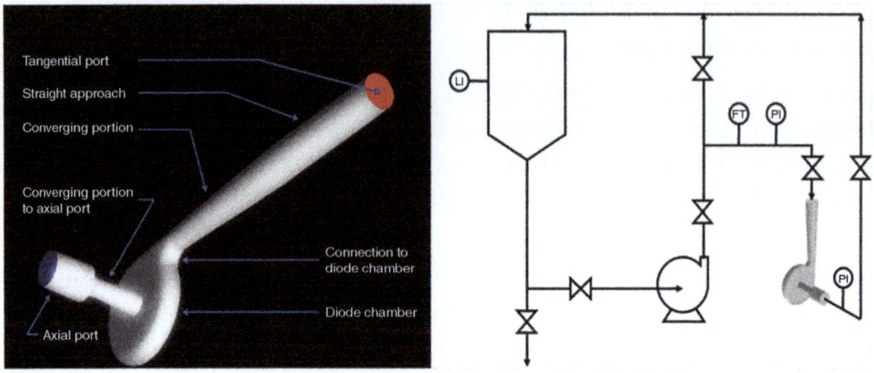

Figure 6.5 Schematic of vortex diode as a cavitating device and its application.

reactors. Considering the efficiency and economics, vortex-based cavitation reactors appear to be promising.

Disinfection studies have been reported for a variety of microorganisms, bacteria, and more commonly for the indicator bacteria such as *E. coli*. Studies on elimination of viruses using hydrodynamic cavitation have been only sparsely reported [60]. Most of these studies have been well reviewed from time to time, with a tabular description of type of the device, the operating conditions, nature of the microorganism, and extent of disinfection, with time and energy requirement, etc.; therefore, the information is not repeated here. However, it is evident that the majority of these studies were carried out using orifice by various research groups [13, 24, 47–50, 57, 61–69], followed by studies using the venturi and vortex diode. The disinfection studies using the vortex diode device were mostly from our research group. In comparison, the studies using rotor elements, devices of rotor-stator/rotating disc or blender/homogenizer types were few [49, 50, 58, 70, 71].

Since the operational and geometrical parameters of cavitation reactors aforementioned and those reported in the literature are quite diverse, it is extremely difficult to conclusively assess the efficacy of different cavitation reactors for disinfection, especially in view of complex coupling of the process parameters, the hydraulic characteristics of a reactor influencing the disinfection rate, and other such factors; greater in-depth investigation in this area is required for complete understanding.

6.3.4 Kinetics of Disinfection in Hydrodynamic Cavitation

For acoustic cavitation, the conventional approach of the chemical reaction rate model can be applied to mathematically predict the performance of the disinfection process. For hydrodynamic cavitation, a model based on per-pass conversion was suggested recently [6, 47, 48] for the schematic of the cavitation process as shown in Figure 6.6. A comparison of the conventional model and the suggested mathematical model is also provided. Since the concentration of bacteria decreases exponentially with time, the kinetic data for disinfection can be fitted using a pseudo first-order equation in terms of k_G (growth rate of microorganisms) and k (disinfection rate

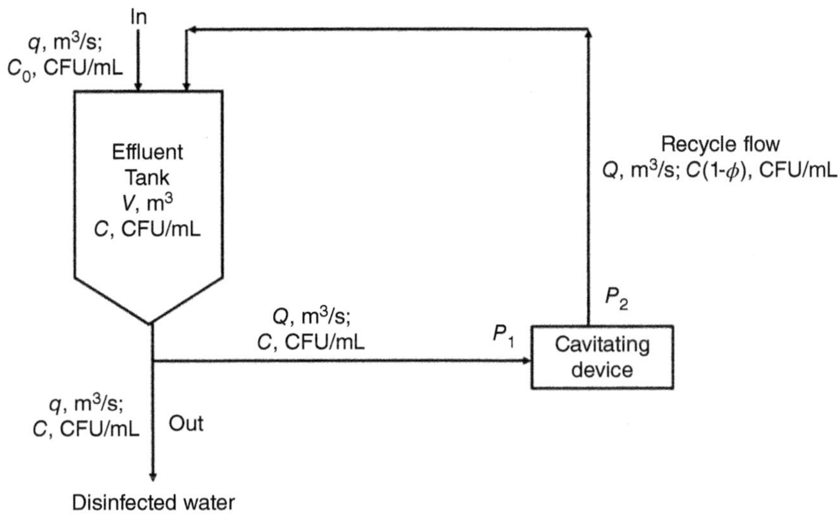

Figure 6.6 Schematic diagram for modeling disinfection kinetics.

of microorganisms). Conventionally, disinfection is also measured in terms of log reduction (3 log or 4 log reduction, etc.; 99.99% reduction is 4 log reduction). For most of the disinfection applications, we need at least 3 log reductions. The representations of different forms – percentage and log reduction – are inter-convertible; however, percentage reduction can provide clear understanding, especially in the low range, while for high removal, log reduction is qualitatively easier to understand.

The concentration of bacteria decreases with respect to time. Therefore, pseudo first-order kinetics for disinfection of water can be applied for both the conventional rate model and the cavitation-based model.

$$\text{Rate of disinfection } \alpha f(\text{C}) \tag{6.3}$$

For conventional rate model, the rate equation is

$$-\frac{dc}{dt} = kC \tag{6.4}$$

For the hydrodynamic cavitation-based model, the rate equation is

$$V\frac{dc}{dt} = V(k_G - k)C \tag{6.5}$$

where C is the concentration of microorganisms in the holding tank, V is a working volume (volume of effluent tank and volume of piping including high-pressure pump), k_G is growth rate of microorganisms, and k is the disinfection rate of microorganisms.

For the conventional rate model, after integrating the rate equation, the concentration of microbes at any time t is

$$C = C_0 e^{-kt} \tag{6.6}$$

For the cavitation-based model, the concentration of microorganisms C can be obtained by assuming the rates to be constant:

$$C = C_0 e^{-(k-k_G)t} \tag{6.7}$$

where, C_0 is the initial concentration of bacteria (CFU/mL) and t is the time for disinfection (min).

In hydrodynamic cavitation, disinfection occurs because of cavitation and it therefore depends on the number of passes through the cavitation device rather than the elapsed time t.

The conventional rate model does not consider the per-pass disinfection (ϕ) or number of passes (n) parameters, which is a disadvantage in this model.

In the cavitation-based rate model, the rate equation can be expressed in terms of the per-pass disinfection factor (ϕ) and the number of passes (n) as:

$$V\frac{dC}{dt} = V\,k_G\,C - Q\phi C;\, n = \frac{Q}{V}\,t \tag{6.8}$$

where Q is the flow rate through the cavitation device, ϕ is the per-pass disinfection factor, and n is the number of passes.

In the conventional rate model, the concentration of microbes at any time t can be written as

$$C = C_0 e^{-kt} \tag{6.9}$$

In the cavitation-based rate model, by the assumption of constant ϕ and k_G, the concentration of microorganism at time t is estimated for the residence time of τ (V/Q) as:

$$C = C_{in} e^{-(\phi-k_G\;\tau)n} \tag{6.10}$$

Correlating the growth rate constant to doubling time, t_D as $k_G = \frac{\ln 2}{t_D}$ and assuming negligible growth of microorganism during the treatment time, the disinfection of water can be described as:

$$C = C_{in} e^{-kt} = C_{in} e^{-\phi n} \tag{6.11}$$

For the conventional rate model, the effective disinfection rate constant (k) is:

$$k = \frac{\ln(C_0/C)}{t} \tag{6.12}$$

For the cavitation-based model, the effective disinfection rate constant (k) may be related to residence time (τ) and ϕ as:

$$k = \frac{\phi n}{t} = \frac{\phi}{\tau} \tag{6.13}$$

The effective disinfection rate constant used in the studies on hydrodynamic cavitation and the cavitation-based model is dependent on the residence time of the holding tank. This cavitation-based rate model is more realistic than the conventional rate model as it is based on more realistic parameters and has also been proven in our previous studies.

The overall cavitation yield Y, CFU/J for both the conventional rate model and the cavitation-based rate model can be obtained as:

$$Y = \frac{V\,(C_{in} - C)}{\Delta P\,Q\,t}\ \text{CFU/J} \tag{6.14}$$

In the conventional rate model, the cavitation yield cannot be expressed in terms of the per-pass disinfection factor (ϕ).

For the cavitation-based model, for small values of ϕ, the overall cavitation yield **Y** may be approximated as:

$$Y = \left(\frac{\phi C}{\Delta P}\right)\ \text{CFU/J} \tag{6.15}$$

The conventional rate model equation in terms of average concentration C_{avg} is:

$$R = kC_{avg}\ \text{(CFU/mL)/s} \tag{6.16}$$

where C_{avg} is average concentration, $C_{avg} = \frac{C_0+C}{2}$

In the cavitation-based model, the average disinfection rate over time "t_{op}" is:

$$R_{avg} = \frac{C_0(1 - e^{-kt})}{t_{op}} = \frac{C_0(1 - e^{-\phi n})}{t_{op}}\ \frac{\text{(CFU/mL)}}{s} \tag{6.17}$$

It should be noted that, in principle, both the mathematical models yield the same result; however, the per-pass conversion-based cavitation for hydrodynamic cavitation is closer to the physical description of the cavitation process. The required data for solving can be experimentally obtained (flow rate (Q), volume (V), and concentration–time data). The number of passes describe the ease of operation; the easier the operation, the lower the value of n; for difficult disinfection, n is large. The number of passes also, therefore, directly reflects the cost of operation. The value of "n" also reflects the effect of reactor geometry. The mathematical treatment of this type can bring out the dependence of disinfection on residence time and is less dependent on initial concentration compared to the conventional reaction rate model.

Table 6.1 summarizes some of the reported data on disinfection using different methods and hydrodynamic cavitation for an analysis on the rates of disinfection in this regard, using the rate model. The performance of disinfection has been compared using "percentage disinfection" for convenience of comparison of the results and better understanding, though as mentioned earlier, many reports also use log reduction terminology for disinfection (disinfection applications require 3 or 4 log reduction, and therefore, it may often be necessary to combine cavitation with other techniques to achieve the desired log reduction).

6.4 Hybrid Methods of Disinfection Involving Cavitation

A large number of hybrid methods have been reported in disinfection with different levels of success and it would be beyond the scope of this chapter to fully discuss

Table 6.1 Comparison of rates of disinfection for different processes.

Sr. no.	Processes	Bacteria	Operating parameters	Percentage disinfection (%)	Rate of disinfection	References
1	Cavitation	*E. coli*	Vortex diode, $\Delta P = 1$ bar, 15 min	44	91 (CFU/mL)/s	[48]
2	Cavitation hybrid	*E. coli*	Vortex diode, Clove oil (0.1%), $\Delta P = 1$ bar, 15 min	86	132 (CFU/mL)/s	[48]
3	Cavitation	*P. aeruginosa*	$\Delta P = 1$ bar, 15 min	28	55.56 (CFU/mL)/s	[6]
4	Cavitation hybrid	*P. aeruginosa*	Peppermint oil (0.1%), VD $\Delta P = 1$ bar, 15 min	99	144.33 (CFU/mL)/s	[6]
5	Vortex diode	*E. coli*	VD $\Delta P = 1$ bar, 60 min	98.7	4.8 (CFU/mL)/s	[47]
6	Orifice	*E. coli*	Orifice $\Delta P = 10$ bar, 60 min	99.5	20.2 (CFU/mL)/s	[47]
7	US bath	Total coliform	Peak operating frequency of 20.5 kHz, 15 min	75	3.85 CFU/s	[23]
8	US bath + 2 mg/L O_3	Total coliform	Peak operating frequency of 20.5 kHz, 2 mg/L O_3, 15 min	98.3	5.54 CFU/s	[23]
9	US horn + 2 mg/L O_3	Total coliform	Frequency of 22 kHz and electrical power rating of 240 W, 2 mg/L O_3, 15 min	99.6	1.53 CFU/s	[23]
10	HC at 5.17 bar	Total coliform	Cavitating device (Ball valve), $P_{in} = 5.17$ bar, 15 min	60		[23]
11	HC at 5.17 bar + 2 mg/L O_3	Total coliform	Cavitating device (Ball valve), $P_{in} = 5.17$ bar + 2 mg/L O_3, 15 min	80	10.72 CFU/s	[23]
12	Orifice (25 holes with a diameter of 1 mm) 2 bar	*E. coli*	Cavitating device (orifice-25 holes with a diameter of 1 mm), $P_{in} = 2$ bar, 120 min	32.7	1.25 (CFU/mL)/s	[13]
13	Venturi (1×10^{-5} m^2) 1 bar	*E. coli*	minimum cross-section (gorge) 1×10^{-5} m^2, $P_{in} = 1$ bar, 120 min		3 (CFU/mL)/s	[13]

14	High-pressure jet device	*E. coli*	Volume treated 7.5–10 L, Operating pressure ~6 MPa	23.7	NA	[66]
15	High-pressure jet device	*B. subtilis*	Volume treated 7.5–10 L, Operating pressure ~6 MPa	28.5	NA	[66]
16	Liquid whistle reactor (LWR)	*E. coli*	P_{in} = 10.34 Mpa, 180 min	71.2	NA	[72]
17	Orifice	*E. coli*	P_{in} = 7.5 bar, 360 min	99.4	NA	[24]
18	Orifice	*E. coli*	P_{in} = 12 MPa, 30 min	100	NA	[69]
19	Venturi	*E. coli*	P_{in} = 600 000 Pa, 120 min	75.4	NA	[73]
20	UV+Per acetic acid(PAA)	*E. coli*	PAA dosage: 2–8 mg/L *t*: 10–30 min UV: 100–300 mJ/cm^2, 10–30 min	100	NA	[74]
21	Sonodisinfection	*E. coli*	US: 200 W Irrad. time: 60 min	90	NA	[75]
22	Electro disinfection	*E. coli*	Current density: 11.46 A m^2	99.9	NA	[75]
23	Sono-electrodisinfection	*E. coli*	US: 200 W Current density: 8.91 A m^2	99.9	NA	[75]
24	Solar photolysis	*E. coli*	Irrad. time: 40 min	90	NA	[76]
25	Solar photocatalytic (absence of HA)	*E. coli*	Irrad. time: 60 min Dosage of dopants (Se-doped TiO_2, N-doped TiO_2 and Se-N-co-doped TiO_2)	90	NA	[76]
26	Solar photocatalytic (presence of HA)	*E. coli*	Irrad. time: 60 min Dosage of dopants (Se-doped TiO_2 and Se-N-co-doped TiO_2); Irrad. time: 90 min (N-doped TiO_2)	99.99	NA	[76]
27	Combination-ultrasound (US) and UV	*E. coli*	US: 1400W UV: 1656 mJ cm^2 *t*: 15 min	99.99	NA	[77]

the various hybrid methods with all the details. Therefore, only a few reasonably successful hybrid methods of disinfection have been briefly discussed subsequently that have shown potential in real-life applications. A very recent development in this regard, hybrid hydrodynamic cavitation using vortex diode and natural oils, is discussed in more detail, not only because of its relevance in terms of hydrodynamic cavitation, but mainly for the reason that this can certainly be a new practical alternative to not just conventional hybrid methods but also to the most widely used chemical methods such as chlorination.

6.4.1 Process Integration—Conventional

The conventional hybrid techniques generally employ a combination of two or more processes to intensify the disinfection operation. The combination can be of different types involving physical processes, chemical processes, and physicochemical processes. Hydrodynamic cavitation processes in principle involve the contribution of hydroxyl radicals ($OH^{\bullet}$), for the oxidation and in situ generation of the hydroxyl radicals is accomplished during the cavitation process itself. Though the hydroxyl radical is a strong oxidizing agent, the efficiency of the process is seldom satisfactory, possibly due to low concentrations. Process intensification or integration therefore largely involves a combination of different oxidizing agents such as hydrogen peroxide and ozone. If the oxidation potentials are to be compared, hydroxyl radicals are the strongest agents (2.8 V), second only to fluorine (3.0 V). Ozone is the third most powerful oxidizing agent, with oxidation potential of 2.1 V, followed by hydrogen peroxide (1.8 V), chlorine dioxide (1.5 V), and chlorine (1.4 V). The reasons for the improvements due to process intensification of this nature are mainly that the small increase in the concentration of oxidants is believed to accelerate free-radical production, and the oxidant concentration effect. The enhancement of this type is however limited and therefore, in spite of various such conventional intensification methodologies, the level of improvements in the disinfection efficiency were not very satisfactory; in many cases, the costs were higher, obviously due to the addition of different processes. Further, the combination of various oxidants is reported more for wastewater treatments (oxidation of organic pollutants) compared to disinfection of water.

6.4.2 Cavitation with Hydrogen Peroxide Addition

Hydrogen peroxide is one of the most powerful known oxidizing agents and is used extensively in different chemistries. It is a known disinfectant and has antiviral and anti-bacterial activity through oxidizing power that produces free radicals, which leads to oxidative damage to proteins and membrane lipids, and consequently cell death. However, it is unstable and decomposes readily to oxygen and water with release of heat. Therefore, though it is miscible with water, it requires specific handling precautions and there are concerns about its concentration and usage.

The addition of hydrogen peroxide is believed to generate a synergistic effect with the conventional cavitation process and additionally generates hydroxyl radicals that contribute to the disinfection and increase the efficiency of disinfection due to increased concentration of oxidizing agents. Jyoti and Pandit [54] reported significant improvements by adding 5 mg/L of H_2O_2 in removal of total coliforms in borewell water and suggested the hybrid method to be more productive compared to conventional hydrodynamic cavitation. Similarly, improvements were also reported by Li et al. [65] for removal of *M. Aeruginosa* by about 8% at various concentrations. However, H_2O_2 addition was not very effective for algae compared to bacteria.

6.4.3 Cavitation with Ozone Addition

Ozone is again a strong oxidizing agent, and intensification using ozone is considered to be better than that using hydrogen peroxide, possibly due to the higher decomposition rate of O_3 molecules into O_2 and $^{\bullet}O$ resulting in the production of $^{\bullet}OH$ radicals, again contributing to increased concentration of the oxidizing species required for disinfection. Moreover, O_3 can be decomposed by hydroxide ions or the conjugate base of H_2O_2 to yield $HO_2^{\bullet}$ and $^{\bullet}OH$ radicals for oxidation.

The positive impact and the synergic effect of ozone addition to hydrodynamic cavitation was discussed by Jyoti and Pandit [53] in detail, where a small addition of ozone in disinfection of borewell water resulted in significantly higher disinfection (>30%) as compared to hydrodynamic cavitation alone. Ozone and other oxidizing species oxidize the organic compounds in the cell membranes of the microorganisms, consequently rupturing the membranes and thereby adversely impacting the cell viability. The final result is disinfection. The dose of ozone and time of treatment play important role – the larger the dose, the shorter the time of treatment for similar disinfection levels. Also, there is a significant effect of ozone addition in improving the efficiency of the hydrodynamic cavitation. The authors felt that with the hybrid cavitation process of this type, the concentration of ozone required for disinfection can be substantially reduced to half or one-third depending upon the type of microorganism. The advantage of ozone addition was also confirmed later in the studies of Chand et al. [72], where an approximately two-fold enhancement in the disinfection of *E. Coli* was obtained. Wu et al. [64] reported comparison of hydrodynamic cavitation, ozone oxidation, and ozone-intensified hydrodynamic cavitation processes in disinfection and found significant improvements in the hybrid hydrodynamic cavitation process using ozone. Increased rates of disinfection and reduced treatment time were observed, implying substantially reduced costs for the ozone-assisted hydrodynamic cavitation.

6.4.4 Cavitation with Aeration or Oxygen

Process intensification using simple aeration and oxygen has not been extensively studied so far, although it was first reported for wastewater treatment by Ranade et al. [78] and recently by Joshi and Gogate [32], and Pravin Patil et al. [35]. The technique typically involves bubbling of air in the liquid during the conventional hydrodynamic cavitation process. Aeration of this type improves the performance, partly due to increase in the number density of the cavities and also partly due to supply of the oxygen necessary for the oxidation. Mane et al. [6] showed, for the first time, quantitative improvements in the efficiency of disinfection due to simple aeration in the hybrid hydrodynamic cavitation process using natural oils and felt that process intensification of this type can improve the economics of the process.

6.5 Hybrid Hydrodynamic Cavitation Technology Using Natural Oils

Natural oils derived from specialty plants are believed to offer additional benefits for increasing the efficiency of disinfection as well as increased rates of disinfection. Medicinal plants are a source of antimicrobial active compounds that can be exploited in the disinfection process through the specific components or through natural oils containing antimicrobial compounds of different types. Different types of antimicrobial compounds are expected to offer an unpredictable capacity to eliminate different microorganisms in different ways and with varying effectiveness, though combinations of such materials are still not reported so far. Integrating biological aspects with existing physicochemical methods can be practical, if the intricacies can be resolved for the various parameters of disinfection, for example, nature of microorganism, concentration, and physical and chemical attributes of the process. More research in this regard to connect traditional knowledge with modern science can provide possible technological alternatives to the existing methods. Cavitation, a physicochemical method of treatment as discussed previously, can be effectively combined with natural oils. Such newer hybrid combinations require inter-related synergy for their effective implementation. For both large- and small-scale water treatment, hybrid technology of hydrodynamic cavitation using natural oils appears to be a highly promising option.

Natural oils or plant extracts can contain a large number of biological substances, largely known and unknown complex compounds, some of which can have antimicrobial activity [79]. The known antimicrobial compounds can assist in the disinfection process; however, when used alone, their efficiency is limited, with practical application being difficult in most cases. The knowledge of the different antimicrobial compounds is essential to facilitate their extraction from the plant; their origin can be from any part of the plant such as flowers, stem, roots, or fruits/pods. The selection therefore depends on the type and composition of the antimicrobial compounds, and a large number of plants/natural oils/essential oils are known in this regard. Recently, Mane et al. [6, 48, 80], for the first time, reported the use of natural oils in the cavitation process, both in acoustic as well as in hydrodynamic cavitation, to increase the efficacy of disinfection by increasing the rates of disinfection with a subsequent positive impact on the techno-economic feasibility of the hybrid technology (**SWASTIIK – Safe Water and Sustainable Technology Initiative from Indian Knowledgebase**). The hybrid form of hydrodynamic cavitation was found most suitable compared to cavitation alone and only a small fraction of natural oil was required (0.1% v/v). It was observed that the disinfection efficiency depends on the type of the oil and the nature of the complex compounds present in the oil that contribute to antimicrobial property. Although this work studied only a few common natural oils such as clove oil, peppermint oil, castor oil, etc., it was evident that by selecting the appropriate oil or plant extract, it should be possible to increase the efficiency of disinfection to acceptable levels for commercial operations.

Another interesting aspect with the hybrid hydrodynamic cavitation technology using natural oils is that the process can be altered (more specifically, the rates can be drastically increased from the practical benefit perspective) or manipulated by selecting the natural oil, by adjusting the process parameters pertaining to the hydrodynamic cavitation, and most importantly, intensification. Again as discussed previously, these process parameters range from the type of hydrodynamic cavitation reactors to operating parameters such as pressure, temperature, etc. Since the literature has so far reported natural oils such as eucalyptus oil, peppermint oil, clove oil, cinnamon oil, etc., it would be prudent to evaluate many other natural oils, hundreds of these, in isolation and in combination. This could be a promising direction for future research.

What creates the disinfection action? This was already discussed with reference to only cavitation and also more particularly for hydrodynamic cavitation using different types of cavitating devices. This discussion of the hybrid process using natural oils is therefore over and above that of hydrodynamic cavitation alone in disinfection. Many of the natural oils have compounds such as phenolics, alcohols, aldehydes, acids, etc. which are considered to contribute to disinfection in different ways. Mane et al. [48] attributed the disinfection by eucalyptus oil to the presence of 1,8-cineole active compound, which is capable of destroying the permeability of bacterial membranes, leading to loss of electrolytes such as K^+, Na^+, and Ca^{2+}. The disinfection in presence of clove oil was presumed to be due to high levels of eugenol which can react with the phospholipids of the cell membrane. A partial list of some common natural oils and the possible contributing antimicrobial compounds contained in these oils is given in Figure 6.7.

The results, which are some of the compiled best results of various processes from literature, are presented in Figure 6.8. It is evident that most of the conventional processes require longer durations for the treatment (average 60 min or even more). Further, the efficiencies are low. Since the mandate is to completely eliminate harmful microorganisms, such options are of practically no value when it comes to commercialization of the process. Hybrid cavitation processes, conventional or otherwise, are certainly able to cut the processing time for disinfection; however, in many cases, at the cost of treatment and ease of operation. In this regard, the most recent hybrid cavitation process using a vortex diode as the cavitating device and natural oils such as peppermint oil clearly seems to provide the most suitable alternative.

The hybrid hydrodynamic cavitation process using natural oils was quite effective in increasing the rates of disinfection drastically – practically complete disinfection in just less than 10 minutes as compared to close to 1–2 hours in most cavitation processes. Thus, the interesting aspect of the hybrid process using natural oils is that whereas the previous and conventional processes desperately explored various configurations of cavitation reactors and that their possible use in disinfection have largely failed, the hybrid process offers a complete ease of operation using any

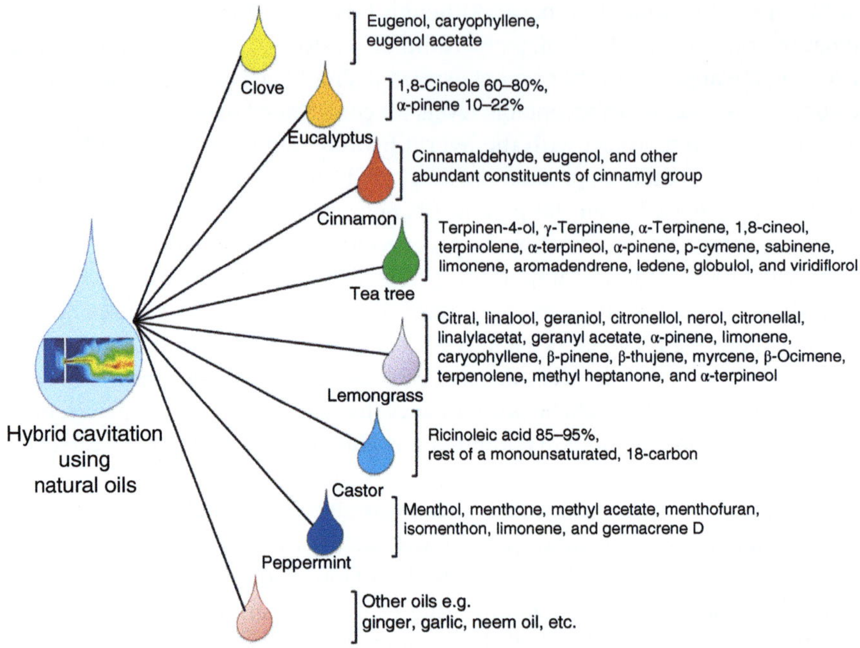

Figure 6.7 Hybrid Process using natural oils and possible options (Source: Adapted from [6, 48, 81–85]).

form of cavitating device with increased efficiency. Thus, although the nature of cavitating device is still important, it is the best combination of different parameters and natural oils which should now dictate the overall process.

In continuation of their investigation on the use of natural oils in hydrodynamic cavitation, Mane et al. [6] also found that exceptionally high rates can be obtained for practically complete removal of the antimicrobial-resistant (AMR) Gram-negative opportunistic pathogen *Pseudomonas aeruginosa* and the Gram-positive methicillin-resistant *Staphylococcus aureus*. Among various natural oils, the authors concluded that peppermint oil gives >99% disinfection in less than 10 minutes, using a vortex diode with operating pressure drop of 1 bar and low dose of 0.1% peppermint oil. A comparison of two different cavitating reactors was provided by employing vortex flow (vortex diode) and linear flow (orifice) for hydrodynamic cavitation; vortex diode giving the highest performance. A very high synergistic index (6 and 5 for vortex diode and orifice, respectively) for *P. aeruginosa*/peppermint oil, values well above 1 clearly indicating the synergistic effect of hydrodynamic cavitation combined with natural oil as an additive.

The implication of hybrid hydrodynamic cavitation using natural oils is particularly relevant in the present context as antimicrobial resistance (AMR) is now considered a global crisis, threatening the future of drugs and of millions of lives worldwide. The presence of AMR bacteria is reported in almost all common sources of drinking water – wells, rivers, and lakes, and astonishingly, even in tap/bottled water. The relevance of hybrid technology is more prominent also because of the fact that disinfection studies pertaining to AMR microorganisms

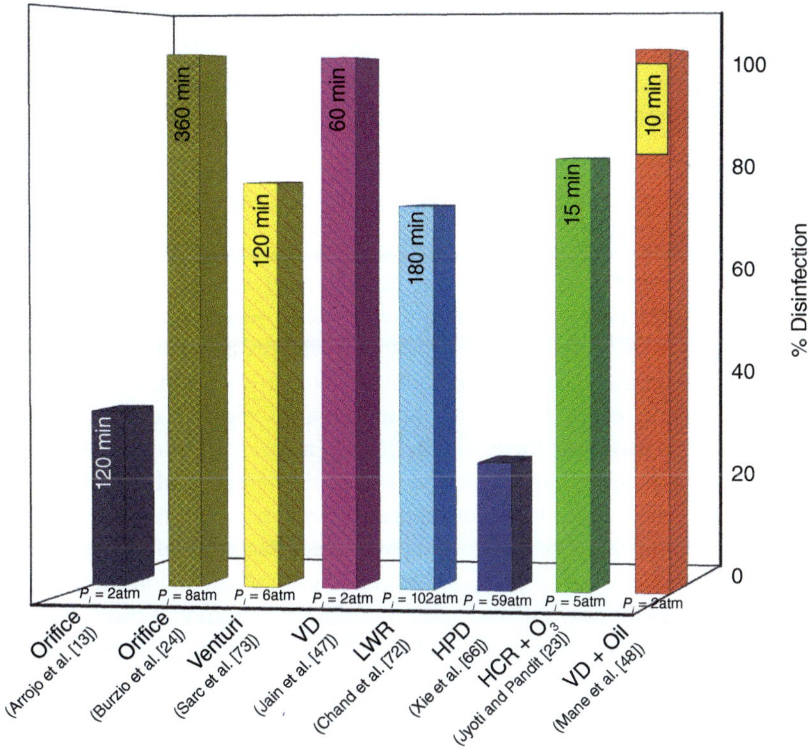

Figure 6.8 Comparison of the disinfection process for various forms-conventional and hybrid.

have been sparsely reported, largely due to difficulty in their removal and the recent emergence of this subject. Thus, the use of natural oils having antimicrobial property in the hybrid hydrodynamic cavitation process provides a unique, practical, and techno-economic strategy to address issues of high efficiency and increased rates of disinfection.

Since the quantity of oil is envisaged to be quite small, it can be easily separated and reused using filters or such designs for oil separations. In view of the low cost of natural oils coupled with the low cost of the hydrodynamic process for reactor types such as vortex diode that require very low pressures compared to conventional devices, the hybrid cavitation process can certainly transform the conventional process into a techno-economically viable process. Further, the health benefits associated with many of these natural oils as mentioned in the ancient Hindu literature, Ayurveda, the process can possibly transfer properties of natural ingredients to the treated water to boost immunity, providing such additional health benefits to humans. The aspects of the nature of oil, the effect of cavitation on the oil/breaking of oil molecule or formation of by-products, apart from health effect, are however not yet confirmed and require further detailed investigations.

As of today, contrary to the opinion expressed in most studies/reviews on the negativity of applying hydrodynamic cavitation in disinfection, now it can be suggested with certainty that the hybrid process using natural oils can be considered as a green and techno-economical process to eliminate common AMR bacteria. It is felt that the technology can also be effective in eliminating viruses and other such harmful microorganisms not reported so far.

6.5.1 Mechanism of Disinfection in Hydrodynamic Cavitation-Conventional vs. Hybrid Processes

The disinfection of water depends on a large number of process parameters, according to the nature of the process. It includes the physical processes such as impingement of a high-velocity jet, generation of shockwaves/pressure impulse, and the physical effect is directed toward accomplishing cell death – again through different mechanisms involving the rupture of cell membrane, oxidation of organics, and so on. The mechanism can get further complicated in various forms of hybrid processes that propose the use of different types of additives in general, and especially for the addition of natural oils having antimicrobial properties. It is no wonder, then, that the exact mechanism of disinfection has remained practically elusive in spite of various attempts at explaining it through the elucidation of cavitation number, different reactor configurations, different rate models, computational fluid dynamic studies, nature of additives, nature of microorganisms, and so on. Sawant et al. [51] suggested that the cavitation number can predict the relative intensity of cavitation in various cavitation devices, mainly linear flow-based, and therefore, can be used as a preliminary tool to compare the relative performance. However, different reactor configurations with wide variations in cavitation numbers have been experimentally shown to provide similar disinfection efficiency, indicating a severe limitation of such models for application to real-life situations.

For hydrodynamic cavitation, the important parameters affecting the disinfection process are mainly pressure drop, pH of the water, temperature, and nature of the cavitating device, apart from the nature of the microorganism. The oxidants, in general, used in various AOPs and also generated in situ during the hydrodynamic cavitation process, oxidize the organic compounds in the cell membranes of the microorganisms, that contributes to the rupture of the membranes, destroying the cell viability and thus resulting in disinfection.

The objective of the disinfection process is to kill or eliminate the bacteria either through the physical removal process (as in the case of adsorption) or by causing the death of the cells.

Acoustic cavitation affects the destruction of the microorganisms largely through mechanical stresses generated by the shockwaves as a consequence of the cavities' implosion and the turbulence generated in the process. It was suggested that sound pressures of 1–8 bar can provide the cavitation threshold in this regard, requiring 8–20 kHz frequency range for generating such pressures.

In hydrodynamic cavitation, because of the integrated nature of the physico-chemical operation, both the physical and chemical forces contribute to the process causing cell death, though the contribution varies mainly according to the oxidation mechanism. The role of the oxidant in disinfection has been known for a long time and earlier studies have reported oxidants such as hydrogen peroxide for generating hydroxyl radicals. However, the use of hydrogen peroxide as a disinfectant was limited due to its instability and the limited effect on the removal of bacteria; other combinations such as ozonation were also attempted with limited success [7, 23, 53, 86, 87].

The conventional hydrodynamic cavitation process (no external addition of chemicals) is believed to proceed through physical and chemical damage to live cells, leading to cell death. As the cavities implode, generating extreme temperatures and pressures at the point of implosion, the temperature and shockwaves of implosion can physically rupture the cell membrane. Further, the hydroxyl radicals generated in the implosion process lead to oxidative damage/rupture of the bonds of the cells, which increases the permeability of the oxidant, further leading to the oxidation of protein and lipoproteins, glycooxidation, and DNA [47, 88–90]. In effect, both physical and chemical oxidation processes in hydrodynamic cavitation contribute to cell death and therefore to disinfection. The efficiency of the disinfection therefore depends on the process parameters such as pressure drop, temperature, and cavitation reactor geometry, apart from the nature of microorganism and its concentration. A significant difference in the rates of destruction of different microorganisms clearly points to the effect of cell wall strength, which is another important parameter controlling the extent of disruption of cell; information on the cell wall or membrane wall strength is far from being satisfactorily reported. There are some other parameters such as the pollutants present in the water and the nature of pollutants, apart from the concentration of dissolved salts and suspended solids, which not only make the mechanism of the disinfection complex but also highly unpredictable.

The recent hybrid process of hydrodynamic cavitation coupled with natural oil is a similarly complex process to define in terms of its mechanism for disinfection of water. The mechanism due to the oil effect itself is not well understood, similar to that of hydrodynamic cavitation alone for water disinfection. Therefore, the combination of different natural oils and hydrodynamic cavitation, different reactor geometries, and different microorganisms with other process parameters, make it practically impossible, a Himalayan task, to understand the mechanism and consequently the prediction of the disinfection behavior. It is felt that the synergistic effect of the antimicrobial property of the natural oil and that of hydrodynamic cavitation can provide an effective and economical pathway for practical application. The mechanism in this hybrid process is believed to follow the similar oxidative pathway as outlined previously for the cavitation and also to contribute due to the antimicrobial compounds present in the natural oils reacting with the phospholipids of the cell membrane, which subsequently increases the permeability of the membrane and in effect denatures the cell protein, leading to cell death (Figure 6.9) [48]. The disinfection ability of eucalyptus oil was mainly believed to be due to the presence of the active compound 1,8-cineole that can destroy the permeability of bacterial membranes, leading to loss of electrolytes such as K^+, Na^+, and Ca^{2+} and disinfection ability of clove oil to high levels of eugenol which can react with the phospholipids of the cell membrane. When the antimicrobial compounds react with the lipids of the cell membrane, the impact is in the form of altering the permeability of the membrane, rupture, leakage of the vital cell components, and denaturing of the cell protein, resulting in cell death.

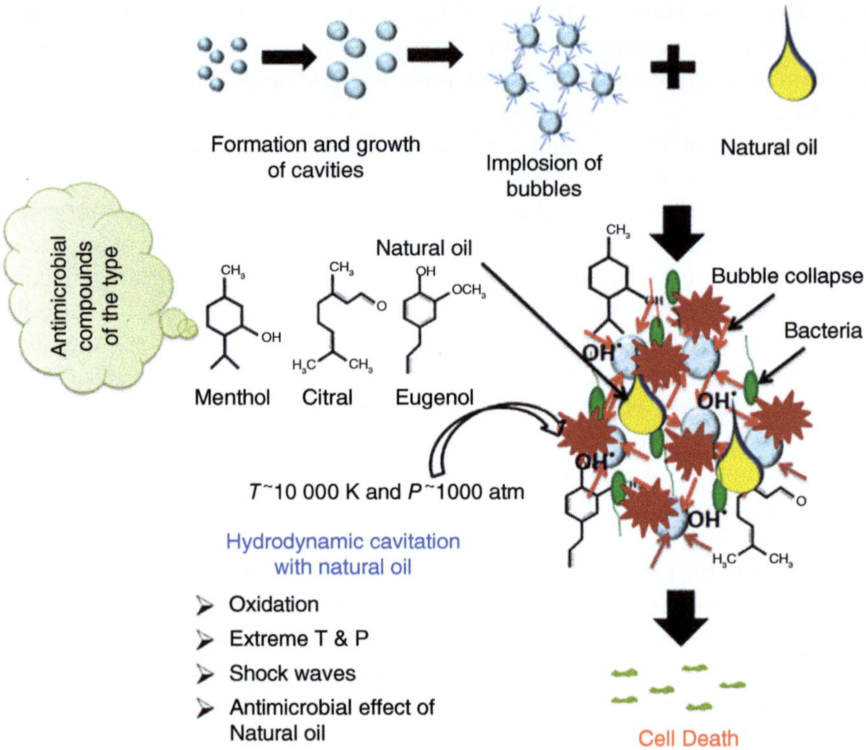

Figure 6.9 Postulated mechanism of hybrid cavitation process using natural oils (Source: Adapted from [48]).

6.5.2 Effect of Temperature

One interesting feature of the hydrodynamic cavitation process is that it also increases the temperature of the bulk liquid due to energy dissipation; as the temperature increases, the liquid properties (viscosity, surface tension, etc.) are also affected, which possibly has an adverse impact on the formation of cavities and also on the cavitation intensity. It is well known that temperature, in itself, has a positive impact on disinfection. An increase of temperature is expected to promote the vapor pressure inside the bubble, resulting in less violent bubble collapse, consequently reducing the chemical effects due to implosion. Another process impact can be in the form of generation of more free radicals at higher temperature. Thus, many such conflicting and contributing parameters/effects, many of which are beyond any means of measurement at this point of time, make the overall effect highly unpredictable. As compared to the conventional hydrodynamic cavitation, the impact could be much more pronounced in the hybrid processes in general, and those using natural oils, in particular. In general, the favorable parameters for enhancing cavitation effects include low liquid vapor pressure, low viscosity, and high surface tension. In the hybrid process using natural oils, the surface tension and viscosity of oil can play an important role in altering the efficiency of the overall process (not investigated in detail so far). The collapse of the cavities is more violent as the surface tension increases and the resultant final collapse temperature and pressure are also expected to be higher due to this violent implosion. In view of this, it is essential that the temperature effect in disinfection, especially in different types of hybrid cavitation, needs to be clearly elucidated for satisfactory and a priori prediction of the disinfection behavior and thereby for tailor-made process modifications in this regard.

6.6 Process Economics

It is well established that among various cavitation processes, only acoustic cavitation and hydrodynamic cavitation processes can be presently considered to be useful as far as water disinfection is concerned. It is also proved by various researchers that acoustic cavitation – because of various issues such as lower efficiency in disinfection, scale-up, and operational problem in large scale – is certainly not techno-economically viable. Thus, the choice at present day is largely restricted to hydrodynamic cavitation for water disinfection, and here, the type of cavitation reactors, process parameters, and usage of cavitation alone or hybrid cavitation are the most important considerations from the point of view of process economics. In view of large energy requirements for cavitation reactors using moving elements such as rotor, it would be useful to consider reactors without moving elements, from an economics point of view. Again, from the conventional cavitating devices such as orifice, venturi of different designs, and reactors using vortex flow such as vortex diode could prove more economical due to lower pressure drop requirements. Among different hydrodynamic cavitation processes using different reactors, the combination of physical or chemical process intensifications of various types again impacts drastically the overall process economics. The operational costs of conventionally reported hydrodynamic cavitation in disinfection can be typically two to three times higher compared to chemical disinfection methods [9, 53, 91]. The addition of natural oils having antimicrobial properties, as an additive for drastic improvements in process efficacy will certainly add a new dimension to conventional hybrid hydrodynamic cavitation techniques. All these aspects need to be considered in detail for the overall assessment of process economics of hydrodynamic cavitation for disinfection of water, either used alone or in hybrid form.

6.6.1 Cost Comparison of Different Processes

A tentative cost comparison of some of the useful methods using the best data from those reported in the literature is presented here in Table 6.2. As per the EPA [94], UV could be considered as a cheaper disinfection method as compared to chlorine and ozonation for a small-scale unit. Chlorination, in general, is the cheapest method and is used worldwide. From the following analysis, the hybrid hydrodynamic cavitation technology using natural oil is also seen as one of the most competitive methods of disinfection.

Table 6.2 Cost comparison for different disinfection processes.

Disinfection process	Cost ($/m^3)	References
Hybrid process with natural oils HC (Vortex diode ΔP,1 bar + 0.1% Peppermint oil)	0.036	[6]
HC (Vortex diode ΔP, 1 bar)	0.22	[6]
Chlorination	0.0042	[52]
Membrane	1	[92]
Solar photo Fenton	0.85	[93]
Ultrasonic bath	0.56	[23, 54]
Ultrasonic bath + 2 mg/L O_3	1	[23, 91]
hydrodynamic cavitation at $\Delta P = 5.17$ bar	8	[91]
Hydrodynamic cavitation (Orifice, ΔP, 5.17 bar) + 5 mg/L H_2O_2	4.6	[54]
Hydrodynamic cavitation (Orifice, ΔP,1.72 bar) + US flow cell, 40 kHz	2	[54]
UV for 40 mJ/cm^2 dose	0.05	[94]
Ozonation(1 mg/L dose)	0.92	[94]
Ultrasonic horn	160	[23]

More recent disinfection techniques include high-pressure homogenizer, high-speed homogenizer, ultrasonication (Ultrasonic horn, Ultrasonic bath), membrane filtration, and hydrodynamic cavitation with or without process integration. The cost of high-speed and high-pressure homogenizers ranges from 0.48 to 3.9 $/m^3, while for hydrodynamic cavitation the costs are comparatively and significantly lower. Cavitation with ozone, hydrogen peroxide, oxygen, etc., can further improve performance and reduce the cost of operation; however at the cost of increased complexity of the process and difficulties in operation. Moreover, there is increased difficulty in prediction of the disinfection behavior and no reports exist of commercial implementation of any of such processes. From Table 6.2, it is evident that the hybrid cavitation process using natural oils for disinfection is low-cost and can be considered as a green alternative to conventional disinfection processes. The low cost of just 0.22 $/m^3 for the hydrodynamic cavitation using vortex diode and the very low cost of only 0.036 $/m^3 in the case of process integration by addition of peppermint oil (0.1%) or with aeration, can certainly make the claims much stronger.

6.6.2 Typical Cost Calculation for Vortex Diode as Reactor in Hybrid Process Using Natural Oils

As explained previously, hybrid hydrodynamic cavitation using natural oils appears to be one of the most economical processes compared to hydrodynamic cavitation alone or other conventional hybrid combinations. A sample cost calculation is explained subsequently for the new hybrid process using the following representative data [6]:

Bacteria: *P. aeruginosa*
Volume: 20 L
Reactor: Vortex diode,
Flow rate: 730 LPH, ΔP: 1.0 bar
Natural oil: 0.1% peppermint oil (20 mL)
Initial concentration of bacteria = 130 000 CFU/mL and in 15 min, 99.9% removal/disinfection
Cavitation yield/volume treated (CFU/mL/J) = $\frac{(C_0-C)}{(\Delta P \times Q \times t)}$ = **7.02** CFU/mL/J
The number of passes can be obtained as: $n = \frac{(Q \times t)}{V}$
The number of passes needed for treatment (n) is 9.13

$$\text{Cost of treatment/m}^3 \text{ of water} = \frac{n \Delta P P_E}{36\eta}$$

Assuming the cost of electricity as 10 Rs/kWh and efficiency of the multistage pump (η) as 0.66, the cost of treatment per m^3 of water is **3.84** Rs/m^3 (~\$0.05/m^3). The cost can vary for different microorganisms and for different process conditions. Mane et al. [48] reported a very low cost, less than **2.5** Rs/m^3 (0.036 \$/m^3), for near-complete elimination of bacteria using peppermint oil and vortex diode in their hybrid hydrodynamic cavitation process. The cost can be optimized using simple process intensification of the aeration type and by further reducing the requirement of natural oils in disinfection.

Overall, techno-economic evaluation of different processes indicates cost of the hybrid methodology to be significantly lower by an order of magnitude compared to the most conventional processes, barring chlorination. In view of the current trend of avoiding chlorination due to the formation of harmful disinfection by-products, the hybrid hydrodynamic cavitation technology using natural oils points to practical solutions and is possibly most suitable alternative by going back to Nature.

6.7 New Developments and Future Potential

6.7.1 Applications in Drinking Water Treatment

Most water treatment plants have water available at high pressure, which is subsequently reduced using pressure reduction stations for classical chemical treatment. Therefore, the designs can be modified for integrating the existing system and hydrodynamic cavitation, especially the hybrid process using natural oils, without needing additional energy. This further facilitates implementation of the hybrid methodology, reduces the treatment cost, and also practically eliminates the use of chemicals and consequently the cost of the chemicals used in the existing treatment. A typical flow diagram of water treatment plant incorporating hybrid hydrodynamic cavitation process using natural oils is shown in Figure 6.10. If any other conventional

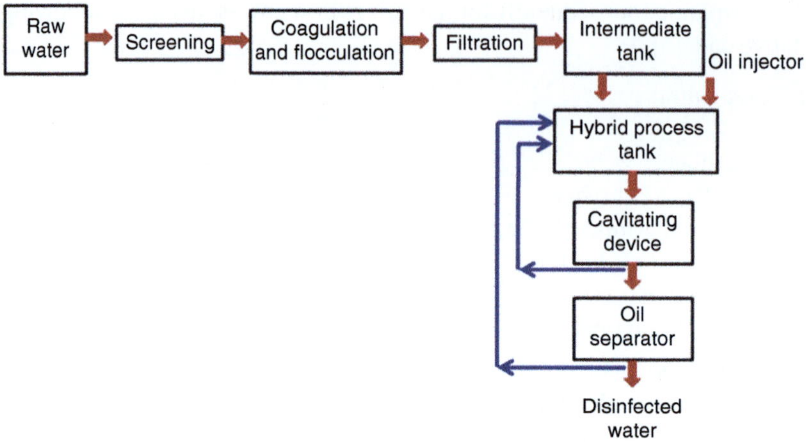

Figure 6.10 Schematic of water treatment plant using a hybrid cavitation process for disinfection.

hybrid process such as hydrogen peroxide or ozone addition is to be used, the oil injector shown in the figure needs to be replaced by an appropriate addition and the oil separator is not required. Since the hydrodynamic cavitation alone is not satisfactory in disinfection, in general, the most appropriate form of the hybrid hydrodynamic cavitation process in shown integrated with the existing water treatment plant from a practical point of view.

It is evident that hydrodynamic cavitation of any type of cavitating reactor can be easily integrated with the existing water treatment plant for disinfecting water. However, it is to be understood that many of the cavitating reactor designs are not robust and also not reliable for large-scale operations. Similarly, though many hybrid processes have shown promising improvements over and above HC alone, consistency and reproducibility in design can limit their application in real life. To the best of our knowledge, the vortex diode that employs vortex flow and the hybrid process employing natural oils having antimicrobial properties are the most suitable methods for water disinfection, both from high efficiency point of view and also for cost effectiveness. They have been shown to be reliable and easy to scale-up, paving way for its application in real-life water treatment.

6.7.2 Applications in Sewage Water Treatment

Sewage wastewater treatment is a routine operation worldwide, especially in the urban areas, for the removal of organic matter/pollutants. The organic matter is mainly from the household water waste, that is, from sewage generated from many different sources comprising washing. This sewage has reasonable amounts of organic matter, mainly comprising BOD (typically COD below 1000 ppm). The sewage wastewater treatment is conventionally carried out using biological processes such as Activated Sludge Process (ASP) involving bacteria, protozoa, metazoan, and other microlife. Wastewater treatment here, as is the case with most biological processes, requires maintaining the right conditions for the right type of microorganisms for best results. For effective degradation and the satisfactory progress of the biological treatment process, the presence of interfering/wrong microorganisms must be avoided.

Barring the conventional sewage wastewater treatment, the other useful application can be found in the sewage water treatment area where the organic matter in sewage water can be fed as nutrient to the microorganisms for some novel applications such as algal pond. The algal ponds/algae are considered to have main applications in deriving oil from algae and require huge amount of water. Since any wastewater– domestic or industrial – is required to be treated satisfactorily before its discharge to surface waters, it would be wise to make use of domestic sewage wastewater that has organic content to feed algae, thereby eliminating the need for separate treatment of such sewage wastewater. However, the application in such areas may not be straightforward, since protozoa, representing zooplanktons, are required to be eliminated as these cells actually eat algae thereby adversely affecting the algal growth. Thus, the removal of zooplanktons is a specific problem that needs to be addressed satisfactorily before the sewage water can be used as a source of food in the algal ponds.

As already elaborated, similar to disinfection, the microorganisms, zooplanktons can be eliminated by various methods such as chemical disinfection, microbial degradation, filtrations, etc. However, hydrodynamic cavitation technology along with a suitable cavitation reactor can provide a useful alternative to existing processes for eliminating zooplanktons. Again, as explained earlier, different reactors such as orifice, venturi, and vortex diode can be used. The devices have been modified over the years by many researchers and designers to improve their performance.

Ranade and Bhandari [95] disclosed a system for controlled destruction of protozoan cells, zooplankton, and dormant cysts from any water stream using a vortex diode. The contaminated water is fed tangentially to the vortex diode so as to form a vortex and effect cavitation which effectively disinfects the water.

6.7.3 Applications in Ballast Water Treatment

Worldwide, shipping is an important pillar of the economy, contributing via transport of large quantities (close to 90%) of commodities. Along with the commodities transported, several billion tons of ballast water get transported every year, and as a consequence, bio-invasion in the form of transport of different microorganisms is a serious concern [51]. Ballast water (BW) is water carried by ships to ensure stability and structural integrity. When a ship loads cargo, the BW is discharged into the sea. Untreated ballast waste water is released directly into the sea, which could potentially introduce new invasive aquatic species as BW contains many microorganisms, phytoplankton, and zooplankton which are native to the ecology of the zone [96].

A global treaty, The Ballast Water and Sediments (BWM) Convention is a key international measure for environmental protection which targets prevention of the spread of potentially invasive aquatic species in ships' BW, in force from September

2017 [97]. This enforces that all ships in international trade manage and treat their ballast waste water prior to releasing it into the sea.

To achieve the requisite BW discharge standards, different water treatment technologies are used. In general, the treatment processes involve three stages, that is, pre-treatment, treatment, and residual control. Though there are a large number of techniques that can be employed for ballast wastewater treatment, there is no single BW treatment technique which can effectively remove all invasive species at a time [98]. The major technologies available for ballast waste water treatment systems are filtration, UV, chemical disinfection, oxidation, and cavitation, some of which are approved and some yet to be approved by the International Maritime Organization (IMO) [99–102]. These technologies are currently employed and evaluated in ballast water treatment (BWT).

Filtration technology in BWT has been reported to be inefficient to eliminate zooplankton, phytoplankton, and pathogenic strains of *Vibrio cholera* [103]. Though improvements could be seen using a combination of filtration and UV radiation, it was not efficient enough to achieve "zero organisms" as suggested by USA-EPA [104]. The spore of bacteria and resting eggs of other organisms are also difficult to exclude by many of the approved technologies like UV radiation, deoxygenation, etc. High heating is considered to be efficient against spore and resting eggs; however, it is highly energy consuming and the heat of the outgoing BW should be tolerated by the vertebrates and invertebrates present in the ecosystem. Elimination of bacteria requires more robust treatment because of their microscopic size in comparison to zooplanktons, regeneration potential, and pliability to adverse environment. Deoxygenation in combination with other techniques shows good results for *E. coli;* however, the process requires relatively longer reaction time which may not be suitable for short voyages [103]. Variation in temperature also affects the disinfection performance of the process and varying atmospheric temperature may lead to an unstable system. A chemical process, alone or combined with other processes, may provide advantages over the physical processes, for example, by lower power consumption and increased efficiency in eliminating bacteria as well as phytoplanktons [105]. Conversely, the toxic response of the treated water due to the persistence of chemical has been validated to be fatal for fish and some other aquatic animals. The use of such hazardous substances on a large scale may create an additional environmental problem. The intermediates generated from oxidation chemicals react with organic matter and result in undesirable by-products which also have a detrimental effect on the environment as well as on human health. Apart from residual toxicity, the use of chemicals for treatment may also cause corrosion of vessels. Advanced oxidation using TiO_2 aid electrolysis has been adopted, which facilitates the generation of free radicals but the formation of hazardous by-products like trihalomethanes, halo-acetic acids have been observed in the process. The persistence and effect of these toxic by-products in environment and aquatic life are yet to be evaluated.

In most cases, process integration of methods was preferred over individual technologies [96]. As already discussed for bacterial killing, process integration is more effective for elimination of bacteria and zooplankton. Filtration can be combined with UV treatment, advanced oxidation by nanoparticles, and chemical disinfection. Coagulation and cavitation processes are also studied individually as well as in combination with other processes. Cavitation is seen as an environmentally and economically sound mechanism, but when used individually, it is slow to work in flow-through systems. The integration of UV and filtration yielded a reduction of >70% of phytoplankton, 60% of zooplankton and only 13.7% of *Artemia* [103]. As is evident, the removal efficiency is not satisfactory and the use of UV will add up to the process cost and complexity. It is important to maintain a balance between process efficiency, cost, and ease of operation when combining different processes. Other than technological and environmental complications, the high operational cost of the available BW treatment solutions is one of the major concerns [98]. The systems must comply with the criteria of being safe for the operating personnel, environmentally acceptable, practical, cost effective, and biologically efficient. Therefore, it is of utmost importance to pursue newer technologies which will be economically feasible and hassle-free in operation and installation.

Hydrodynamic cavitation has now shown significant potential for the disinfection of water. Along the similar lines, it can also be effectively applied in the treatment of ships' ballast water. There are only a few reports suggesting the application of hydrodynamic cavitation in destruction of zooplankton that can be particularly useful in ballast water treatment [51, 95]. The elimination of harmful protozoa zooplanktons was briefly discussed in the section on sewage water treatment.

The pumped seawater can be fed to a heat exchanger having a standard source such as steam or engine exhaust gas, or another suitable method. Control of the vortex diode may be through a regulator connected in series with the water intake means that is connected to the cavitation chamber. The treated water expelled from the vortex diode is connected to a storage tank(s). The water collected from the bleed holes can be recycled back to the water intake means or discarded into the sea.

The water treatment should ensure the vessel's water quality at the time of intake and also during discharge. A re-circulation mechanism can be devised to further treat water, and the need for such a mechanism may be determined by appropriately monitoring the parameters for quality. It is possible to devise sensors for detecting the presence of specific pathogens. If a re-circulation mechanism needs to be activated, the shutdown operation may be controlled from a control panel. Thus, the method of water treatment can be employed without the use of chemicals for efficiently treating water.

6.8 Summary

Disinfection of water is the most important, critical, and essential operation in water treatment to eliminate harmful microorganisms such as *E. coli*, and harmful, pathogenic, difficult-to-treat, and antimicrobial-resistant bacteria (AMR). Cavitation, in general, and hydrodynamic cavitation, in particular, can certainly be presented as promising alternatives to the existing methods of treatment such as chemical disinfection, or physical methods such as UV radiation or membrane separation. However, most of the existing practices of cavitation, in its conventional format, may not completely eliminate the harmful bacteria; moreover, they are largely not economical. Further, hydrodynamic cavitation appears to be the most effective compared to the other forms of cavitation such as acoustic cavitation.

The severe temperature and pressure conditions generated during the cavitation, formation of pressure-waves and micro-jets releasing large amounts of energy, generation of intense shear fluid stresses, oxidation due to consequently generated oxidizing species (e.g., $OH^{\bullet}$), and many such factors damage the cell membrane resulting in microorganism death or inactivation, thereby accomplishing disinfection of water.

Reactor configuration in the hydrodynamic cavitation process is critical and a judicious choice between cavitation devices, with or without moving elements, is required. In view of the large number of cavitation devices reported in the literature, ranging from a simple single-hole orifice to rotating impeller designs, the rotational flow device seems to be the most efficient. In this category, the vortex diode, a cavitating device that has no moving parts and employs a vortex for generating cavitation, can be considered as the best choice as of today.

Hybrid hydrodynamic cavitation technology using natural oils with antimicrobial properties appears to provide not just a techno-economical alternative, but the process is simple, requires less space, does not involve the use of harmful chemicals – thereby avoiding harmful disinfection by-products, is easy to operate, makes use of natural oils that can assist in producing healthy water, and is easy to scale-up – most crucial for commercial applications for implementation in large-scale water treatment plants.

The overall hydrodynamic cavitation process – the reactor configuration in hydrodynamic cavitation, fluid dynamics within the reactors, effect of various process parameters, and synergistic effects of additives for water treatment – is still in its infancy and not fully understood. The process optimization is largely empirical, by experimentally evaluating the governing parameters. However, it is essential to understand the theory and practice – the qualitative as well as quantitative roles of various parameters in the process of disinfection.

It is essential to drive research in this area so as to develop a techno-economical process using cavitation that is capable of completely eliminating the harmful microorganisms.

References

1 WHO (2012). *Potable Water Specification*, 2e. Kigali: Rwanda Bureau of Standard.
2 WHO (2017) *2017 Who Guidelines for Drinking Water Quality: First Addendum to the Fourth Edition.*
3 WHO (2016). *Quantitative Microbial Risk Assessment: Application for Water Safety Management*, vol. 187. WHO Press.
4 Pfaller, S. (2014). *Innovative Methods for Determining the Microbiological Quality of Drinking Water*. Elsevier Ltd.

5 WHO (2008). WHO guidelines for drinking-water quality. 1: 1–668. https://doi.org/10.1248/jhs1956.35.307.

6 Mane, M., Bhandari, V., Balapure, K., and Ranade, V. (2020). Ultrasonics – sonochemistry destroying antimicrobial resistant bacteria (AMR) and difficult, opportunistic pathogen using cavitation and natural oils/plant extract. *Ultrasonics Sonochemistry* 69: 105272.

7 Gogate, P.R. (2007). Application of cavitational reactors for water disinfection: current status and path forward. *Journal of Environmental Management* 85 (4): 801–815.

8 Dular, M., Griessler-Bulc, T., Gutierrez-Aguirre, I. et al. (2016). Use of hydrodynamic cavitation in (waste)water treatment. *Ultrasonics Sonochemistry* 29: 577–588.

9 Sun, X., Liu, J., Ji, L. et al. (2020). A review on hydrodynamic cavitation disinfection: the current state of knowledge. *Science of the Total Environment* 737: 139606.

10 Collivignarelli, M.C., Abbà, A., Benigna, I. et al. (2018). Overview of the main disinfection processes for wastewater and drinking water treatment plants. *Sustainability* 10 (1): 1–21.

11 Krasner, S.W., Weinberg, H.S., Richardson, S.D. et al. (2006). Occurrence of a new generation of disinfection byproducts. *Environmental Science & Technology* 40 (23): 7175–7185.

12 Lee, O.M., Kim, H.Y., Park, W. et al. (2015). A comparative study of disinfection efficiency and regrowth control of microorganism in secondary wastewater effluent using UV, ozone, and ionizing irradiation process. *Journal of Hazardous Materials* 295: 201–208.

13 Arrojo, S., Benito, Y., and Martínez Tarifa, A. (2008). A parametrical study of disinfection with hydrodynamic cavitation. *Ultrasonics Sonochemistry* 15 (5): 903–908.

14 Perugu, S., Nagati, V., and Bhanoori, M. (2016). Green synthesis of silver nanoparticles using leaf extract of medicinally potent plant Saraca indica: a novel study. *Applied Nanoscience* 6 (5): 747–753.

15 Nithya, R. and Ragunathan, R. (2009). Synthesis of silver nanoparticle using Pleurotus sajor caju. *Digest Journal of Nanomaterials and Biostructures* 4 (4): 623–629.

16 Kiruba Daniel, S.C.G., Abirami, J., Kumaran, S., and Sivakumar, M. (2015). Microbicidal tissue paper using green synthesized silver nanoparticles. *Current Nanoscience* 11 (1): 64–68.

17 Gondal, M.A., Dastageer, M.A., Khalil, A. et al. (2011). Nanostructured ZnO synthesis and its application for effective disinfection of Escherichia coli micro organism in water. *Journal of Nanoparticle Research* 13: 3423–3430.

18 Asuncion, E., Dimapilis, S., Hsu, C. et al. (2018). Zinc oxide nanoparticles for water disinfection. *Sustainable Environment Research* 28 (2): 47–56.

19 Bondarenko, O., Juganson, K., Ivask, A. et al. (2013). Toxicity of Ag, CuO and ZnO nanoparticles to selected environmentally relevant test organisms and mammalian cells in vitro: a critical review. *Archives of Toxicology* 87: 1181–1200.

20 Huang, J., Zhan, G., Zheng, B. et al. (2011). Biogenic silver nanoparticles by Cacumen Platycladi extract: synthesis, formation mechanism, and antibacterial activity. *Industrial and Engineering Chemistry Research* 50 (15): 9095–9106.

21 Piccirillo, C., Demiray, S., Ferreira, A.C.S. et al. (2013). Chemical composition and antibacterial properties of stem and leaf extracts from Ginja cherry plant. *Industrial Crops and Products* 43: 562–569.

22 Kirti, S., Bhandari, V.M., Jena, J., and Bhattacharyya, A.S. (2018). Elucidating efficacy of biomass derived nanocomposites in water and wastewater treatment. *Journal of Environmental Management* 226 (November): 95–105.

23 Jyoti, K.K. and Pandit, A.B. (2004). Ozone and cavitation for water disinfection. *Biochemical Engineering Journal* 18 (1): 9–19.

24 Burzio, E., Bersani, F., Caridi, G.C.A. et al. (2020). Water disinfection by orifice-induced hydrodynamic cavitation. *Ultrasonics Sonochemistry* 60 (September 2019): 104740.

25 Suslick, K.S. (1990). Sonochemistry. *Science* 247: 1439–1445.

26 Ranade, V.V. and Bhandari, V.M. (ed.) (2014). Industrial wastewater treatment, recycling, and reuse. In: *Industrial Wastewater Treatment, Recycling and Reuse*, 1–80. Elsevier.

27 Chakinala, A.G., Gogate, P.R., Burgess, A.E., and Bremner, D.H. (2009). Industrial wastewater treatment using hydrodynamic cavitation and heterogeneous advanced Fenton processing. *Chemical Engineering Journal* 152 (2–3): 498–502.

28 Gore, M.M., Saharan, V.K., Pinjari, D.V. et al. (2014). Degradation of reactive orange 4 dye using hydrodynamic cavitation based hybrid techniques. *Ultrasonics Sonochemistry* 21 (3): 1075–1082.

29 Suryawanshi, P.G., Bhandari, V.M., Sorokhaibam, L.G. et al. (2018). Solvent degradation studies using hydrodynamic cavitation. *Environmental Progress & Sustainable Energy* 37 (1): 295–304.

30 Tao, Y., Cai, J., Huai, X. et al. (2016). Application of hydrodynamic cavitation to wastewater treatment. *Chemical Engineering and Technology* 39 (8): 1363–1376.

31 Choi, J., Khim, J., Neppolian, B., and Son, Y. (2019). Enhancement of sonochemical oxidation reactions using air sparging in a 36 kHz sonoreactor. *Ultrasonics Sonochemistry* 51 (July 2018): 412–418.

32 Joshi, S.M. and Gogate, P.R. (2019). Intensification of industrial wastewater treatment using hydrodynamic cavitation combined with advanced oxidation at operating capacity of 70 L. *Ultrasonics Sonochemistry* 52 (December 2018): 375–381.

33 Panda, D. and Manickam, S. (2019). Hydrodynamic cavitation assisted degradation of persistent endocrine-disrupting organochlorine pesticide Dicofol: optimization of operating parameters and investigations on the mechanism of intensification. *Ultrasonics Sonochemistry* 51 (May 2018): 526–532.

34 Thanekar, P. and Gogate, P. (2018). Application of hydrodynamic cavitation reactors for treatment of wastewater containing organic pollutants: intensification using hybrid approaches. *Fluids* 3 (4): 98.

35 Patil, P.B., Bhandari, V.M., and Ranade, V.V. (2021). Improving efficiency for removal of ammoniacal nitrogen from wastewaters using hydrodynamic cavitation. *Ultrasonics Sonochemistry* 70 (July 2020): 105306.

36 Kim, H., Sun, X., Koo, B., and Yoon, J.Y. (2019). Experimental investigation of sludge treatment using a rotor-stator type hydrodynamic cavitation reactor and an ultrasonic bath. *Processes* 7 (11): 790.

37 Suryawanshi, N.B., Bhandari, V.M., Sorokhaibam, L.G., and Ranade, V.V. (2016). A non-catalytic deep desulphurization process using hydrodynamic cavitation. *Scientific Reports* 6 (September): 1–8.

38 Suryawanshi, N.B., Bhandari, V.M., Sorokhaibam, L.G., and Ranade, V.V. (2017). Developing techno-economically sustainable methodologies for deep desulfurization using hydrodynamic cavitation. *Fuel* 210 (September): 482–490.

39 Gogate, P.R. (2011). Hydrodynamic cavitation for food and water processing. *Food and Bioprocess Technology* 4 (6): 996–1011.

40 Mohod, A.V., Gogate, P.R., Viel, G. et al. (2017). Intensification of biodiesel production using hydrodynamic cavitation based on high speed homogenizer. *Chemical Engineering Journal* 316: 751–757.

41 Chipurici, P., Vlaicu, A., Calinescu, I. et al. (2019). Ultrasonic, hydrodynamic and microwave biodiesel synthesis – a comparative study for continuous process. *Ultrasonics Sonochemistry* 57 (May): 38–47.

42 Badve, M.P., Gogate, P.R., Pandit, A.B., and Csoka, L. (2014). Hydrodynamic cavitation as a novel approach for delignification of wheat straw for paper manufacturing. *Ultrasonics Sonochemistry* 21 (1): 162–168.

43 Gogate, P.R. and Pandit, A.B. (2000). Engineering design methods for cavitation reactors II: hydrodynamic cavitation. *AIChE Journal* 46 (8): 1641–1649.

44 Vichare, N.P., Gogate, P.R., and Pandit, A.B. (2000). Optimization of hydrodynamic cavitation using a model reaction. *Chemical Engineering and Technology* 23 (8): 683–690.

45 Yan, Y. and Thorpe, R.B. (1990). Flow regime transitions due to cavitation in the flow through an orifice. *International Journal of Multiphase Flow* 16 (6): 1023–1045.

46 Braeutigam, P., Franke, M., and Wu, Z. (2010). Role of different parameters in the optimization of hydrodynamic cavitation. *Chemical Engineering and Technology* 33 (6): 932–940.

47 Jain, P., Bhandari, V.M., Balapure, K. et al. (2019). Hydrodynamic cavitation using vortex diode: an efficient approach for elimination of pathogenic bacteria from water. *Journal of Environmental Management* 242 (September 2018): 210–219.

48 Mane, M., Bhandari, V., Balapure, K., and Ranade, V. (2020). A novel hybrid cavitation process for enhancing and altering rate of disinfection by use of natural oils derived from plants. *Ultrasonics Sonochemistry* 61 (October 2019): 104820.

49 Save, S.S., Pandit, A.B., and Joshi, J.B. (1997). Use of hydrodynamic cavitation for large scale microbial cell disruption. *Food and Bioproducts Processing* 75 (1): 41–49.

50 Save, S.S., Pandit, A.B., and Joshi, J.B. (1994). Microbial cell disruption: role of cavitation. *Chemical Engineering Journal and the Biochemical Engineering Journal* 55 (3): 67–72.

51 Sawant et al. (2008). Effect of hydrodynamic cavitation on zooplankton: a tool for disinfection. 42: 320–328.

52 Jyoti, K.K. and Pandit, A.B. (2001). Water disinfection by acoustic and hydrodynamic cavitation. *Biochemical Engineering Journal* 7 (3): 201–212.

53 Jyoti, K.K. and Pandit, A.B. (2004). Effect of cavitation on chemical disinfection efficiency. *Water Research* 38 (9): 2249–2258.

54 Jyoti, K.K. and Pandit, A.B. (2003). Hybrid cavitation methods for water disinfection. *Biochemical Engineering Journal* 14 (1): 9–17.

55 Liu, Z., Zhu, M., Deng, C. et al. (2016). Pollutant and microorganism removal from water by hydrodynamic cavitation. *Open Biotechnology Journal* 10 (1): 258–264.

56 Balasundaram, B. and Harrison, S.T.L. (2011). Optimising orifice geometry for selective release of periplasmic products during cell disruption by hydrodynamic cavitation. *Biochemical Engineering Journal* 54 (3): 207–209.

57 Loraine, G., Chahine, G., Hsiao, C.T. et al. (2012). Disinfection of gram-negative and gram-positive bacteria using DynaJets® hydrodynamic cavitating jets. *Ultrasonics Sonochemistry* 19 (3): 710–717.

58 Mezule, L., Tsyfansky, S., Yakushevich, V., and Juhna, T. (2009). A simple technique for water disinfection with hydrodynamic cavitation: effect on survival of Escherichia coli. *Desalination* 248 (1–3): 152–159.

59 Bhandari, V.M., Sorokhaibam, L.G., and Ranade, V.V. (2016). Industrial wastewater treatment for fertilizer industry – a case study. *Desalination and Water Treatment* 57 (57): 27934–27944.

60 Kosel, J., Gutiérrez-Aguirre, I., Rački, N. et al. (2017). Efficient inactivation of MS-2 virus in water by hydrodynamic cavitation. *Water Research* 124 (July): 465–471.

61 Balasundaram, B. and Pandit, A.B. (2001). Selective release of invertase by hydrodynamic cavitation. *Biochemical Engineering Journal* 8 (3): 251–256.

62 Xu, Y., Yang, J., Wang, Y. et al. (2006). The effects of jet cavitation on the growth of Microcystis aeruginosa. *Journal of Environmental Science and Health, Part A: Toxic/Hazardous Substances and Environmental Engineering* 41 (10): 2345–2358.

63 Balasundaram, B. and Harrison, S.T.L. (2006). Disruption of brewers ' yeast by hydrodynamic cavitation: process variables and their influence on selective release. *Biotechnology and Bioengineering* 94 (2): 303–311.

64 Wu, Z., Shen, H., Ondruschka, B. et al. (2012). Removal of blue-green algae using the hybrid method of hydrodynamic cavitation and ozonation. *Journal of Hazardous Materials* 235–236: 152–158.

65 Li, P., Song, Y., and Yu, S. (2014). Removal of Microcystis aeruginosa using hydrodynamic cavitation: performance and mechanisms. *Water Research* 62: 241–248.

66 Xie, L., Terada, A., and Hosomi, M. (2017). Disentangling the multiple effects of a novel high pressure jet device upon bacterial cell disruption. *Chemical Engineering Journal* 323: 105–113.

67 Waghmare, A., Nagula, K., Pandit, A., and Arya, S. (2019). Hydrodynamic cavitation for energy efficient and scalable process of microalgae cell disruption. *Algal Research* 40 (March): 101496.

68 Mevada, J., Devi, S., and Pandit, A. (2019). Large scale microbial cell disruption using hydrodynamic cavitation: energy saving options. *Biochemical Engineering Journal* 143 (August 2018): 151–160.

69 Dalfré Filho, J.G., Assis, M.P., and Genovez, A.I.B. (2015). Bacterial inactivation in artificially and naturally contaminated water using a cavitating jet apparatus. *Journal of Hydro-Environment Research* 9 (2): 259–267.

70 Shirgaonkar, I.Z., Lothe, R.R., and Pandit, A.B. (1998). Comments on the mechanism of microbial cell disruption in high-pressure and high-speed devices. *Biotechnology Progress* 14 (4): 657–660.

71 Sun, X., Park, J.J., Kim, H.S. et al. (2018). Experimental investigation of the thermal and disinfection performances of a novel hydrodynamic cavitation reactor. *Ultrasonics Sonochemistry* 49 (February): 13–23.

72 Chand, R., Bremner, D.H., Namkung, K.C. et al. (2007). Water disinfection using the novel approach of ozone and a liquid whistle reactor. *Biochemical Engineering Journal* 35 (3): 357–364.

73 Šarc, A., Kosel, J., Stopar, D. et al. (2018). Removal of bacteria *Legionella pneumophila, Escherichia coli,* and *Bacillus subtilis* by (super)cavitation. *Ultrasonics Sonochemistry* 42 (September 2017): 228–236.

74 Caretti, C. and Lubello, C. (2003). Wastewater disinfection with PAA and UV combined treatment: a pilot plant study. *Water Research* 37 (10): 2365–2371.

75 Cotillas, S., Llanos, J., Castro-Ríos, K. et al. (2016). Synergistic integration of sonochemical and electrochemical disinfection with DSA anodes. *Chemosphere* 163: 562–568.

76 Birben, N.C., Paganini, M.C., Calza, P., and Bekbolet, M. (2017). Photocatalytic degradation of humic acid using a novel photocatalyst: Ce-doped ZnO. *Photochemical & Photobiological Sciences* 16 (1): 24–30.

77 Naddeo, V., Landi, M., Belgiorno, V., and Napoli, R.M.A. (2009). Wastewater disinfection by combination of ultrasound and ultraviolet irradiation. *Journal of Hazardous Materials* 168 (2–3): 925–929.

78 Ranade, V.V., Kulkarni, A.A., Bhandari, V.M., and Data, P. (2016). Vor. DODES AS Effl. Treat. DEVICES. (12) United States Patent 9,422,952 (B2).

79 de Oliveira, K.Á.R., Berger, L.R.R., de Araújo, S.A. et al. (2017). Synergistic mixtures of chitosan and Mentha piperita L. essential oil to inhibit Colletotrichum species and anthracnose development in mango cultivar Tommy Atkins. *Food Microbiology* 66: 96–103.

80 Mane, M.B., Bhandari, V.M., and Ranade, V.V. (2021). Safe water and technology initiative for water disinfection: application of natural plant derived materials. *Journal of Water Process Engineering* 43 (August): 1–11.

81 Riley, T.V., Carson, C.F., and Hammer, K.A. (2006). Melaleuca alternifolia (Tea Tree) oil: a review of antimicrobial and other medicinal properties. *Clinical Microbiology Reviews* 19 (1): 50–62.

82 Siddiqui, N. and Ahmad, A. (2013). A study on viscosity, surface tension and volume flow rate of some edible and medicinal oils. *International Journal of Environmental Science and Technology* 2 (6): 1318–1326.

83 Baker, B.P. and Grant, J.A. (2018). Cinnamon & cinnamon oil profile. 1–16. https://hdl.handle.net/1813/56117.

84 Tajidin, N.E. (2012). Chemical composition and citral content in lemongrass (Cymbopogon citratus) essential oil at three maturity stages. *African Journal of Biotechnology* 11 (11): 2685–2693.

85 Baker, B., Grant, J.A., and Malakar-Kuenen, R. (2018). Peppermint & peppermint oil profile. 1–18. https://hdl.handle.net/1813/56135.

86 Dahl, E. (1976). Physicochemical aspects of disinfection of water by means of ultrasound and ozone. *Water Research* 10 (8): 677–684.

87 Yoshpe-Purer, Y. and Eylan, E. (1968). Disinfection of water by hydrogen peroxide. *Health Laboratory Science* 5 (4): 233–238.

88 Burleson, G.R., Murray, T.M., and Pollard, M. (1975). Inactivation of viruses and bacteria by ozone, with and without sonication. *Applied Microbiology* 29 (3): 340–344.

89 Bandyopadhyay, U., Das, D., and Banerjee, R. (1999). Reactive oxygen species: oxidative damage and pathogenesis. *Current Science* 77 (5): 658–666.

90 Phaniendra, A., Jestadi, D.B., and Periyasamy, L. (2015). Free radicals: properties, sources, targets, and their implication in various diseases. *Indian Journal of Clinical Biochemistry* 30 (1): 11–26.

91 Jyoti, K.K. and Pandit, A.B. (2003). Hybrid cavitation methods for water disinfection: simultaneous use of chemicals with cavitation. *Ultrasonics Sonochemistry* 10 (4–5): 255–264.

92 Pooi, C.K. and Ng, H.Y. (2018). Review of low-cost point-of-use water treatment systems for developing communities. *npj Clean Water* 1 (1): https://doi.org/10.1038/s41545-018-0011-0.

93 Michael, I., Hapeshi, E., Michael, C. et al. (2012). Solar photo-Fenton process on the abatement of antibiotics at a pilot scale: degradation kinetics, ecotoxicity and phytotoxicity assessment and removal of antibiotic resistant enterococci. *Water Research* 46 (17): 5621–5634.

94 EPA (1999). *Wastewater Technology Fact Sheet, Ultar-Violet Disinfection*. EPA.

95 Ranade, V.V. and Bhandari, V.M. (2014). Apparatus and method for eliminating or reducing protozoa-zoo planktons, algal predators from wastewaters using cavitation. 2014-INV-0083.

96 David, M. and Gollasch, S. (ed.) (2015). *Global Maritime Transport and Ballast Water Management*, Invading Nature - Springer Series in Invasion Ecology 8, 133–169. Springer.

97 IMO (2016). Global treaty to halt invasive aquatic species to enter into force in 2017. International Maritime Organisation Website, 1–3. http://www.imo.org/

en/MediaCentre/PressBriefings/Pages/22-BWM-.aspx (accessed 5 November 2020).

98 Tsolaki, E. and Diamadopoulos, E. (2010). Technologies for ballast water treatment: a review. *Journal of Chemical Technology and Biotechnology* 85 (1): 19–32.

99 Liang, J., Bing Chen, B.Z., and Peng, H. (2012). A review of ballast water management practices and challenges in harsh and arctic environments. *Environmental Reviews* 20 (2): 83–108.

100 Herwig, R.P., Cordell, J.R., Perrins, J.C. et al. (2006). Ozone treatment of ballast water on the oil tanker S/T Tonsina: chemistry, biology and toxicity. *Marine Ecology Progress Series* 324: 37–55.

101 Davidson, I.C., Minton, M.S., Carney, K.J. et al. (2017). Pioneering patterns of ballast treatment in the emerging era of marine vector management. *Marine Policy* 78 (January): 158–162.

102 Bai, M., Zheng, Q., Tian, Y. et al. (2016). Inactivation of invasive marine species in the process of conveying ballast water using OH based on a strong ionization discharge. *Water Research* 96: 217–224.

103 Balaji, R., Yaakob, O., and Koh, K.K. (2014). A review of developments in ballast water management. *Environmental Reviews* 22 (3): 298–310.

104 Lee, H., Reusser, D., Frazier, M.R., and Ruiz, G. (2010) Density matters: review of approaches to setting organism-based ballast water discharge standards. EPA/600/R-10/031 (NTIS PB2012-110754). Washington, DC: U.S. Environmental Protection Agency.

105 Stehouwer, P.P., Buma, A., and Peperzak, L. (2015). A comparison of six different ballast water treatment systems based on UV radiation, electrochlorination and chlorine dioxide. *Environmental Technology (United Kingdom)* 36 (16): 2094–2104.

7

Wastewater Treatment

7.1 Introduction

Water is crucial to life and is a resource that is widely used for a variety of human activities ranging from domestic use and agriculture to industrial consumption. The unique physical and chemical properties of water make it ideal as a universal solvent [1]. Water is an irreplaceable commodity used in agricultural, food, and chemical processing [2–4], and domestic purposes. More than 70% of the total consumption is taken up by the agricultural sector in developing countries [5–7]. Key societal requirements such as food security and human sustenance are served by the presence of water in our lives, and it is impossible to imagine daily life without the use of water. With increasing population and rapid industrialization, the availability of water per capita has reduced by 40% in less than a century (Figure 7.1). The need for water is expected to increase by 20–30% of current consumption by the year 2050 [8] and many regions in the world are expected to be water-stressed in the coming years [9]. The availability of freshwater is not geographically uniform, implying that some regions are more water-stressed. By 2050, more than 80 countries are expected to have less than 1700 m^3 per capita, which is 30% of the current available limit [10]. Due to impending factors such as increased withdrawal of freshwater and decreased per-capita availability, the treatment strategies for contaminated water become not just pertinent, but imperative for sustainability.

Wastewater streams could result from various natural or man-made activities and can be grouped broadly as agricultural, agro-industrial, domestic/sewage, and industrial processing. In general, these processes contaminate water with harmful pollutants and generate wastewater. Such wastewater needs to be treated before being discharged back to the environment, to avoid damage to aquatic ecosystems and threatening human livelihood. Ideally, wastewater needs to be treated, reused, or recycled wherever possible. Reuse/recycling of wastewater is quite important and relevant from the point of reducing the stress on freshwater resources. For example, the US Environmental Protection Agency reported a reuse of nearly 6.4 gigaliters per day, which is 15% of their daily requirement, and the Australian government reported the use of recycled water to irrigate vineyards, with significant economic benefits [11]. Recycled water mainly finds its use in irrigation for agricultural and non-potable purposes. There is still a significant resistance to recycle and reuse wastewater for direct

Hydrodynamic Cavitation: Devices, Design, and Applications, First Edition.
Vivek V. Ranade, Vinay M. Bhandari, Sanjay Nagarajan, Varaha P. Sarvothaman and Alister T. Simpson.

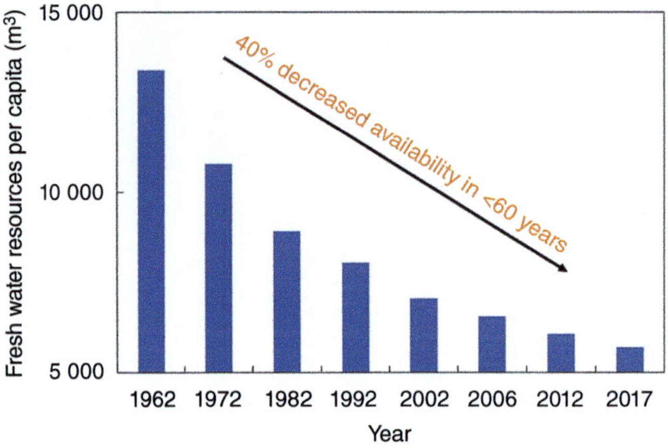

Figure 7.1 Availability of water per capita worldwide (Source: Data from [8]).

consumption [12–14]. A drastic change in usage philosophy and better treatment technologies are needed to overcome such resistance, in addition to public awareness and planning.

Several different physical, chemical, and biological treatment methods are being used for wastewater treatment. The book by Ranade and Bhandari [9] has presented key methods used for the treatment, recycle, and reuse of industrial wastewater. The book by Ranade and Bhandari also highlights the large spatial footprint of typical wastewater treatment processes and emphasizes the need for developing intensified wastewater treatment technologies. It also points out the fact that more often than not, the simultaneous use of multiple treatment technologies is needed for adequate treatment of wastewater.

The technologies available for treating wastewater are summarized in Figure 7.2. These are broadly classified into two groups: with and without reactions. The primary stage in water treatment focuses on removal of solids by settling and filtration. Flocculation and coagulation are often used as well. Secondary treatment typically involves more advanced processes, comprising physical, chemical, and physicochemical treatments, apart from possible biological treatments such as aerobic or anaerobic processes. Tertiary treatment, which aims to generate water fit for either recycle/reuse or discharge to environment, uses multiple specific techniques to achieve the desired specifications, generally referred to as a polishing step. These may again include advanced treatments such as oxidation, membrane filtration, and disinfection processes.

The conventional wastewater treatment techniques such as coagulation, precipitation, flotation, and biodegradation suffer from disadvantages such as the requirement of additional chemicals, secondary effluent formation (sludge formation), and large spatial footprint. These conventional technologies are also not very effective against specific pollutants. For example, biodegradation might not be suitable for decolorization and hence the technique might not be suitable for wastewater containing dyes and colorants. Incineration or wet-air oxidation mineralizes

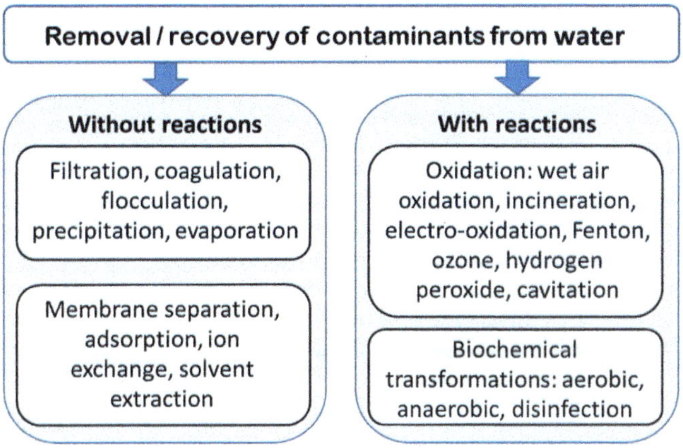

Figure 7.2 Treatment technologies for wastewater treatment.

pollutants without the formation of any secondary solid or liquid effluents. However, these are typically useful only for relatively high concentrations of organic pollutants (COD > 10^5 ppm). The other option for such high COD streams is valorization using anaerobic digestion, which is discussed in Chapter 8. Membrane or solvent extraction-based processes are generally expensive and also generate secondary effluents. The disinfection of water has already been discussed in Chapter 6. The focus of this chapter is on advanced oxidation processes targeted toward mineralization of organic pollutants with minimal or no formation of secondary effluents. Such advanced oxidation methods can be effectively tailored to target the removal of specific pollutants, as well as to intensify the overall treatment process.

The advanced oxidation processes (AOPs) use highly oxidizing and reactive hydroxyl and other radicals to treat organic pollutants in water. Hydroxyl $OH^{\cdot}$ radicals, which are the main oxidant species formed in AOPs, have very high reactivity toward organic contaminants in wastewater, with rate constants (second-order rate constant) greater than 10^6 L/mol.s [15]. By virtue of such highly oxidizing radicals, AOPs can potentially mineralize a wide range of contaminants—such as amino acids and antibiotics—and therefore are suitable for treatment of industrial wastewater such as phenolic wastewater, tannery wastewater, paper mill effluent, pesticide-contaminated water, and wastewater containing endocrine disruptors [16], specifically for the removal of pollutants which are otherwise difficult to remove from wastewater. Several different types of AOPs have been developed based on different methods used for generating oxidizing radicals [16, 17]. These are schematically summarized in Figure 7.3.

AOPs based on microwave, plasma, or ultrasound are effective in treating small quantities of effluents containing recalcitrant species. However, these methods are usually cost-intensive and are difficult to scale-up for treating large quantities of wastewater. Photocatalysis-based techniques are restricted to transparent effluents. The scale-up is difficult as well. Chemical-based AOPs have been scaled up and used in practice. The required chemical agents like ozone or hydrogen peroxide

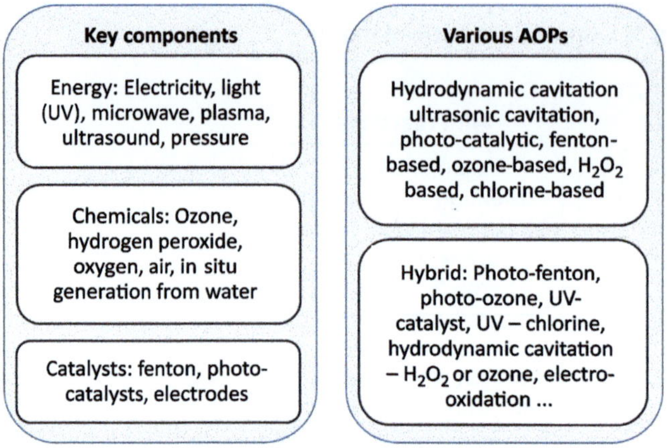

Figure 7.3 Advanced oxidation processes (AOPs).

are directly proportional to the COD of effluents. Therefore, chemical-based AOPs are typically used when the COD is less than 10^4 ppm, to contain costs associated with the chemicals [15]. The application of chemicals and catalyst-based AOPs to wastewater treatment has many undesirable features [18]. For example, the use of hazardous chemicals like chlorine, mineral acid, and alkali or sludge formation (Fenton-based processes). Ozone is known to cause respiratory problems in humans, and thus needs a lot of precaution in handling. UV-based processes often require a pre-treatment step. Photocatalysts have issues of agglomeration, mass transfer limitations and difficulties of separation post treatment. Unlike these methods, hydrodynamic cavitation offers an attractive platform for implementing AOP since it generates hydroxyl radicals in situ without the requirement of any externally added chemical. It is also intrinsically amenable for scale-up. Hydrodynamic cavitation-based AOP, therefore, is a "green" process (which does not generate harmful by-products, does not require additional chemicals, etc.). It is easy to retrofit in the existing treatment plants, relatively inexpensive, and can be used in combination with other AOPs and water-treatment technologies. This chapter is focused on the application of hydrodynamic cavitation in wastewater treatment. Some discussion on how hydrodynamic cavitation may be combined with other techniques such as Fenton, ozonation, hydrogen peroxide, and photo catalysis is also included.

7.2 Hydrodynamic Cavitation for Wastewater Treatment

Hydrodynamic cavitation has been applied for wastewater treatment for more than two decades [18, 19]. Studies have indicated its successful application in the destruction of various types of organic compounds encountered in industrial wastewater [20–23] and industrial effluents [24, 25]. These processes rely on the very high pressure and temperatures generated during the collapse of cavities. Various physicochemical transformation processes occurring in and around collapsing cavities have been discussed in Chapter 5 and shown schematically in Figure 5.3. These intense physicochemical processes during cavity collapse are instrumental in treating pollutants in water. This primarily happens in two modes: (i) via in situ generation of hydroxyl radicals and their subsequent reactions with pollutants in water, and (ii) via pyrolysis of pollutants (see discussion on cavity dynamics in Chapter 5). The extreme temperature and pressure generated during cavity collapse leads to various reactions in the vaporous cavity, leading to the formation of various oxidizing radicals. Some of the key reactions are listed as (R7.1)–(R7.8) [9]:

$$H_2O \rightarrow H^{\cdot} + OH^{\cdot} \tag{R7.1}$$

$$O_2 \rightarrow 2\,O^{\cdot} \tag{R7.2}$$

$$O^{\cdot} + H_2O \rightarrow 2\,OH^{\cdot} \tag{R7.3}$$

$$HO^{\cdot} + H^{\cdot} \rightarrow H_2O \tag{R7.4}$$

$$2\,HO^{\cdot} \rightarrow O^{\cdot} + H_2O \tag{R7.5}$$

$$H^{\cdot} + O_2 \rightarrow HOO^{\cdot} \tag{R7.6}$$

$$2\,HO^{\cdot} \rightarrow H_2O_2 \tag{R7.7}$$

$$2\,HOO^{\cdot} \rightarrow H_2O_2 + O_2 \tag{R7.8}$$

Such in situ generation of strongly oxidizing species without the addition of external chemicals makes hydrodynamic cavitation attractive. The generated radicals react with other gaseous species within the cavity (see for example, Pichat et al. [26]) or diffuse into the surrounding liquid and react either in the interfacial zone or in the bulk zone (for example, Flint and Suslick [27]; Ashokkumar et al. [28]). If the liquid contains volatile species, these volatile species will be present in the gaseous cavity. In such a case, either pyrolysis or thermal cracking may occur within the cavity (for example, Ashokkumar et al. [28]). It can thus be seen that various chemical transformations in and around the cavity will be influenced by the characteristics of the chemical species in the considered system (volatility, hydrophobicity, molecular weight, reactivity towards radicals, presence of radical scavengers, propensity to cracking, and so on), as well as a variety of other macroscopic factors such as design of cavitation device and operating conditions (pH, temperature, pressure drop, and so on).

The implementation of hydrodynamic cavitation for water treatment is quite straightforward, requiring essentially a pump, a cavitation device, and at the most, a holding tank. Various cavitation devices, the typical experimental setup used for evaluating efficacy of hydrodynamic cavitation, and models for design and scale-up are discussed in Chapters 2–5. Considering simplicity of implementation and amenability to scale-up, hydrodynamic cavitation has been tested and used for a

Table 7.1 Some examples of application of hydrodynamic cavitation for treating real effluents.

Effluent type	Device type	Working volume (L)	COD of influent (ppm)	Pressure drop (kPa)	Number of passes (–)	Cavitational yield (mg COD/J)	Percentage reduction in COD (%)	Reference
Effluent containing ammoniacal nitrogen/fertilizer Industry	Vortex diode	15	125 000	50	100	NA	85	[24]
Industry not specified	Orifice	70	100 000	600	75	NA	3	[29]
Pharmaceutical industry	Venturi	34	69 960 (TOC)	800	7	0.065	21	[30]
Varnish type, wood industry	Rotor-stator	4	38 000	NA	195	0.0025	42	[31]
Distillery	Venturi	15	35 000	500	85	NA	31	[32]
Vegetable oil refinery wastewater	Orifice	15	22 300	600	NA	NA	31	[33]
Organics, industry not specified	Venturi	4	20 800	400	205	0.0724	20	[34]
Pesticide industry	Venturi	6	17 000–18 000	600	100	NA	5	[35]
Phenolic derivatives, resins and sodium acetate	Venturi	4	14 400	400	NA	0.023	8[a)]	[36]
Aromatics, industry not specified	Liquid whistle reactor	5	4 200	10 340	NA	NA	30[a)]	[37]
Textile industry	Venturi	6	4 000	500	200	NA	12	[38]
Aromatics, industry not specified	Orifice	8	1 680	10 340	NA	NA	60	[39]
Laundry industry	Venturi	4	680	500	170	0.0064	25	[40]
Textile industry	Venturi	40	54–68	40	NA	0.0002	45	[41]

a) For diluted effluent.

wide range of effluents. Some examples of application of hydrodynamic cavitation to real effluents are listed in Table 7.1. It can be seen that the efficacy of hydrodynamic cavitation in degrading different pollutants varies over a wide range. In this section, overall implementation of hydrodynamic cavitation for wastewater treatment is briefly discussed. The influence of various design and operating parameters on performance or effectiveness of treatment is discussed in a subsequent section. The potential for augmenting hydrodynamic cavitation by combining it with other complementary techniques is discussed at the end. Collectively, this chapter provides useful information for designing and implementing hydrodynamic cavitation for wastewater treatment.

The typical arrangement while using hydrodynamic cavitation for wastewater treatment is shown schematically in Figure 7.4. It comprises a holding tank to which the hydrodynamic cavitation device is connected. Depending on the effective number of passes needed through the cavitation device, the flow rate of effluent through the cavitation device (Q) and the net flow rate (q) can be established using the models discussed in Chapter 5.

The overall behavior of such a water treatment system may be represented as (see Chapter 5 for more details):

$$V\frac{dC}{dt} = q(C_{in} - C) - Q\emptyset C \tag{7.1}$$

Here, C is the concentration of mixed pollutants (or COD), V is the holding tank volume, Q is the flow rate through the cavitation device, and $\emptyset$ is the per-pass degradation factor. The per-pass degradation factor $\emptyset$ can be obtained by carrying out batch experiments (without net flow or $q = 0$). For such batch experiments, the measured concentration of pollutant as a function of time and the following Eq. (7.2) may be used to calculate $\emptyset$:

$$C = C_{in}e^{-\beta\emptyset t} \tag{7.2}$$

where β is a ratio of flow rate through the cavitation device and the holding tank volume (Q/V, s^{-1}). The product of β and time indicates the number of passes, n, through the cavitation device for a batch system:

$$n = \beta t \tag{7.3}$$

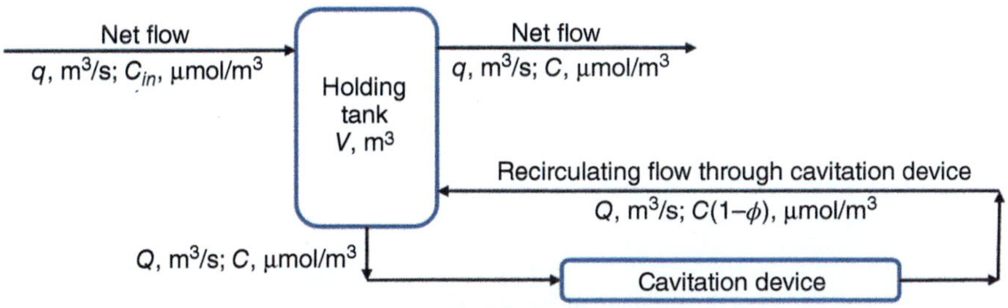

Figure 7.4 Schematic of wastewater treatment for application of hydrodynamic cavitation (Source: From [42]/with permission of American Chemical Society).

For a continuous mode of operation, at a steady state, the outlet concentration of pollutant may be related to the inlet concentration, β and $\emptyset$ as:

$$C = \frac{q}{q + Q\emptyset} C_{in} = \frac{C_{in}}{1 + \beta\tau\emptyset} \tag{7.4}$$

where τ is the residence time in holding tank based on net flow (V/q). Flow rate through the cavitation device can be estimated from the desired degradation and the system-specific value of per-pass degradation (for the given device and operating conditions) as:

$$\beta\tau = \frac{Q}{q} = \frac{1}{\emptyset}\left(\frac{C_{in}}{C} - 1\right) \tag{7.5}$$

With Eq. (7.5), knowing the range of desired concentration reduction and the per-pass degradation coefficient, it will be possible to determine the value of $\beta\tau$—the ratio between flow rate through the cavitation device (Q) and net flow rate (q). Equation (7.5) may also be used to generate a chart such as that in Figure 7.5, which may be used for design purposes.

For a specified inlet concentration of pollutant (C_{in}), desired outlet concentration (C), and net flow rate (q), Figure 7.5 may be used to estimate flow rate though the cavitation device (Q), provided the per-pass degradation factor for the considered device is known. Once this flow rate is known, power (or cost) required for the desired reduction in pollutant concentration can be calculated as:

$$\text{Power requirement, } W = \frac{\Delta Pq}{\eta}\left(\frac{Q}{q}\right) \tag{7.6}$$

where η is pump efficiency and ΔP is pressure drop across the cavitation device (Pa) at a flow rate of Q (m^3/s). With the assumption of efficiency of the multistage pump (η) as 0.66 and the pressure drop across the device as 2.4 bar, the energy requirement for one pass through the cavitation device per m^3 of water is 0.1 kWh. The net cost of treatment is dependent on the required ratio $\left(\frac{Q}{q}\right)$.

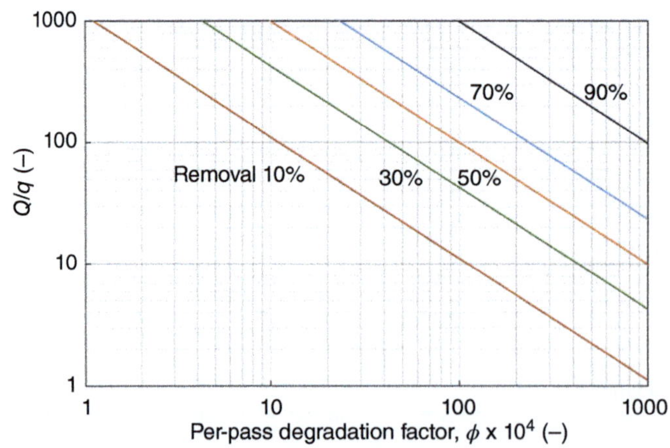

Figure 7.5 Cavitation device sizing for a given pollutant mixture with known degradation characteristics.

It is obvious from Figure 7.5 that the value of required $\left(\frac{Q}{q}\right)$ depends on the desired reduction and per-pass degradation factor. Since the per-pass degradation factor is key in the design of treatment capacities, it is important to understand the influence of device (type, design, and scale) and operating parameters on the per-pass degradation factor, and the optimization of these parameters. These aspects are discussed in Section 7.3.

7.3 Performance of Hydrodynamic Cavitation-based Wastewater Treatment

The overall performance of hydrodynamic cavitation-based water treatment may be quantified by using the per-pass degradation factor. Many of the earlier studies interpreted batch experiments of water treatment based on hydrodynamic cavitation using an apparent first-order degradation rate constant [38, 43, 44]. However, as discussed in Chapter 5, the per-pass degradation factor is more appropriate for systematic comparison of performance of different devices and operating conditions. This is particularly true for the continuous operation of hydrodynamic cavitation-based treatment.

Hydrodynamic cavitation processes are typically operated at ambient temperature and pressure conditions. As discussed earlier, the collapse of cavities generates localized very high pressure and temperature regions. These hot spots contribute to pollutant degradation either via generating oxidizing radicals or via high-temperature cracking and pyrolysis. The temperature around the collapsed cavity is shown schematically in Figure 7.6, based on the conceptual model proposed by Suslick [45].

Successive studies employing cavitation-based degradation have qualitatively explained the observed results based on the three-zone theory depicted in Figure 7.6. For example, Sivasankar and Moholkar [46] have analyzed degradation of five aromatic compounds and have proposed the mechanism and the location of degradation. Their results are summarized in Table 7.2. Pollutants such as phenol or nitrobenzene with relatively high hydrophilic character do not undergo pyrolysis-based attack. While the compounds such as chlorobenzene (or toluene, although not part of the study by Sivasankar and Moholkar [46]) might undergo pyrolytic decomposition. The per-pass degradation approach proposed by Ranade et al. [43] is useful to quantify observed degradation performance. However, it should be noted that this approach does not distinguish between two modes of degradation (oxidation via radicals or pyrolysis) and lumps overall degradation in one parameter.

It may also be noted that when hydrodynamic cavitation is used for degrading complex organic molecules, the process of mineralization proceeds through several intermediate steps and products (see for example, Sarvothaman [47] and references cited therein; Carpenter et al. [48] and references cited therein). These intermediates may influence the per-pass degradation factor, leading to a variable per-pass degradation factor instead of a constant one. Recently, Ranade et al. [43] have included

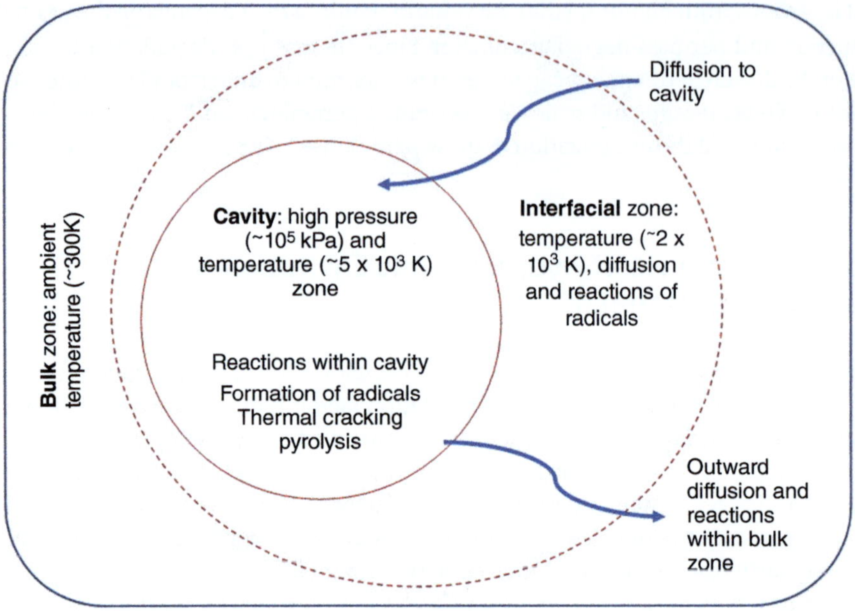

Figure 7.6 Schematic of hot-spot theory of cavitation (Source: Adapted from [45]/with permission of Elsevier).

Table 7.2 Predominant mechanism and location of degradation for organic pollutants.

Pollutant	Mechanism of degradation	Location of degradation
Phenol	Hydroxylation	Bulk medium
Chlorobenzene	Pyrolytic decomposition	Inside the cavitation bubble
Nitrobenzene	Hydroxylation	Bubble–bulk interfacial region
p-Nitrophenol	Thermal decomposition followed by hydroxylation	Bubble–bulk interfacial region (thermal decomposition) and Bulk medium (hydroxylation)
2,4-Dichlorophenol	Hydroxylation	Bubble–bulk interfacial region

Source: From [46]/with permission of Elsevier.

a varying per-pass degradation factor for interpreting the performance of hydrodynamic cavitation-based water treatment. Though some attempts have been made to develop computational models for a priori prediction of the per-pass degradation factor, the commonly used methods for quantifying the performance of hydrodynamic cavitation-based treatment processes rely on experimental measurements. The overall performance depends on the design of cavitation devices used in the process as well as on a variety of operating process parameters. The influence of these factors on overall performance is discussed in the following section.

7.3.1 Influence of Device Design

Several different types of hydrodynamic cavitation devices are available (see recent review by Wang et al. [18] and Chapters 2–3). The per-pass degradation factor realized by these different devices is a complex function of device design, operating conditions, and characteristics of the wastewater under consideration. A systematic comparison of different devices, however, is not frequently reported in the open literature. Specific pollutant degradation per unit power consumption may be used to evaluate different devices and processes. Using Eqs. (7.4)–(7.6), specific cavitation yield for continuous processes may be calculated as (concentration of pollutant is in kg/m³):

$$\text{Cavitation yield} = \frac{C_{in}\, q\, \emptyset\, \eta}{\Delta P\, Q\left(\emptyset + \frac{q}{Q}\right)}\ \frac{\text{kg}}{\text{J}} \tag{7.7}$$

For batch systems, cavitation yield may be calculated as:

$$\text{Cavitation yield} = \frac{C_{in}\,(1 - e^{-\emptyset n})\,\eta}{\Delta P\, n}\ \frac{\text{kg}}{\text{J}} \tag{7.8}$$

It can be seen that cavitation yield is directly proportional to inlet pollutant concentration and pump efficiency. It is inversely proportional to operating pressure drop across the cavitation device. For small values (up to 0.1) of $\emptyset n$ for batch systems or $\emptyset\left(\frac{Q}{q}\right)$ for continuous systems, the cavitation yield is directly proportional to the per-pass degradation factor and is independent of the number of passes $\left(n \text{ or } \left(\frac{Q}{q}\right)\right)$. For higher values of $\emptyset n$ or $\emptyset\left(\frac{Q}{q}\right)$, cavitation yield is inversely proportional to the number of passes $\left(n \text{ or } \left(\frac{Q}{q}\right)\right)$ and independent of the per-pass degradation factor.

Sivakumar and Pandit [49] compared the degradation of rhodamine B dye for six different orifice plates (with a varying number of holes—thereby varying the flow area and perimeter) at a fixed pressure drop of nearly 200 kPa. This was one of the first studies to compare the performance of different devices for wastewater treatment. Their study showed that the orifice with 33 holes had a cavitational yield of 3×10^{-8} g of rhodamine B/J, which was nearly two times that of the next best orifice plate and the ultrasound bath used for comparison. In another study using Rhodamine B dye as a pollutant, Mishra and Gogate [50] compared the performance of an orifice and a venturi. The venturi device showed a marginal increase in degradation from 20 to 25% when the pressure drop was increased from 290 to 580 kPa, whereas the orifice device used showed a negligible influence of pressure drop (~22% for all cases in the considered range). For both the devices considered in this work, the flow rate at a pressure drop of 484 kPa was more or less the same (400 LPH for venturi and 380 LPH for orifice), indicating nearly the same power consumption. Thus, the cavitation yield (mg of pollutant degraded/W) was found to be nearly the same for both the devices.

In another study of dye degradation, Orange G dye was degraded using two venturi devices and an orifice device [51]. The venturis used were of the slit and the circular type. The slit venturi was operated between 200 and 400 kPa pressure drop.

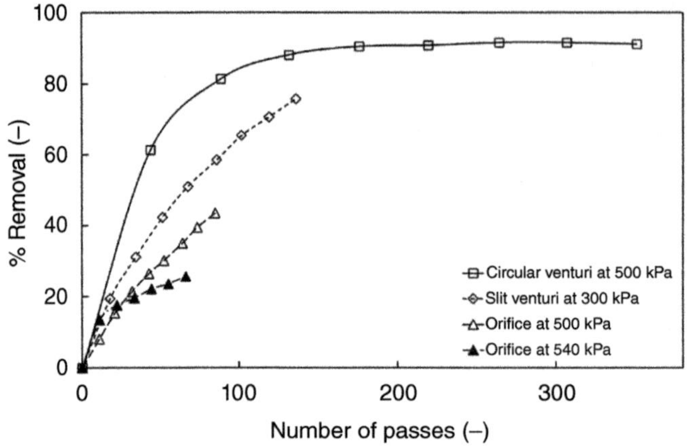

Figure 7.7 Influence of cavitation device on degradation performance for Orange G dye (Source: Data for devices: Circular venturi (500 kPa), slit venturi (300 kPa), and orifice (500 kPa), from [51] and Orifice at 540 kPa from [52]).

Other devices in the study were operated between 300 and 700 kPa pressure drop. The optimal pressure drop for dye degradation was found to be 300 and 500 kPa, for the slit venturi and the circular venturi, respectively. The decolorization profiles from this study are presented in Figure 7.7. The performance of orifice was not found to be dependent on pressure drop within the considered range. The slit venturi at 300 kPa was found to be 1.65 times superior to that of the circular venturi at 500 kPa and 10 times superior to the orifice device. In a more recent study, Doltade et al. [53] used two venturis and an orifice device for the degradation of a petroleum industry effluent. The orifice device used in their study was useful in removing nearly 40% of COD at pressure drops of 500 and 700 kPa across the orifice. No significant removal was observed at lower pressure drop (300 kPa) for 50 passes through the device. The performance observed in these different studies could not be directly compared because of lack of systematic basis (different pollutants, initial concentrations, pressure drops, and device designs).

Sarvothaman et al. [54] presented the influence of different device types—namely the orifice, orifice with swirl, venturi, venturi with swirl, and a vortex-diode on degradation of a model pollutant (dichloroaniline). These devices were designed (by Simpson and Ranade [55]) to ensure that all of them exhibit similar pressure drop versus flow characteristics. This ensures that when operated at the same pressure drop, the energy consumption of all of these devices is the same, lending a uniform basis for comparison. Sarvothaman et al. [54] evaluated the degradation of dichloroaniline for these five devices at two (100 and 200 kPa) pressure-drop conditions, and their results showed a negligible influence of device type on degradation performance (see Figure 7.8).

Though there are numerous published studies on the applications of hydrodynamic cavitation, information on scale-up of cavitation devices is conspicuous by its absence. There are almost no reports presenting data and performance of geometrically similar hydrodynamic cavitation devices on different scales

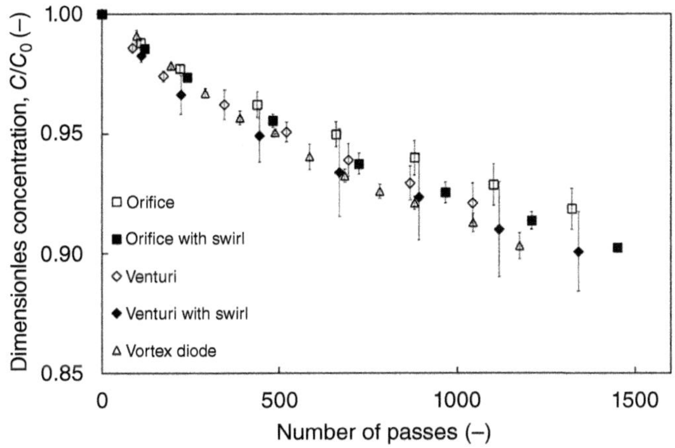

Figure 7.8 Influence of device type on degradation of 2,4-dichloroaniline at 200 kPa, $C_0 = 35$ ppm, pH = 7, $T = 18\,°C$ (Orifice and Orifice with swirl: $d_t = 5$ mm; Venturi and Venturi with swirl of $d_t = 4$ mm; Vortex diode: $d_t = 12$ mm; Flow rate of devices = 1200 LPH at 250 kPa (Source: Data from [54]).

in open literature. Recently, Ranade et al. [43] reported experimental data on performance of geometrically similar vortex-based cavitation devices for degradation of dichloroaniline in water (up to 200 times scale-up in capacity). Their results (shown in Figure 7.9) showed strong influence of device scale on degradation performance. They were able to correlate the per-pass degradation factor as a function of device scale as:

$$\emptyset = \emptyset_{\infty} \exp\left(\frac{\gamma}{d_t}\right) \tag{7.9}$$

where $\emptyset_{\infty}$ is the value of the per-pass degradation factor at infinite size, d_t is the characteristic length scale of the cavitation device (throat diameter in case

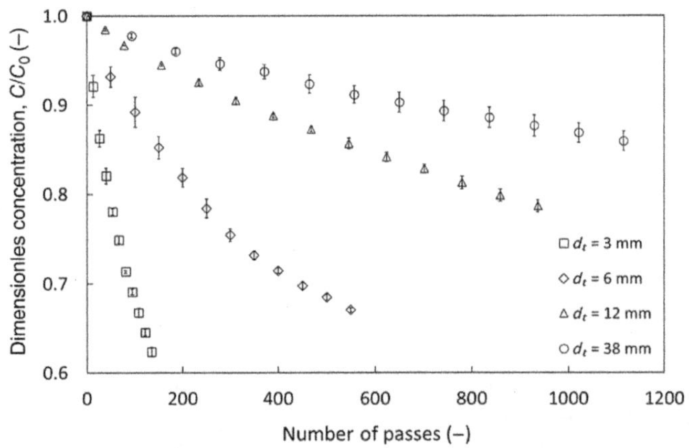

Figure 7.9 Influence of device scale on degradation of 2,4-dichloroaniline for vortex diode type cavitation devices at 280 kPa, $C_0 = 35$ ppm, pH 7, $T = 18\,°C$ (Source: From [43]/with permission of Elsevier).

of vortex-based devices considered by Ranade et al. [43]), and γ is an empirical parameter. Further research is needed on quantitatively estimating the influence of scale-up. The models discussed in Chapter 5 may be used as a starting point for such work. It is also very important to devise new strategies and internals for mitigating adverse influence of device scale on degradation performance. The other alternative is to use the numbering-up approach instead of scaling up. A judicious combination of scale-up (with appropriate internals) and number-up will allow overall optimization and scale-up needed for large-scale industrial applications of hydrodynamic cavitation-based wastewater treatment processes.

7.3.2 Influence of Operating Parameters

Several operating process parameters influence the overall performance of hydrodynamic cavitation-based water treatment. These may range from inlet concentration of pollutant, operating pressures, temperature, pH, and so on. The key aspects and some guidelines on selecting the appropriate process parameters are briefly discussed in the following section.

7.3.2.1 *Inlet Concentration of Pollutant*

When pollutant degradation occurs primarily through oxidizing reactions with hydroxyl radicals, the per-pass degradation factor depends on the rate of generation of hydroxyl radicals, concentration of pollutants, concentration of other radical scavengers, and relative reactivity of pollutants and scavengers with hydroxyl radicals. Sarvothaman et al. [56] have formulated the expression for per-pass degradation factor as:

$$\emptyset = \frac{\left(\frac{G}{Q\,C_S}\right)}{\left(\frac{k_S}{k_2} + \frac{C}{C_S}\right)} \tag{7.10}$$

For relatively low concentrations of pollutant ($C \ll C_S$), the value of the per-pass degradation factor will be independent of pollutant concentration. It will only depend on the effective generation rate of hydroxyl radicals, G, and the ratio of k_S to k_2 (assuming high and constant concentration of scavengers, C_S, at a specified flow rate, Q). At the other extreme, when $C \gg C_S$, the per-pass degradation coefficient will be inversely proportional to C. In both these cases, the dimensionless profiles of C/C_0 versus number of passes is not expected to vary with inlet/initial concentration of pollutant, C_0.

Some studies investigating the influence of initial concentration studied by hydrodynamic cavitation are listed in Table 7.3. All of these studies are batch experiments carried out in laboratory-scale setups. The data from some of the studies included in Table 7.3 are shown in Figure 7.10. It can be seen from Figure 7.10 that the experimental data on degradation of dicofol reported by Panda and Manickam [58] exhibit significant influence of initial concentration. When the initial concentration was increased from 20 to 100 ppm, the proportion of removal decreased from 97 to 67%. Similar behavior was also reported by Patil et al. [63] for ammoniacal

Table 7.3 Studies which report influence of pollutant concentration using hydrodynamic cavitation.

Pollutant	Range of concentration (ppm)	Cavitation device type	Reference
Acid Red 18 dye	15–30	Orifice	[33]
p-Aminophenol	100–500	Vortex diode	[57]
Dicofol	20–100	Orifice and blade assembly	[58]
Acetone	100 and 500	Vortex diode	[59]
Toluene	100 and 500	Vortex diode	[59]
2,4-Dichloroaniline	17.5–70	Vortex diode	[43]
Imidacloprid	20–60	Venturi	[60]
Methyl parathion	20–50	Orifice	[61]
Rhodamine B	5–40	Swirling jet cavitation	[62]
Industrial effluent	840–4200	Orifice and blade assembly	[37]
Industrial effluent	1440–14 400	Venturi	[34]
Textile industry effluent	690–4000	Venturi	[38]

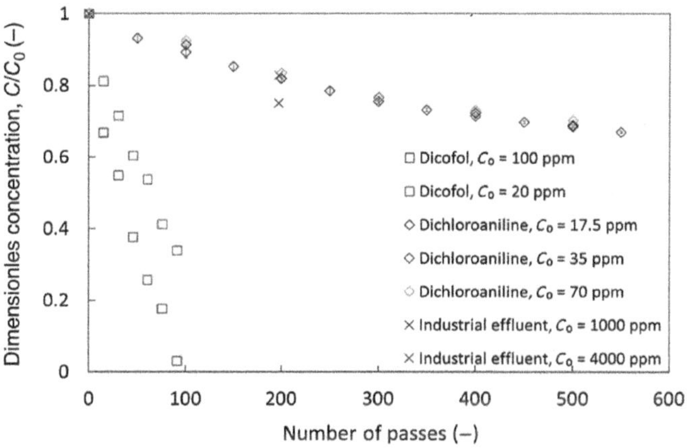

Figure 7.10 Influence of initial concentration on degradation (Source: Data for Dicofol: [58]; Dichloroaniline: [43]; Industrial effluent: [38]).

nitrogen. Unlike these, the data of degradation of an industrial effluent, reported by Rajoriya et al. [38], and of the degradation of dichloroaniline, reported by Ranade et al. [43], exhibit negligible influence of the initial concentration of pollutant (profiles of dimensionless pollutant concentration with respect to number of passes are independent of initial concentration). Further work is needed on developing better understanding of the influence of initial pollutant concentration on treatment performance. In general, overall cavitation yield increases with initial concentration (see Eqs. 7.7 and 7.8).

7.3.2.2 *Pressure Drop Across Cavitation Device*

For a specified pollutant concentration and a given cavitation device, the flow rate through the device, or in other words, the pressure drop across the device is a primary operating parameter. It is important to operate the process at a flow rate/pressure drop value beyond the cavitation inception point. The identification of inception was discussed in Chapter 4. Typically, the overall per-pass degradation factor shows a maxima with respect to the operating pressure drop across the cavitation device. See for example, the degradation of four organic pollutants from published studies, shown in Figure 7.11.

At the inception condition, the cavities start to emerge and turn the otherwise single phase into a multiphase flow. The overall performance or the per-pass degradation factor depends on the number of collapsing cavities (proportional to the generated vapor volume fraction) and intensity of collapse (generated hydroxyl radicals or extent of pyrolysis occurring in cavities). Beyond the inception condition, as the flow rate (pressure drop) increases and the extent of cavitation increases, indicated by increase in vapor volume. However, as vapor volume fraction increases, effective compressibility of the gas–liquid medium increases and therefore the intensity of cavity collapse (and the corresponding degradation per collapsing cavity) decreases. Because of these two competing effects, the overall performance exhibits a maxima with respect to flow rate/pressure drop as shown in Figure 7.11. It should also be noted that the pressure drop condition at which maximum degradation (or per-pass degradation factor) is observed may not be an optimum condition. For example, consider a case discussed by Sarvothaman et al. [42] where performance of two cavitation devices was evaluated over a range of operating pressure drops. A part of their data is shown in Figure 7.12. It can be seen from this figure that the highest degradation was observed at pressure drop of 250 kPa. However, the overall cavitation yield or cost is proportional to the ratio

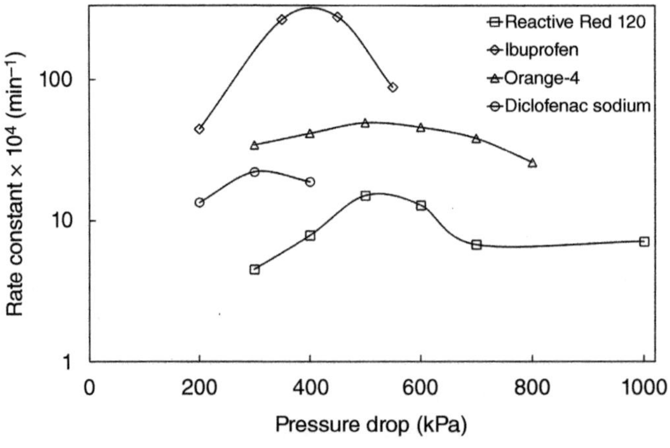

Figure 7.11 Influence of operating pressure drop on rate constant for different pollutants in hydrodynamic cavitation (Source: Data for Reactive-Red 120– [64]; for Ibuprofen– [23]; Orange–4: [65]; Diclofenac sodium: [66]).

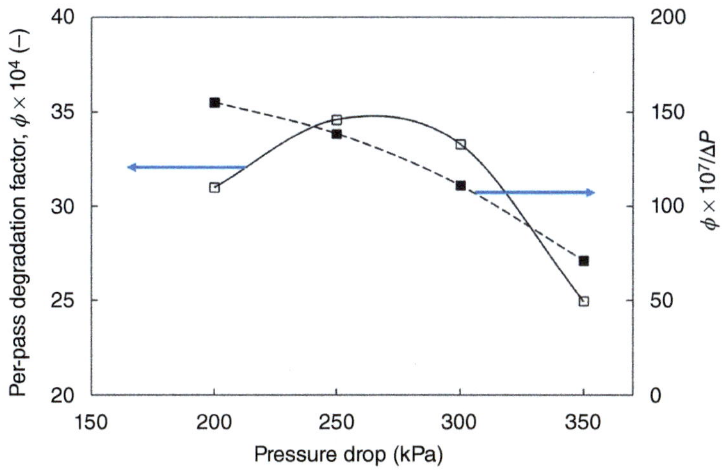

Figure 7.12 Existence of maxima and optima in degradation performance across the cavitation device (Source: Data from [42]).

of pressure drop and per-pass degradation factor. The inverse of this ratio, that is, the ratio of per-pass degradation factor and pressure drop as a function of operating pressure drop is also shown in Figure 7.12.

It can be seen from this figure that the operation at pressure drop of 200 kPa will be more favorable in terms of cavitational yield. Appropriate consideration of capital costs and electricity costs is required while selecting optimal pressure drop for a given effluent and hydrodynamic cavitation device.

7.3.2.3 *Downstream Pressure*

It is possible to improve the degradation/per-pass degradation factor by elevating the downstream pressure. When the cavities generated in the device are exposed to an elevated downstream pressure, they are expected to experience more intense collapse in a shorter timescale and thereby improve the per-pass degradation factor. Sarvothaman et al. [56] used a vortex-based device and showed that the per-pass degradation factor can be enhanced by 20% by increasing downstream pressure by 100 kPa (gauge) as against the no-downstream-pressure condition. Angaji and Ghiaee [67] have also shown the beneficial influence of downstream pressure (see Figure 7.13). Their results show an existence of maxima with respect to enhanced downstream pressure, observed at 300 kPa downstream pressure. In an application unrelated to water treatment, Soyama and Hoshino [68] had presented their results of increased cavitational intensity for elevated downstream pressure at a given inlet pressure for a venturi-based cavitation device. The concept of using elevated downstream pressure has not been adequately explored, evaluated and used in the studies of hydrodynamic cavitation-based wastewater treatment. However, the limited results discussed here indicate that the strategy looks quite promising and deserves further investigations and applications in practice.

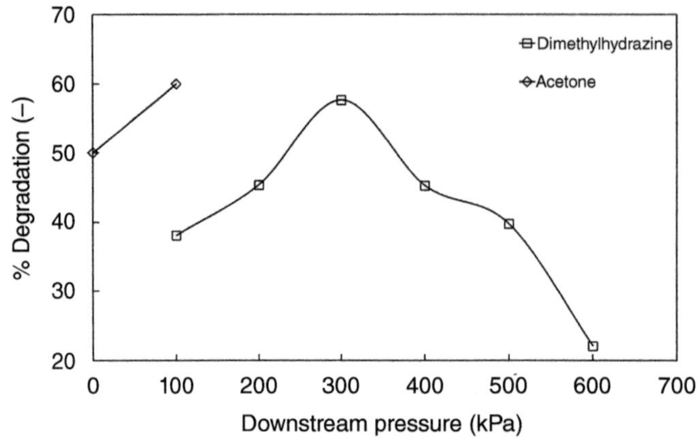

Figure 7.13 Influence of downstream pressure on degradation performance—inlet pressure for (i) dimethylhydrazine: 600, 600, 650, 700, 740, and 780 kPa, and (ii) acetone: 300 kPa (Source: Data—Dimethylhydrazine from [67], and Acetone from [42]).

7.3.2.4 *Operating pH*

The operating pH of cavitation influences the efficacy of contact between the pollutant molecules and the generated oxidizing radicals, or the extent of pyrolysis, since it influences transport of pollutant molecules near and inside the cavities (see Figure 7.6). The prevailing pH influences whether pollutant molecules preferentially reside in the hydrophobic interior of the cavity, the interface region, or the bulk region. The specific nature of the pollutant molecules and the prevailing pH control this. There has been considerable interest in investigating the influence of pH on sonochemical and hydrodynamic cavitation-based degradation [48, 69–71]. These studies indicate that generally, the operating pH should be less than the pKa of the pollutant (the pKa of the compound is the pH at which the pollutant will exist in 50% dissociated state) for better degradation of pollutants. For example, Capocelli et al. [72] observed that the degradation of *p*-nitrophenol dramatically increased by decreasing pH from 8 to 5.5 (see Figure 7.14).

When pH < pKa, the pollutant species are present predominantly in molecular form, whereas when pH > pKa, they are predominantly in ionic form. The ionic form of the compound will tend to be present in bulk-phase, rather limiting its exposure to the radicals being formed from cavitation. When pollutant species exist predominantly in molecular forms, there will be enhanced contact with oxidizing radicals and enhanced degradation. Studies which report the variation of solution pH with hydrodynamic cavitation are listed in Table 7.4. From the information available on pKa and the optimal pH observed for degradation, it can be said that acidic pH is favorable for the degradation of organic pollutants.

For the degradation of methomyl [76], the pKa of the compound is ~13, while the investigated pH conditions were less than or equal to 4. The degradation continually increased on decreasing the pH from 4 to 1.5. This decrease in pH, increased the per-pass constant by nearly 35 times, suggesting that pH played a significant

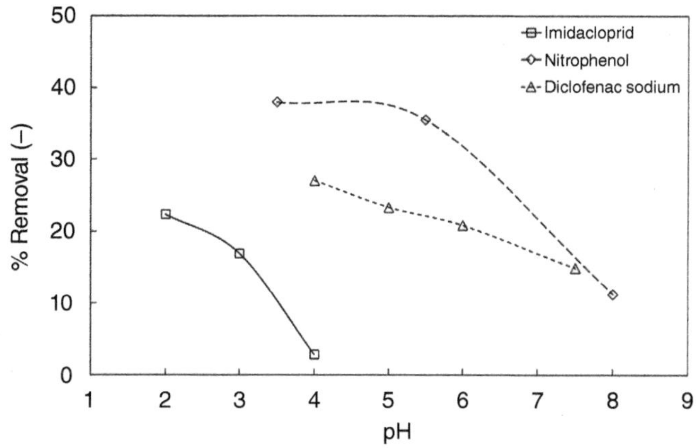

Figure 7.14 Influence of pH on degradation (Source: Data – imidacloprid [60], *p*-nitrophenol [72], and reactive orange-4 dye [65]).

Table 7.4 Studies which report influence of pH using hydrodynamic cavitation.

Pollutant	pH range	pKa of pollutant	Optimal pH for degradation	Reference
p-Nitrophenol	3.5–8	7.15	3.5	[72]
Rhodamine B	2–10	4.2	3	[73]
Dichlorvos	3 and 5.7	Not reported	3	[74]
Methyl parathion	2.2–8.2	Not reported	3	[61]
Acid Red 88 Dye	2–10	10.7	2	[75]
Imidacloprid	2–7.5	1.6	2	[35]
Ibuprofen	2–9	4.9	–	[23]
Methomyl	2–6	13.3	2.5	[76]
Orange-4	2–10	2	2	[65]
Diclofenac	4–7.5	4.2	4	[66]
Dicofol	3–7	10.7	3	[58]
Rhodamine B	2.5–11	4.2	2.5	[50]

role in improving the degradation performance for this compound. Whereas in the case of ibuprofen, the degradation was not observed to be a function of pH [23]. Apparently, the presence of multiple functional groups may complicate the general guideline of having pH < pKa (for example, in the case of ibuprofen degradation). For example, consider the degradation of the pharmaceutical pollutant, tetracycline. The pollutant has three contributing functional groups to its pKa (Figure 7.15a). The degradation obtained at the basic pH 10—72%—was superior to that of acidic pH (2.1)—46%—for a treatment time of 20 minutes (Figure 7.15b). The three groups contributing to the pKa of the compound make it difficult to determine an optimal

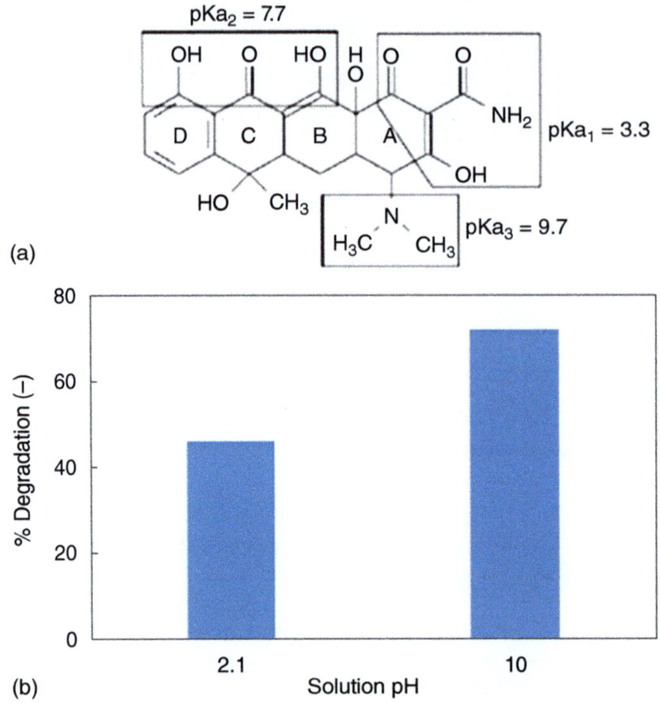

Figure 7.15 Structure and influence of pH on degradation of tetracycline using hydrodynamic cavitation. (a) Chemical structure of tetracycline (Source: From [77]/with permission of Elsevier). (b). Influence of pH on degradation of tetracycline (Source: Data from [77]).

pH. In such cases, the determination of the dissociated states alone cannot help in predicting the degradation performance.

Lowering the pH below 2 has been reported to favor the recombination of OH radicals to form H_2O_2 and alter the effectiveness of cavitation [9]. While these studies are of individual pollutants and these compounds are favored at acidic conditions, for the treatment of a mixed stream of effluents, it is best to evaluate degradation in acidic (3), neutral (7), and basic (10) conditions for determining the optimal treatment pH. Gągol et al. [78] investigated the degradation of commonly encountered organic pollutants in industrial effluents in the presence of three mineral acids to manipulate the pH of the solution (Figure 7.16).

It was noted that the degradation was not consistent at similar pH conditions on using different mineral acids, as Cl^-, SO_4^{2-}, and NO_3^-. From Figure 7.16, it can be seen that the degradation is consistently near 80% for H_2SO_4. Whereas in the case of hydrochloric acid, the degradation was not consistent and only 2–3 out of 15 compounds exhibited nearly 80% degradation. It was hypothesized that in the case of sulfuric acid, the sulfate radicals have a longer lifetime than OH radicals and also a remarkable oxidation potential. In the case of hydrochloric acid, the chloride radicals would scavenge the formed OH radicals. Based on this study, the

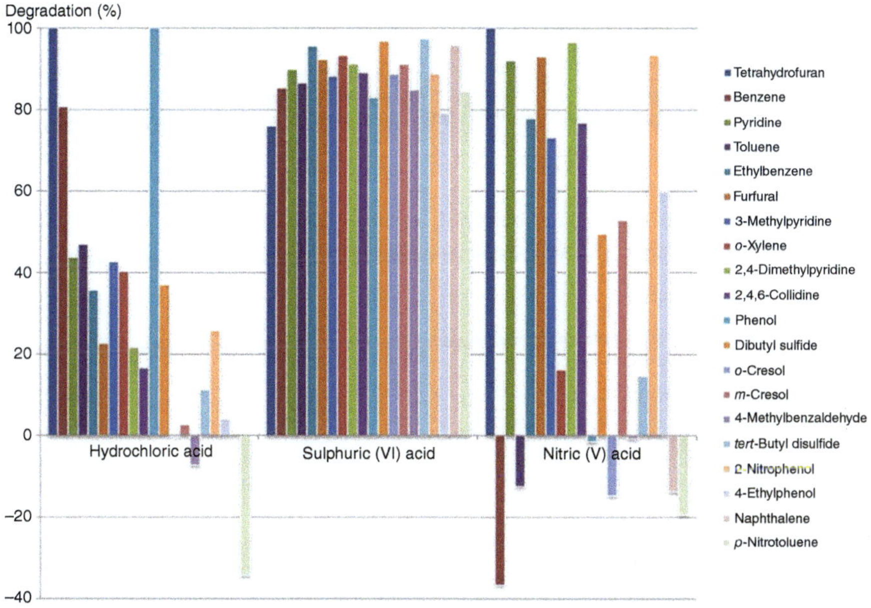

Figure 7.16 Degradation of organic pollutants by hydrodynamic cavitation process in the presence of different mineral acids (Source: From [78]/with permission of Elsevier).

beneficial effect can be ranked as $H_2SO_4 > HNO_3 > HCl$ for the use of mineral acids for hydrodynamic cavitation-based studies.

In general, pH is an important process parameter for wastewater treatment via hydrodynamic cavitation. Careful evaluation of the role of prevailing pH on degradation performance is needed, particularly for real-life effluents containing myriad organic species.

7.3.2.5 *Operating Temperature*

The operating temperature will change the vapor pressure of liquid streams (wastewater). It is therefore expected to have significant influence on the extent and effect of hydrodynamic cavitation. The degradation of a variety of pollutants such as dyes and phenolics, and the influence of hydrodynamic cavitation at different bulk liquid temperatures are summarized in Table 7.6. These studies often use isothermal conditions for the degradation. A study with ultrasonic cavitation systems reported the effect of temperature on the cavitation intensity of a number of liquids. A cavitation intensity meter was used for this purpose Niemczewski [82]. Niemczewski reported that cavitation intensity was highest with water (among the 38 liquids used) at $T = 35\,°C$. The results suggest that the highest degradation performance would also be under similar operating temperature in water. However, not all the studies (in Table 7.5) report $T = 35\,°C$ as the temperature for highest degradation.

In a study reporting the degradation of two dyes (Brilliant Green and Orange Acid II) for the same cavitation device [86], the reaction rate was found to decrease with increase in temperature from 20 to 40 °C: a reduction by 0.23 times for Orange Acid II

Table 7.5 Studies reporting the effect of temperature on hydrodynamic cavitation performance.

Pollutant	Cavitation device type	Temperature range (°C)	Optimal temperature (°C)	Reference
Acid Red-18	Orifice	25–40	35	[33]
Dicofol	Orifice blade assembly	20–40	30	[58]
Rhodamine B	Multiple venturi tubes	15–35	25	[83]
Potassium ferrocyanide	Venturi	20–50	30	[84]
2,4-Dinitrophenol	Orifice	30–40	35	[85]
Orange Acid—II and Brilliant Green	Orifice	20–40	20	[86]
Carbamazepine	Orifice	25 and >25	25	[87]
Rhodamine B	Orifice Venturi	30 and 40	—	[50]
Industrial wastewater	Orifice	20 and 30	30	[39]
Alachlor	Swirling jet-induced cavitation	20–60	40	[88]

and by 0.78 times for Brilliant Green. This indicates that the influence of temperature might be dependent on the pollutant type or its molecular properties. For the degradation of dichlorvos, Joshi and Gogate [74] evaluated three temperatures: 31, 36, and 39 °C and observed that the lowest temperature was found to be beneficial. For the degradation of an industrial effluent, Chakinala et al. [39] observed 50% more degradation at 30 °C, than at 20 °C. A study reporting the degradation of alachlor [88] evaluated its degradation between 30 and 60 °C (Figure 7.17). The optimal temperature was found to be 40 °C for the study and increasing temperature from 40 to 60 °C decreased the degradation rate constant by 0.35 times.

For the degradation of solvent-based contaminants, Sarvothaman et al. [42] reported that degradation monotonically increased with temperature in the range of 20–45 °C. The results from hydrodynamic cavitation indicate that the degradation efficiency decreased on operating at temperatures greater than 40 °C. Overall, the optimal temperature reported by the studies vary from 20 to 40 °C based on the pollutant degradation performance. However, it was consistent that a higher temperature (>40 °C) would be detrimental to the overall degradation performance (Table 7.5). Therefore, an optimal temperature for operation may exist between 20 and 40 °C, and this will be dependent on the type of pollutant used.

7.3.2.6 *Influence of Dissolved Gases/Sparged Gases*

The aeration or modification of dissolved gases is expected to influence per-pass degradation performance in three ways: the first is by modifying the inception; the

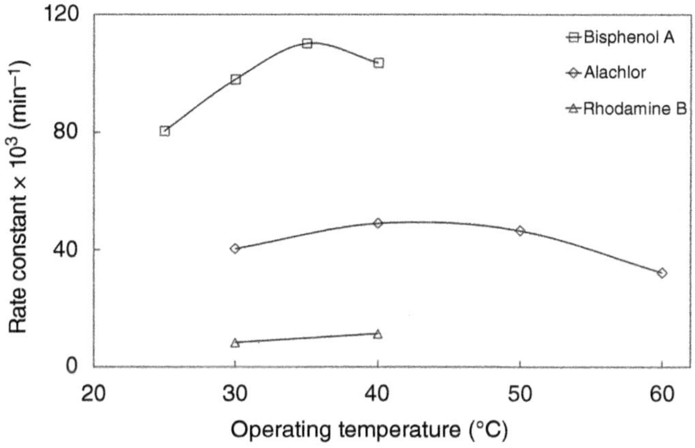

Figure 7.17 Influence of temperature on degradation rate constant for the degradation of Bisphenol A, Alachlor, and Rhodamine B (Source: Data—Bisphenol A from [89], Alachlor from [88], and Rhodamine B from [50]).

second by modification of the intensity of cavity collapse; and the last by influencing the radical reactions. The low-pressure region generated in cavitation devices causes desorption of dissolved gases which may act as a nucleation for further cavitation, i.e. growth by evaporation of liquid components into the gaseous cavity. The extent of dissolved gases will therefore influence the inception of cavitation. Secondly, the dissolved gases will influence composition as well as transport properties (e.g., heat capacity, compressibility) at the cavity collapse, which has a significant impact on generated high temperature, pressure, and hydroxyl radicals [47]. There are relatively fewer studies investigating the role of dissolved gas in hydrodynamic cavitation as compared to acoustic cavitation [70, 90].

While the bubble temperature at moments of collapse is expected to be higher in the case of argon, the gas being inert is unlikely to influence radical formation. However, in the case of oxygen, it is expected to directly participate in the reactions leading to radical formations [91]. The degradation of pollutants such as phenol [25, 46, 92], dichlorophenols [93], and dyes [94, 95] under different dissolved gases has been attempted in acoustic cavitation. Since the underlying mechanisms of degradation of hydrodynamic and acoustic modes of cavitation are the same, it is worth discussing the results from these studies. The introduction of the dissolved gas enables either the production of more radicals or increases the bubble's interior temperature [90]. Consider the effective rate constant for the degradation of 50 ppm 2,4-dichlorophenol under different gases, presented in Figure 7.18. The trend obtained at different pH conditions is uniform, to indicate that: Oxygen > Argon > Air > Nitrogen.

Berlan et al. [92] have investigated the use of three ratios of Air: Argon and conclude that 60 : 40 is the best for phenol degradation. Harada and Kumagai [96] evaluated the production of hydrogen peroxide from cavitation and noticed that 2 : 3 ratio of oxygen : argon was suitable for the highest yield of hydrogen peroxide. The studies such as those of the degradation of phenolics and dyes do point out interesting

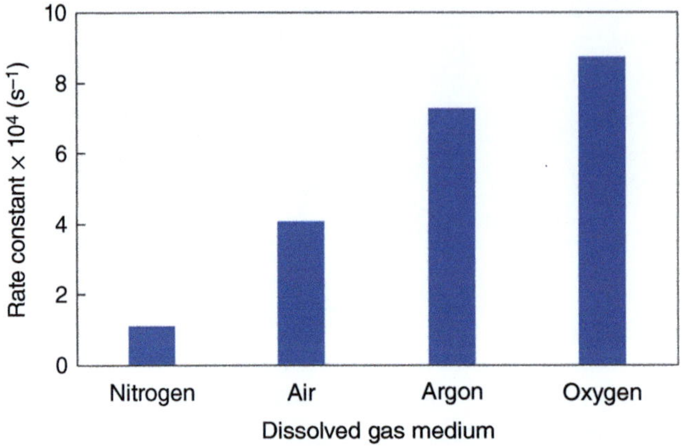

Figure 7.18 Influence of dissolved gas on rate constant for degradation of 2,4-dichlorophenol at pH 6.3, $C_0 = 50$ ppm, $T = 20\,°C$ (Source: Data from [93]).

insights to the degradation under the influence of dissolved gases. In the review by Rooze et al. [90], a comparison of rate constants obtained for argon and air is presented. Sivasankar and Moholkar [46] reported a 1.3-fold increase in degradation of phenol with air as dissolved gas compared to that of argon. In another study for the degradation of phenol, Kidak et al. [97] reported a three-fold increase in degradation in presence of air as medium when compared to argon. This indicates that the influence of dissolved gas on degradation might be pollutant-specific. From a theoretical standpoint, the hotspot temperature and radical formation do differ with different dissolved gases [91, 98]. The use of argon, air, and oxygen as dissolved gases can potentially improve cavitation performance.

There are very few studies which have focused on the role of dissolved gas on degradation performance for hydrodynamic cavitation [42, 57, 99, 100]. The study by Suslick et al. [100] used different ratios of argon and helium gas. The effect of the use of these gases was quantified by the oxidation reaction of potassium iodide (the so-called Weissler reaction). The study showed a nearly five-fold increase in the rate of oxidation of potassium iodide when argon was used, when compared to an equal proportion of argon/helium. In the case of phenol degradation studied by Wu et al. [99], nitrogen, air, and oxygen media were studied. The best degradation was obtained in the case of introduction of oxygen > air (normal conditions) > nitrogen. Aeration thus is a promising means to improve performance of hydrodynamic cavitation. Suryawanshi et al. [59] reported an increase of 20–30% in COD removal with aeration. Sarvothaman et al. [42] used aeration as a means to improve degradation of acetone in water with vortex-based devices. They report an optimal aeration flow rate (about 1 vvm) for acetone degradation. In a recent study, for p-aminophenol degradation of $C_0 = 500$ ppm, the degradation was enhanced by nearly four-fold with aeration, three-fold with oxygen media and a hindered performance of a four-fold reduction with helium as against no aeration [57].

Studies in hydrodynamic cavitation should focus on clarifying the effect of dissolved gas on different pollutant types. For example, a ratio such as 2 : 3 (oxygen : argon) was suggested as optimal for the production of hydrogen peroxide [96]. Such studies should be extended to cover different pollutants. Further research on developing adequate computational models is needed to relate concentration of dissolved gases and overall performance or per-pass degradation factor.

7.4 Enhancing the per-pass Performance: Augmentation by Hybrid Processes

The previous section discussed the effect of various design and operating parameters on the overall degradation performance of hydrodynamic cavitation-based wastewater treatment. The results reported for various pollutants in this section reveal that the extent of degradation for different pollutants differs quite significantly. Particularly, some types of pollutants (e.g., containing –OH groups like phenol) were found to be difficult to degrade adequately using just hydrodynamic cavitation [22, 48, 99]. Investigation into degradation pathways of phenol (shown in Figure 7.19) by Yang et al. [101] indicates the formation of short-chain carboxylic acids and subsequent slowdown of further mineralization. Similar observations are reported in Berlan et al. [92]. For such pollutants, which are difficult to remove via hydrodynamic cavitation alone, the process may be augmented by synergistically combining them with other AOPs. As discussed in the introduction (and shown Figure 7.2), the overall oxidation of pollutants realized by AOPs can be augmented by (a) enhancing energy input, (b) additional oxidizing chemicals, and (c) catalysts. Some options of such hybrid technologies are briefly summarized in this section.

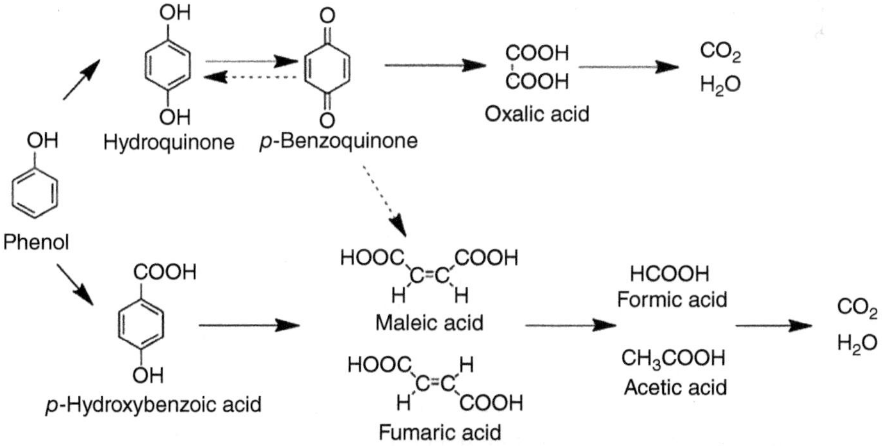

Figure 7.19 Degradation pathway of phenol by advanced oxidation processes (Source: From [101]/with permission of John Wiley & Sons).

Before such combined or hybrid technologies are discussed, it will be useful to define a parameter to evaluate the effectiveness of the hybrid process. It is customary to define the "synergistic index" for this purpose, which may be defined as:

$$\text{Synergistic index (SI)} = \frac{\emptyset_{HC+AOP}}{(\emptyset_{HC} + \emptyset_{AOP})} \tag{7.11}$$

where Ø is the per-pass degradation factor and the subscripts HC, AOP, and HC + AOP denote only hydrodynamic cavitation, other AOP and combined processes of hydrodynamic cavitation, and other AOP. The values of $SI > 1$ indicate synergy between the combined processes which is the desired outcome of the combination of processes.

7.4.1 Coupling of HC with AOPs Using Alternative Energy Sources

Several attempts have been made to combine hydrodynamic cavitation with other novel energy sources like microwave, UV, or plasma to enhance its performance. Microwave-assisted hydrodynamic cavitation has been used for altering carbon number profile—value addition of crude oil [102], liposome destruction [103], and biodiesel synthesis [104]. However, this has not been used as much for wastewater treatment applications. Application of UV and plasma-assisted hydrodynamic cavitation is briefly discussed in the following section.

7.4.1.1 *UV-assisted HC*

Ultraviolet is a form of electromagnetic irradiation which is commonly used with chemical additive-based processes such as peroxide, ozone, and per sulfate-based processes for the disinfection and treatment of wastewater [105–107]. Among the different types of UV sources: UV-A, UV-B, and UV-C, it is UV-C which is mostly used for the treatment of wastewater [108]. UV-C light can induce the cleavage of additives (such as peroxide, ozone, and per sulfate) and improve the reaction with organic pollutants in water. In a similar sense, UV-C-based processes can be incorporated with hydrodynamic cavitation to cleave and additionally provide a means of photo-degradation of organic pollutants. A schematic of an orifice-based hydrodynamic cavitation incorporated with UV-based treatment is presented in Figure 7.20.

Čehovin et al. [109] employed this arrangement for the degradation of natural organic matter. In this study, hydrodynamic cavitation and UV-based treatment were not used in isolation, but with peroxide and ozone-based treatments. In a study to degrade carbamazepine, Thanekar et al. [21] report a detrimental effect of combining hydrodynamic cavitation and UV. Bagal and Gogate [66] employed a combination of hydrodynamic cavitation and UV for the degradation of diclofenac sodium. The extent of degradation was found to be 26.8, 49.8, and 65.6% for only hydrodynamic cavitation, only UV, and their combination, respectively. The authors report a synergistic index of 1.1. Thanekar et al. [110] incorporated a combination of UV illumination in their liquid-holding tank with an 8 W and a 16 W UV source; the rate constants from their study are plotted in Figure 7.21. They reported a synergistic index of 1.25. Studies relating to other pollutants such as dyes and other phenolics are yet to be explored.

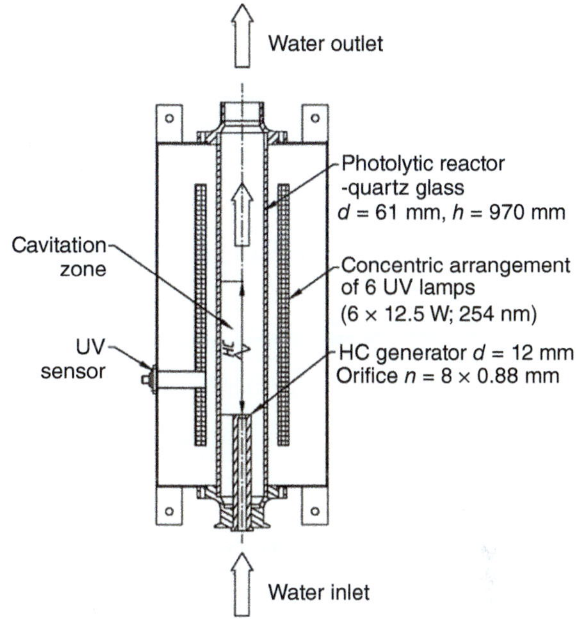

Figure 7.20 Configuration of photolytic UV reactor with hydrodynamic cavitation (Source: From [109]/with permission of Elsevier).

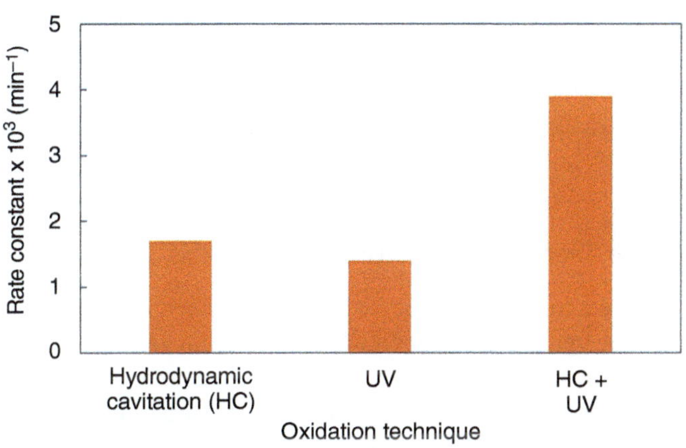

Figure 7.21 First-order rate constants for the application of hydrodynamic cavitation, UV, and their combination in the degradation of dimethoate (Source: Data from [110]).

7.4.1.2 *Plasma-based HC*

Plasma irradiation is divided into thermal and non-thermal plasma. Non thermal plasma offers a high selectivity and energy efficiency making it attractive for water treatment applications [111]. Of the three types of non-thermal plasma, the application of electric discharge to bubbles/vapor in a flowing liquid (which serves as an electrode) is the one coupled with hydrodynamic cavitation. By this coupling, as shown in Figure 7.22, formation and prolonged life of radical species and generation of UV light can be achieved. Very limited work pertaining to the coupling of plasma and hydrodynamic cavitation has been performed in the area of

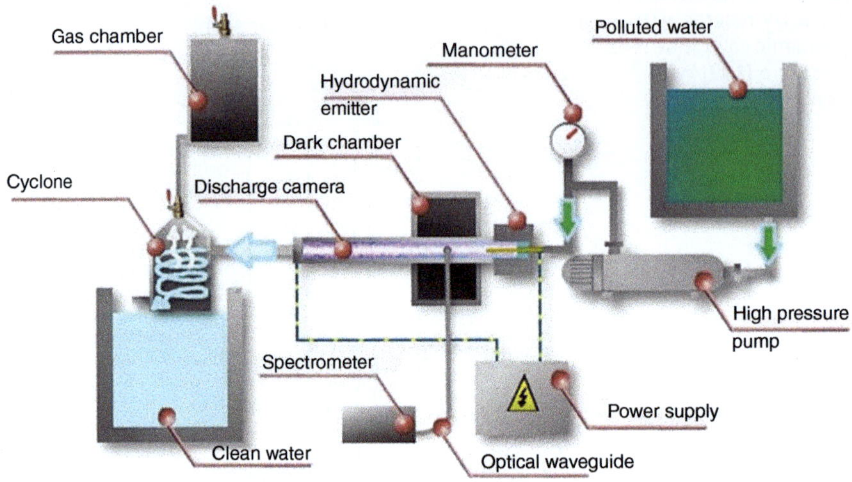

Figure 7.22 Schematic of experimental setup for wastewater treatment incorporating hydrodynamic cavitation and plasma discharge (Source: From [112]/with permission of Elsevier).

wastewater treatment [112, 113]. Abramov et al. [112] employed different electrode materials namely silver, graphite, and brass, primarily for water disinfection and further for methanol degradation. Greater than 99.99% disinfection was reported in three passes. Their results also indicate potential applications for decolorization of wastewater containing dyes.

7.4.2 Coupling of HC with Chemical-based AOPs

Various studies have reported that the use of hydrodynamic cavitation alone is not sufficient for the degradation of organic pollutants [21, 22, 60, 74, 114], which might be due to the recalcitrant nature of the pollutant used and insufficient oxidant production. For example, the degradation of organic compounds such as dyes—Rhodamine-6G, Orange-4; pesticides: dichlorvos, triazophos, and phenolics: 4-chloro-2-aminophenol, 2,4,6-trichlorophenol—with stand alone hydrodynamic cavitation treatment may not yield a satisfactory degradation without a very high number of passes through the hydrodynamic cavitation device. The performance of hydrodynamic cavitation can be augmented by combining it with other chemical-based AOPs such as hydrogen peroxide (H_2O_2) addition, ozonation, or the combination of these two techniques (peroxonation). The rationale behind the coupling is to produce more radicals in the bulk phase and reduce treatment times/number of passes required for the desired degradation. Hydrodynamic cavitation improves mixing, the dissociation of hydrogen peroxide, and contact of oxidizing species with pollutant species in water. Therefore, coupling chemical-based AOPs with hydrodynamic cavitation reduces the consumption of H_2O_2 or ozone (O_3).

7.4.2.1 Hydrogen Peroxide (H_2O_2) Treatment

In the case of hydrogen peroxide addition to conventional stirred reactions, although the oxidation potential is high (1.78 V), the dissociation is poor. However, in the case of hydrogen peroxide treatment in combination with hydrodynamic cavitation, the dissociation occurs at an enhanced rate [71, 109]. The reactions (R7.9)–(R7.16) are expected to take place in such a combined treatment:

$$H_2O_2 \rightarrow 2\,OH^{\cdot} \qquad (R7.9)$$

$$H_2O \rightarrow H^{\cdot} + OH^{\cdot} \qquad (R7.10)$$

$$OH^{\cdot} + OH^{\cdot} \rightarrow H_2O_2 \qquad (R7.11)$$

$$OH^{\cdot} + H_2O_2 \rightarrow HO_2^{\cdot} + H_2O \qquad (R7.12)$$

$$OH^{\cdot} + HO_2^{\cdot} \rightarrow H_2O + O_2 \qquad (R7.13)$$

$$HO_2^{\cdot} + H_2O_2 \rightarrow OH^{\cdot} + H_2O + O_2 \qquad (R7.14)$$

$$\text{Pollutant molecules} + OH^{\cdot} \rightarrow CO_2 + H_2O + \text{other intermediates} \qquad (R7.15)$$

$$\text{Pollutant molecules} + H_2O_2 \rightarrow CO_2 + H_2O + \text{other intermediates} \qquad (R7.16)$$

Several laboratory studies have reported a synergistic index greater than 2 when hydrodynamic cavitation is combined with hydrogen peroxide. For example, in a study to degrade an industrial effluent, Joshi and Gogate [29] report that the treatment by HC alone was effective in 6% COD removal after three hours of operation, despite altering the hydrodynamic cavitation device and operating parameters. In the study, with addition of 15 g/L of H_2O_2, a meagre 8.5% removal was effected. Whereas with the incorporation of hydrodynamic cavitation, the removal was boosted to 30%, with a synergistic index of 2.6. Suresh Kumar et al. [115] report the degradation of a ternary dye by hydrodynamic cavitation treatment alone was only 12.5% in 40 minutes. The degradation with oxidant, by maintaining molar ratio of dye: H_2O_2 of 1 : 40 was 22.5% in 40 minutes by conventional stirring. Whereas with the same oxidant dosage, hydrodynamic cavitation was able to effect a complete decolorization. The synergistic index in both these studies [29, 115] was greater than or equal to 2, indicating that hydrodynamic cavitation played a significant role in aiding the dissociation of additive introduced.

Another example of synergy between hydrodynamic cavitation and hydrogen peroxide treatment is the case of degradation of 2,4,6-trichlorophenol reported by Barik and Gogate [79]. Their results are shown in Figure 7.23. With the hydrogen peroxide treatment (pollutant: $H_2O_2 = 1 : 5$), the degradation achieved was ~20%. Combining it with hydrodynamic cavitation improved the degradation by nearly threefold. It can also be observed that while comparing the treatment of (i) stoichiometric ratio of pollutant to hydrogen peroxide with hydrodynamic cavitation treatment and (ii) hydrogen peroxide treatment (pollutant: $H_2O_2 = 1 : 5$), the case with hydrodynamic

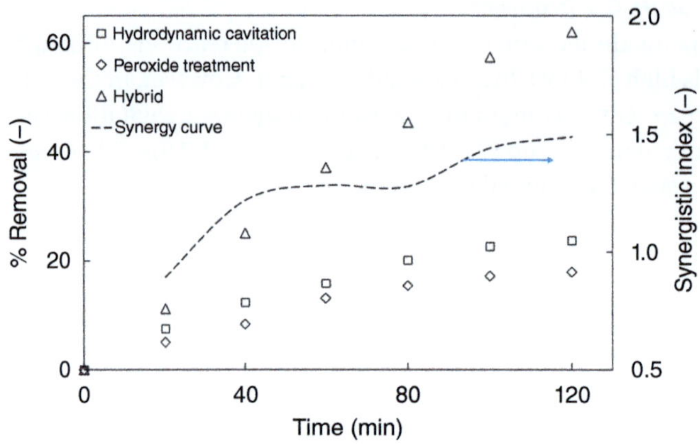

Figure 7.23 Effect of H_2O_2 loading on degradation of 2,4,6-trichlorophenol (TCP) for the combination of hydrodynamic cavitation and H_2O_2 (inlet pressure = 400 kPa, C_0 = 20 ppm, T = 30 °C, pH 7) (Source: Adapted from [79]).

cavitation treatment outperforms by twofold, implying hydrodynamic cavitation treatment can lower the additive requirement. From the figure (Figure 7.23), it can be seen that the synergistic index gradually increases from 1 to 1.5 over the entire treatment duration and there is lowering of treatment time. Similar to the previous report of hydrodynamic cavitation with hydrogen peroxide, several studies observe the synergistic performance of hydrodynamic cavitation with additives. A representative set of studies is presented in Table 7.6.

7.4.2.2 *Ozone (O_3) Treatment*

The ozone treatment is strongly influenced by the operating pH. At pH < 8, ozone reacts directly and selectively via electrophilic attack with electron-rich centers. At pH > 10, an indirect oxidation pathway would dominate. At a neutral pH, both the pathways are prominent [116]. Most of the studies which used hydrodynamic cavitation with ozone used acidic pH conditions. The reactions (R7.17)–(R7.22) are expected to occur for this combined (hydrodynamic cavitation with ozone) treatment. The incorporation of ozone in hydrodynamic cavitation requires some level of sophistication such as a control system and cooling systems (shown in Figure 7.24), as opposed to peroxide treatment:

$$O_3 \rightarrow O_2 + O\,(^3P) \tag{R7.17}$$

$$H_2O \rightarrow H^\cdot + OH^\cdot \tag{R7.18}$$

$$O\,(^3P) + H_2O \rightarrow H^\cdot + OH^\cdot \tag{R7.19}$$

$$OH^\cdot + OH^\cdot \rightarrow H_2O_2 \tag{R7.20}$$

$$\text{Pollutant molecules} + OH^\cdot \rightarrow CO_2 + H_2O + \text{other intermediates} \tag{R7.21}$$

$$\text{Pollutant molecules} + H_2O_2 \rightarrow CO_2 + H_2O + \text{other intermediates} \tag{R7.22}$$

Table 7.6 Representative studies coupling the treatment of hydrogen peroxide and ozone treatment with hydrodynamic cavitation.

Pollutant	Cavitation device flow rate (LPM)	Liquid tank holding volume (L)	Concentration of pollutant (ppm)	Optimal hydrogen peroxide dosage (ratio with pollutant)	Optimal ozone flow rate (mg/h)	Reference
Carbamazepine	—	4	10	1 : 5	400	[21]
2,4,6-Trichlorophenol	—	7	20	1 : 5	400	[79]
Rhodamine-6G	9.9	6	10–50	1 : 30 ($C_0 = 10$ ppm)	700 ($C_0 = 50$ ppm)	[22]
4-Chloro 2-aminophenol	—	7	20	—	400	[80]
Triazophos	—	4	20	—	576	[81]
Orange-4	6.8	5	40	1 : 30 (pH 2)	3000 (pH 2)	[65]
Imidacloprid	—	4	20	1 : 4	—	[60]
Dichlorvos	—	4	20	1 : 0.8	—	[74]

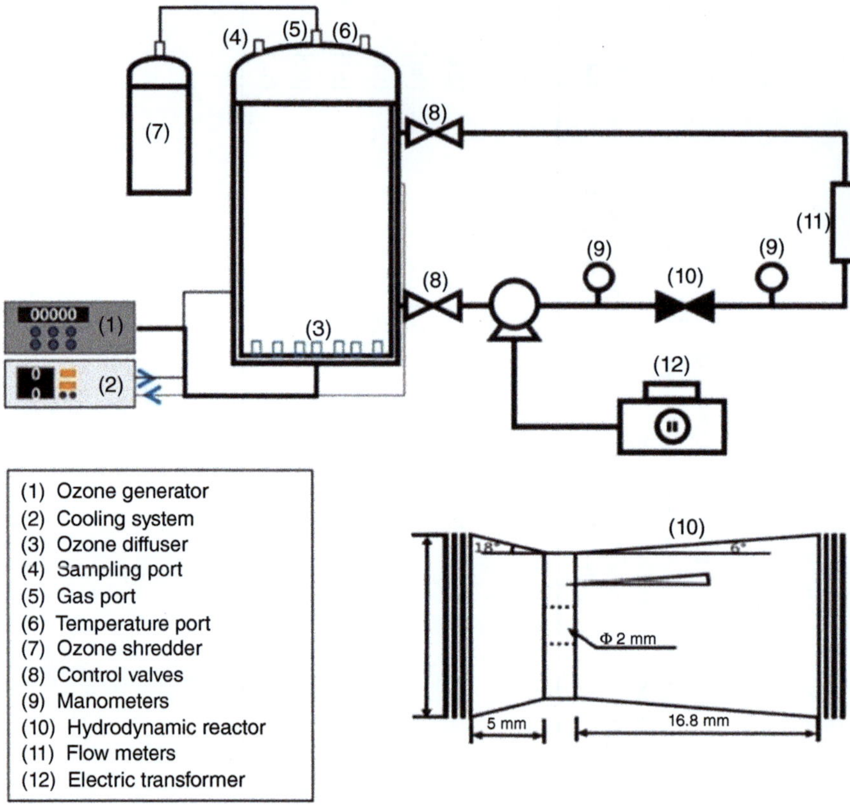

Figure 7.24 Assembly of hydrodynamic cavitation device with ozonator assembly (Source: From [117]/with permission of Elsevier).

Similar to studies which have reported a synergistic effect with peroxide treatment, incorporation of ozone with hydrodynamic cavitation has shown synergistic index greater than 2 [21]. In the case of 2,4,6-trichlorophenol degradation, the authors observed removals of 24 and 86% with hydrodynamic cavitation and ozone, respectively. Whereas the combination of the two resulted in 96% removal within 120 minutes, with a synergistic factor of 1.5 [79], in this study the pollutant used was 2,4,6-trichlorophenol and the rate constants for these techniques are plotted in Figure 7.25. In a recent study to degrade laundry effluents, the use of 400 mg/h of ozone to 5 L of holding tank volume increased the performance of the hydrodynamic cavitation from 24.7 to 33.3% in 90 minutes of operation [40].

7.4.2.3 *Peroxonation (Hydrogen Peroxide–H_2O_2 + Ozone–O_3)*

The simultaneous addition of the two oxidants—hydrogen peroxide and ozone—is termed as peroxonation. Such an approach can be used to accelerate the process and could provide a means to tackle multiple pollutant types simultaneously. The current literature reports the use of this technique for single pollutants and has reported either shortened reaction durations or beneficial synergistic factors. Thanekar et al.

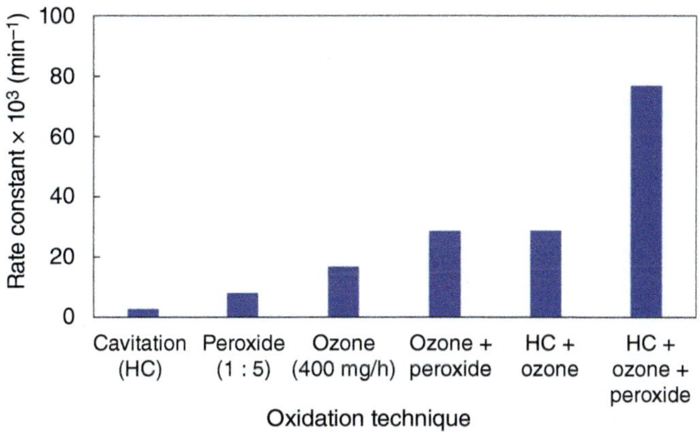

Figure 7.25 Effect of different oxidation techniques on the degradation rate constant for 2,4,6-trichlorophenol, cavitation device pressure drop = 400 kPa, $C_0 = 20$ ppm, $T = 30$ °C and pH = 7 (Source: Adapted from [79]).

[21] reported that the combined treatment of hydrodynamic cavitation with peroxonation resulted in 100% degradation of carbamazepine, the synergistic index for the incorporation of three AOPs (hydrodynamic cavitation—peroxide—ozone) was calculated as slightly higher than 3. The combination of hydrodynamic cavitation with chemical AOPs has found application in decolorizing dye-based wastewater [22, 65] and reducing the toxicity of pharmaceuticals [21, 79], and is indeed quite promising for intensifying wastewater treatment processes. Further work on developing computational models for simulating and optimizing such hybrid AOPs for maximizing synergy is needed.

7.4.3 Augmenting Hydrodynamic Cavitation by Catalyst-based AOPs

Another way of augmenting performance of hydrodynamic cavitation-based wastewater treatment processes is to use catalysts. Such an augmentation may be achieved by using either photocatalysts (which require additional source of energy in the form of light) or a conventional Fenton catalyst (which generates sludge as secondary effluent). These two processes are briefly discussed in the following section.

7.4.3.1 *Fenton's Process*

Fenton's processes are those which are catalyzed by the addition of iron salts [116] —typically iron (II) sulfate, which can accelerate production of ·OH radicals from H_2O_2. The use of iron particles as such is termed Advanced Fenton's processes [118]. The use of iron salts to catalyze the formation of ·OH radicals, referred to as Fenton's reaction, is quite commonly employed to improve the treatment efficacy. The reactions involved for this treatment are shown in reactions (R7.23)–(R7.26) [9]. The process is sensitive to pH, and typically, a pH of 3 is maintained. The ferrous ion is oxidized by hydrogen peroxide to ferric ions. These ferric ions form an intermediate complex, which under cavitating conditions would form ferrous ions and the hydroperoxyl radical. The regeneration of the ferrous ion would enable the iron species to participate in the scheme from (R7.23) again.

$$Fe^{2+} + H_2O_2 \rightarrow Fe^{3+} + OH^- + OH^{\cdot} \qquad (R7.23)$$

$$Fe^{3+} + H_2O_2 \rightarrow Fe - OOH^{2+} + H^+ \qquad (R7.24)$$

$$Fe - OOH^{2+} \rightarrow Fe^{2+} + HO_2^{\cdot} \qquad (R7.25)$$

$$H_2O_2 \rightarrow 2\,OH^{\cdot} \qquad (R7.26)$$

Overall performance is dependent on pH, temperature, the pollutant concentration and the $[H_2O_2]/[Fe^{2+}]$ ratio. The optimal pH is typically found to be 2–4. The temperature would influence the decomposition of H_2O_2 and the dosage of peroxide—iron salt would influence of the sludge formed at the end of the process [116]. Patil et al. [60] report the use of Fenton's reagents for the removal of imidacloprid (Figure 7.26). They report that in the absence of reagents (H_2O_2 or Fenton's), only 20% removal is achieved by hydrodynamic cavitation. With addition of Fenton's reagents, a near-complete removal was achieved within 15 minutes. Similar enhancements for the degradation with different Fenton's dosages are reported for triazophos [81] and dichlorvos [74].

7.4.3.2 *Photocatalysis*

In photocatalytic processes, reactive oxygen species are generated with the help of a photocatalyst and light. The process is shown schematically in Figure 7.27. Incident light with sufficient energy excites electrons in the valence band and transfers them to the conduction band, leaving behind holes. This initiates further reactions with typical electron acceptors such as O_2 and adsorbed water molecules to produce

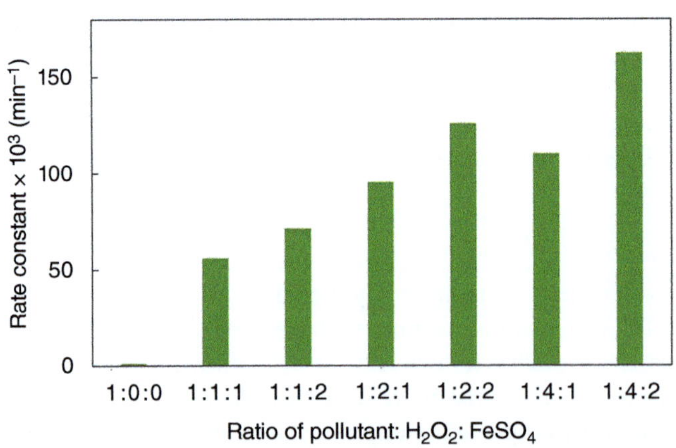

Figure 7.26 Effect of loading of Fenton's reagent on extent of degradation of imidacloprid, cavitation device pressure drop = 400 kPa, C_0 = 20 ppm, T = 34 °C and pH = 3 (Source: Data from [60]).

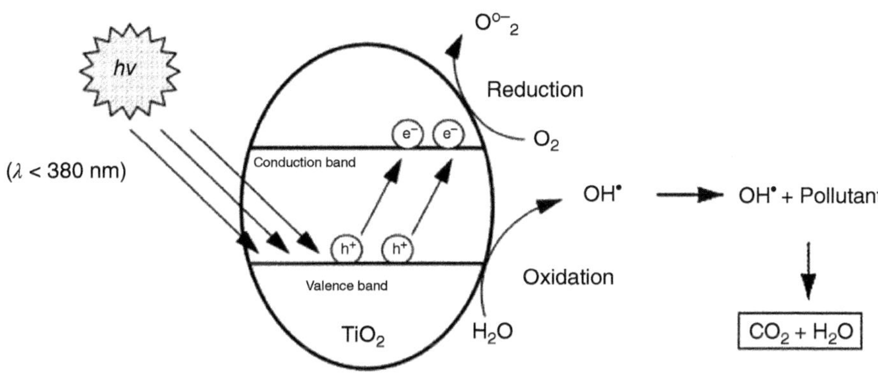

Figure 7.27 Schematic of principle of photocatalyst to wastewater treatment (Source: From [119]/with permission of Elsevier).

reactive oxygen species via reactions, shown as Reactions (R7.27)–(R7.31). The use of photocatalysis as an AOP predates the application of HC to wastewater treatment. One of the first reports for the removal of pollutants in wastewater was in the 1970s, and since then, the interest has grown multifold. Roughly half of the research efforts directed toward photocatalysts has been to treat wastewater [120], and the relevant photochemistry is well understood. Photocatalytic wastewater treatment is quite attractive since it does not require additional chemicals:

$$\text{Photocatalyst} + h\vartheta \rightarrow h^+_{vb} + e^-_{cb} \quad \text{(R7.27)}$$

$$h^+_{vb} + \mathrm{H_2O} \rightarrow \mathrm{OH\cdot} + \mathrm{H^+} \quad \text{(R7.28)}$$

$$h^+_{vb} + \mathrm{OH^-} \rightarrow \mathrm{OH^{\cdot}_{ad}} \quad \text{(R7.29)}$$

$$h^+_{vb} + R_{ads} \rightarrow R^+ \quad \text{(R7.30)}$$

$$e^-_{cb} + \mathrm{O_2} \rightarrow \mathrm{O_2^{\cdot -}} \quad \text{(R7.31)}$$

As a stand alone AOP, photocatalysis was found to be inferior to the processes based on UV/H_2O_2 [121]. However, development of cheaper photocatalysts (such as titanium dioxide) and potential synergy with hydrodynamic cavitation may offer better prospects. The combination of cavitation with photocatalysis is particularly attractive for mineralization of pollutants like phenol which are otherwise difficult to treat by cavitation alone. The use of photocatalysts acts as an additional source of reactive oxygen species, which will augment the performance of hydrodynamic cavitation.

Suspended photocatalysts are the most common form of photocatalytic processes combined with hydrodynamic cavitation, despite concerns about the separation of the catalyst and wastewater. Immobilized photocatalysts, because of lower surface areas, are not as efficient as the suspended form [122]. The studies prior to 2010 typ-

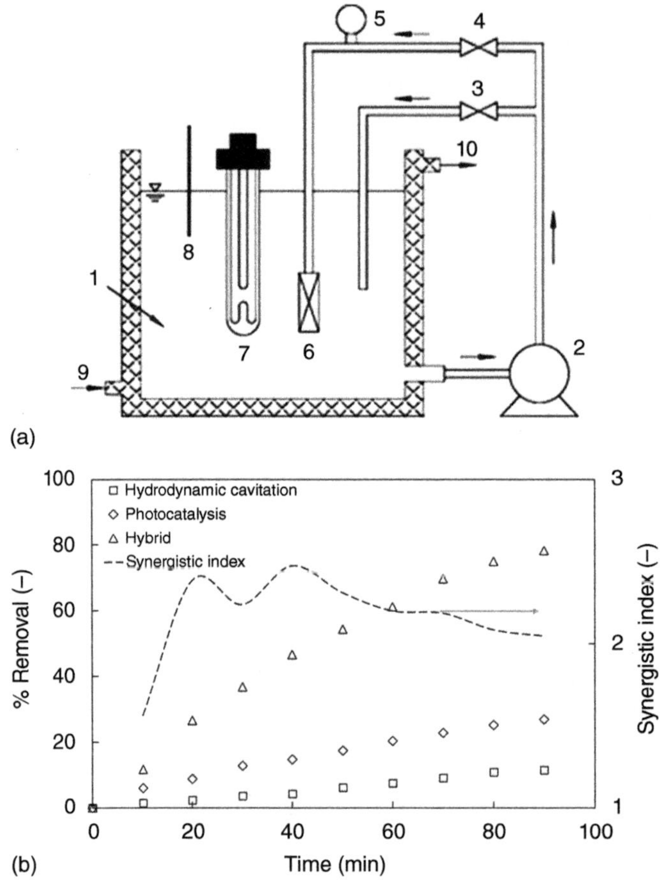

Figure 7.28 Combined hydrodynamic cavitation and photocatalytic process. (a) Schematic of the experimental setup used for the hydrodynamic cavitation and photocatalysis experiments. (1) Reactor; (2) centrifugal pump; (3 and 4) throttle valves; (5) pressure gauge; (6) venturi tube or straight pipe; (7) UV254 lamp; (8) thermometer; (9) cooling water inlet; (10) cooling water outlet. (b) Concentration profile of tetracycline with individual and hybrid process (Source: From [77]/with permission of Elsevier).

ically combined acoustic cavitation with photocatalysis [66]. In the past decade or so, efforts are being made to combine hydrodynamic cavitation with photocatalysis [77, 115, 123]. The typical setup combining hydrodynamic cavitation with photocatalyst is shown in Figure 7.28a. The cavitation prevents agglomeration and cleans the catalyst surface, thereby improving the photoactivity of these catalysts and enabling radical generation abilities [77]. Such an effect of increased destruction of the targeted pollutant—tetracycline—is reported in the study, as shown in Figure 7.28b.

Further research is needed on quantifying the interactions of suspended and immobilized photocatalysts with hydrodynamic cavitation. Particularly, the potential of using visible light and inexpensive photocatalysts to augment the performance of hydrodynamic cavitation needs to be further developed.

7.5 Summary and Outlook

Water scarcity and security have become major concerns, owing to the pressures exerted on water sources; climatic changes are further expected to intensify this crisis. It is therefore essential to develop intensified and effective technologies to promote water treatment, its recycle and reuse. In this chapter, the influence of parameters on hydrodynamic cavitation has been discussed. There have been significant experimental efforts in the understanding of these parameters influencing hydrodynamic cavitation. However, there is a need to unify this understanding using a common framework and consistent computational models. In addition to reviewing these experimental efforts, some suggestions for additional experiments to bridge the gaps in the existing knowledge base have been made. This is expected to strengthen the implementation of stand alone hydrodynamic cavitation-based processes for the degradation of organic pollutants in wastewater. For recalcitrant compounds such as phenol and primary carboxylic acids, often encountered as intermediates during the treatment of pollutants, hydrodynamic cavitation alone might not be suitable as a stand alone process. For such cases, the potential of augmenting hydrodynamic cavitation by combining it with other AOPs has been discussed. This includes a combination of hydrodynamic cavitation with processes based on alternative energy sources (UV, plasma); chemicals (peroxide, ozone), and catalysts (Fenton, photocatalysis). The hybrid processes based on hydrodynamic cavitation and Fenton catalysts have shown significant synergy (with a synergistic index of 3 or so). Further work on optimization and scale-up of hydrodynamic cavitation devices and the processes based on those will facilitate large-scale implementation of hydrodynamic cavitation-based wastewater treatment processes.

Notations

C	Outlet concentration for industrial cavitation setup (μmol/m^3)
C_{in}	Inlet concentration for industrial cavitation setup (μmol/m^3)
C_S	Concentration of scavengers (μmol/m^3)
d_t	Characteristic length scale/throat diameter of hydrodynamic cavitation device (mm)
G	Generation rate of hydroxyl radicals (kmol/s)
k_2	Second-order rate constant (M/s)
k_S	Rate of reactivity between pollutant and scavengers (–)
M	Molarity (mol/L)
V	Volume of holding tank (m^3)
n	Number of passes through cavitation device (–)
q	Net flow in a typical industrial cavitation setup (m^3/s)
Q	Recirculating flow through cavitation device (m^3/s)
W	Power requirement (W)
β	Ratio of flow rate through cavitation device (Q) and holding tank volume (V) (s^{-1})

γ Empirical parameter for scale-up correlation
η Pump efficiency (%)
$\emptyset$ Per-pass degradation factor (–)
$\emptyset_\infty$ Per-pass degradation factor for infinite scale-up (–)
τ Residence time of liquid in holding tank (s)

References

1 Westall, F. and Brack, A. (2018). The importance of water for life. *Space Science Reviews* 214 (2): 50.
2 Ng, W.J. (2006). *Industrial Wastewater Treatment*. World Scientific.
3 Garcia, D.J. and You, F. (2016). The water-energy-food nexus and process systems engineering: a new focus. *Computers & Chemical Engineering* 91: 49–67.
4 Garcia-Segura, S., Qu, X., Alvarez, P.J.J. et al. (2020). Opportunities for nanotechnology to enhance electrochemical treatment of pollutants in potable water and industrial wastewater – a perspective. *Environmental Science: Nano* 7 (8): 2178–2194.
5 Saccon, P. (2018). Water for agriculture, irrigation management. *Applied Soil Ecology* 123: 793–796.
6 Markland, S.M., Ingram, D., Kniel, K.E., and Sharma, M. (2018). Water for agriculture: the convergence of sustainability and safety. *Preharvest Food Safety* 143–157.
7 D'Odorico, P., Chiarelli, D.D., Rosa, L. et al. (2020). The global value of water in agriculture. *Proceedings of the National Academy of Sciences* 117 (36): 21985–21993.
8 UN-Water (2019). World Water Development Report 2019. https://data.worldbank.org/ (accessed 20 May 2021).[
9 Ranade, V.V. and Bhandari, V.M. (2014). *Industrial Wastewater Treatment, Recycling and Reuse*. Butterworth-Heinemann.
10 Baggio, G., Qadir, M., and Smakhtin, V. (2021). Freshwater availability status across countries for human and ecosystem needs. *Science of the Total Environment* 792: 148230.
11 Chen, Z., Ngo, H.H., and Guo, W. (2013). A critical review on the end uses of recycled water. *Critical Reviews in Environmental Science and Technology* 43 (14): 1446–1516.
12 Dolnicar, S. and Schäfer, A.I. (2009). Desalinated versus recycled water: public perceptions and profiles of the accepters. *Journal of Environmental Management* 90 (2): 888–900.
13 Hui, I. and Cain, B.E. (2018). Overcoming psychological resistance toward using recycled water in California. *Water Environment Journal* 32 (1): 17–25.
14 McNabb, D.E. (2017). Managing recycled water. *Water Resources Management* 283–306.

15 Andreozzi, R., Caprio, V., Insola, A., and Marotta, R. (1999). Advanced oxidation processes (AOP) for water purification and recovery. *Catalysis Today* 53 (1): 51–59.

16 Parsons, S. (2004). *Advanced Oxidation Processes for Water and Wastewater Treatment*. IWA Publishing.

17 Dewil, R., Mantzavinos, D., Poulios, I., and Rodrigo, M.A. (2017). New perspectives for advanced oxidation processes. *Journal of Environmental Management* 195: 93–99.

18 Wang, B., Su, H., and Zhang, B. (2021). Hydrodynamic cavitation as a promising route for wastewater treatment – a review. *Chemical Engineering Journal* 412: 128685.

19 Pandit, A.B. and Joshi, J.B. (1993). hydrolysis of fatty oils: effect of cavitation. *Chemical Engineering Science* 48 (19): 3440–3442.

20 Thanekar, P., Murugesan, P., and Gogate, P.R. (2018). Improvement in biological oxidation process for the removal of dichlorvos from aqueous solutions using pretreatment based on Hydrodynamic Cavitation. *Journal of Water Process Engineering* 23: 20–26.

21 Thanekar, P., Panda, M., and Gogate, P.R. (2018). Degradation of carbamazepine using hydrodynamic cavitation combined with advanced oxidation processes. *Ultrasonics Sonochemistry* 40: 567–576.

22 Rajoriya, S., Bargole, S., and Saharan, V.K. (2017). Degradation of a cationic dye (Rhodamine 6G) using hydrodynamic cavitation coupled with other oxidative agents: reaction mechanism and pathway. *Ultrasonics Sonochemistry* 34: 183–194.

23 Musmarra, D., Prisciandaro, M., Capocelli, M. et al. (2016). Degradation of ibuprofen by hydrodynamic cavitation: reaction pathways and effect of operational parameters. *Ultrasonics Sonochemistry* 29: 76–83.

24 Bhandari, V.M., Sorokhaibam, L.G., and Ranade, V.V. (2016). Industrial wastewater treatment for fertilizer industry – a case study. *Desalination and Water Treatment* 57 (57): 27934–27944.

25 Wu, Z., Yuste-Córdoba, F.J., Cintas, P. et al. (2018). Effects of ultrasonic and hydrodynamic cavitation on the treatment of cork wastewater by flocculation and Fenton processes. *Ultrasonics Sonochemistry* 40: 3–8.

26 Pichat, P., Guillard, C., Pe, C., and Chopin, T. (1999). Degradation of phenyltrifluoromethylketone in water by separate or simultaneous use of TiO 2 photocatalysis and 30 or 515 kHz ultrasound. *Physical Chemistry Chemical Physics* 1 (19): 4663–4668.

27 Flint, E.B. and Suslick, K.S. (1991). The temperature of cavitation. *Science* 253 (5026): 1397–1399.

28 Ashokkumar, M., Mulvaney, P., and Grieser, F. (1999). The effect of pH on multibubble sonoluminescence from aqueous solutions containing simple organic weak acids and bases. *Journal of the American Chemical Society* 121 (32): 7355–7359.

29 Joshi, S.M. and Gogate, P.R. (2019). Intensification of industrial wastewater treatment using hydrodynamic cavitation combined with advanced oxidation at operating capacity of 70 L. *Ultrasonics Sonochemistry* 52: 375–381.

30 Lalwani, J., Gupta, A., Thatikonda, S., and Subrahmanyam, C. (2020). Oxidative treatment of crude pharmaceutical industry effluent by hydrodynamic cavitation. *Journal of Environmental Chemical Engineering* 8 (5): 104281.

31 Badve, M., Gogate, P., Pandit, A., and Csoka, L. (2013). Hydrodynamic cavitation as a novel approach for wastewater treatment in wood finishing industry. *Separation and Purification Technology* 106: 15–21.

32 Padoley, K.V., Saharan, V.K., Mudliar, S.N. et al. (2012). Cavitationally induced biodegradability enhancement of a distillery wastewater. *Journal of Hazardous Materials* 219: 69–74.

33 Dhanke, P.B. and Wagh, S.M. (2020). Intensification of the degradation of Acid RED-18 using hydrodynamic cavitation. *Emerging Contaminants* 6: 20–32.

34 Thanekar, P. and Gogate, P.R. (2019). Combined hydrodynamic cavitation based processes as an efficient treatment option for real industrial effluent. *Ultrasonics Sonochemistry* 53: 202–213.

35 Raut-Jadhav, S., Badve, M.P., Pinjari, D.V. et al. (2016). Treatment of the pesticide industry effluent using hydrodynamic cavitation and its combination with process intensifying additives (H_2O_2 and ozone). *Chemical Engineering Journal* 295: 326–335.

36 Thanekar, P. and Gogate, P.R. (2020). Improved processes involving hydrodynamic cavitation and oxidants for treatment of real industrial effluent. *Separation and Purification Technology* 239: 116563.

37 Chakinala, A.G., Gogate, P.R., Burgess, A.E., and Bremner, D.H. (2008). Treatment of industrial wastewater effluents using hydrodynamic cavitation and the advanced Fenton process. *Ultrasonics Sonochemistry* 15 (1): 49–54.

38 Rajoriya, S., Bargole, S., George, S., and Saharan, V.K. (2018). Treatment of textile dyeing industry effluent using hydrodynamic cavitation in combination with advanced oxidation reagents. *Journal of Hazardous Materials* 344: 1109–1115.

39 Chakinala, A.G., Gogate, P.R., Burgess, A.E., and Bremner, D.H. (2009). Industrial wastewater treatment using hydrodynamic cavitation and heterogeneous advanced Fenton processing. *Chemical Engineering Journal* 152 (2–3): 498–502.

40 Patil, V.V., Gogate, P.R., Bhat, A.P., and Ghosh, P.K. (2020). Treatment of laundry wastewater containing residual surfactants using combined approaches based on ozone, catalyst and cavitation. *Separation and Purification Technology* 239: 116594.

41 Wang, J., Chen, H., Yuan, R. et al. (2020). Intensified degradation of textile wastewater using a novel treatment of hydrodynamic cavitation with the combination of ozone. *Journal of Environmental Chemical Engineering* 8 (4): 103959.

42 Sarvothaman, V.P., Nagarajan, S., and Ranade, V.V. (2018). Treatment of solvent-contaminated water using vortex-based cavitation: influence of operating pressure drop, temperature, aeration, and reactor scale. *Industrial & Engineering Chemistry Research* 57 (28): 9292–9304.

43 Ranade, V.V., Prasad Sarvothaman, V., Simpson, A., and Nagarajan, S. (2021). Scale-up of vortex based hydrodynamic cavitation devices: a case of degradation of di-chloro aniline in water. *Ultrasonics Sonochemistry* 70: 105295.

44 Kadam, S.R., Jadhav, N.L., Pandit, A.B., and Pejaver, M.K. (2021). Degradation kinetics and mechanism of hazardous metribuzin herbicide using advanced oxidation processes (HC & HC+ H_2O_2). *Chemical Engineering and Processing: Process Intensification* 166: 108486.

45 Suslick, K.S., Cline, R.E., and Hammerton, D.A. (ed.) (1985). *Determination of Local Temperatures Caused by Acoustic Cavitation*. IEEE.

46 Sivasankar, T. and Moholkar, V.S. (2009). Physical insights into the sonochemical degradation of recalcitrant organic pollutants with cavitation bubble dynamics. *Ultrasonics Sonochemistry* 16 (6): 769–781.

47 Sarvothaman, V.P. (2020). *Hydrodynamic Cavitation for Effluent Treatment: Using Vortex-based Cavitation Devices*. Queen's University Belfast.

48 Carpenter, J., Badve, M., Rajoriya, S. et al. (2017). Hydrodynamic cavitation: an emerging technology for the intensification of various chemical and physical processes in a chemical process industry. *Reviews in Chemical Engineering* 33 (5): 433–468.

49 Sivakumar, M. and Pandit, A.B. (2002). Wastewater treatment: a novel energy efficient hydrodynamic cavitational technique. *Ultrasonics Sonochemistry* 9 (3): 123–131.

50 Mishra, K.P. and Gogate, P.R. (2010). Intensification of degradation of Rhodamine B using hydrodynamic cavitation in the presence of additives. *Separation and Purification Technology* 75 (3): 385–391.

51 Saharan, V.K., Rizwani, M.A., Malani, A.A., and Pandit, A.B. (2013). Effect of geometry of hydrodynamically cavitating device on degradation of orange-G. *Ultrasonics Sonochemistry* 20 (1): 345–353.

52 Cai, M., Su, J., Zhu, Y. et al. (2016). Decolorization of azo dyes Orange G using hydrodynamic cavitation coupled with heterogeneous Fenton process. *Ultrasonics Sonochemistry* 28: 302–310.

53 Doltade, S.B., Dastane, G.G., Jadhav, N.L. et al. (2019). Hydrodynamic cavitation as an imperative technology for the treatment of petroleum refinery effluent. *Journal of Water Process Engineering* 29: 100768.

54 Sarvothaman, V.P., Simpson, A., and Ranade, V.V. (2020). Comparison of hydrodynamic cavitation devices based on linear and swirling flows: degradation of dichloroaniline in water. *Industrial & Engineering Chemistry Research* 59 (30): 13841–13847.

55 Simpson, A. and Ranade, V.V. (2019). 110th anniversary: comparison of cavitation devices based on linear and swirling flows: hydrodynamic characteristics. *Industrial & Engineering Chemistry Research* 58 (31): 14488–14509.

56 Sarvothaman, V.P., Simpson, A.T., and Ranade, V.V. (2019). Modelling of vortex based hydrodynamic cavitation reactors. *Chemical Engineering Journal* 377: 119639.

57 Patil, P.B., Bhandari, V.M., and Ranade, V.V. (2021). Improving efficiency for removal of ammoniacal nitrogen from wastewaters using hydrodynamic cavitation. *Ultrasonics Sonochemistry* 70: 105306.

58 Panda, D. and Manickam, S. (2019). Hydrodynamic cavitation assisted degradation of persistent endocrine-disrupting organochlorine pesticide Dicofol: optimization of operating parameters and investigations on the mechanism of intensification. *Ultrasonics Sonochemistry* 51: 526–532.

59 Suryawanshi, P.G., Bhandari, V.M., Sorokhaibam, L.G. et al. (2018). Solvent degradation studies using hydrodynamic cavitation. *Environmental Progress & Sustainable Energy* 37 (1): 295–304.

60 Patil, P.N., Bote, S.D., and Gogate, P.R. (2014). Degradation of imidacloprid using combined advanced oxidation processes based on hydrodynamic cavitation. *Ultrasonics Sonochemistry* 21 (5): 1770–1777.

61 Patil, P.N. and Gogate, P.R. (2012). Degradation of methyl parathion using hydrodynamic cavitation: effect of operating parameters and intensification using additives. *Separation and Purification Technology* 95: 172–179.

62 Wang, X., Wang, J., Guo, P. et al. (2008). Chemical effect of swirling jet-induced cavitation: degradation of rhodamine B in aqueous solution. *Ultrasonics Sonochemistry* 15 (4): 357–363.

63 Patil, P.B., Bhandari, V.M., and Ranadec, V.V. (2021). Wastewater treatment and process intensification for degradation of solvents using hydrodynamic cavitation. *Chemical Engineering and Processing: Process Intensification* 166: 108485.

64 Saharan, V.K., Badve, M.P., and Pandit, A.B. (2011). Degradation of Reactive Red 120 dye using hydrodynamic cavitation. *Chemical Engineering Journal* 178: 100–107.

65 Gore, M.M., Saharan, V.K., Pinjari, D.V. et al. (2014). Degradation of reactive orange 4 dye using hydrodynamic cavitation based hybrid techniques. *Ultrasonics Sonochemistry* 21 (3): 1075–1082.

66 Bagal, M.V. and Gogate, P.R. (2014). Degradation of diclofenac sodium using combined processes based on hydrodynamic cavitation and heterogeneous photocatalysis. *Ultrasonics Sonochemistry* 21 (3): 1035–1043.

67 Angaji, M.T. and Ghiaee, R. (2015). Decontamination of unsymmetrical dimethylhydrazine waste water by hydrodynamic cavitation-induced advanced Fenton process. *Ultrasonics Sonochemistry* 23: 257–265.

68 Soyama, H. and Hoshino, J. (2016). Enhancing the aggressive intensity of hydrodynamic cavitation through a Venturi tube by increasing the pressure in the region where the bubbles collapse. *AIP Advances* 6 (4): 045113.

69 Mancuso, G., Langone, M., and Andreottola, G. (2020). A critical review of the current technologies in wastewater treatment plants by using hydrodynamic cavitation process: principles and applications. *Journal of Environmental Health Science and Engineering* 18 (1): 311–333.

70 Holkar, C.R., Jadhav, A.J., Pinjari, D.V., and Pandit, A.B. (2019). Cavitationally driven transformations: a technique of process intensification. *Industrial & Engineering Chemistry Research* 58 (15): 5797–5819.

71 Gągol, M., Przyjazny, A., and Boczkaj, G. (2018). Wastewater treatment by means of advanced oxidation processes based on cavitation – a review. *Chemical Engineering Journal* 338: 599–627.

72 Capocelli, M., Prisciandaro, M., Lancia, A., and Musmarra, D. (2014). Hydrodynamic cavitation of p-nitrophenol: a theoretical and experimental insight. *Chemical Engineering Journal* 254: 1–8.

73 Parsa, J.B. and Zonouzian, S.A.E. (2013). Optimization of a heterogeneous catalytic hydrodynamic cavitation reactor performance in decolorization of Rhodamine B: Application of scrap iron sheets. *Ultrasonics Sonochemistry* 20 (6): 1442–1449.

74 Joshi, R.K. and Gogate, P.R. (2012). Degradation of dichlorvos using hydrodynamic cavitation based treatment strategies. *Ultrasonics Sonochemistry* 19 (3): 532–539.

75 Saharan, V.K., Pandit, A.B., Satish Kumar, P.S., and Anandan, S. (2012). Hydrodynamic cavitation as an advanced oxidation technique for the degradation of acid red 88 dye. *Industrial & Engineering Chemistry Research* 51 (4): 1981–1989.

76 Raut-Jadhav, S., Saini, D., Sonawane, S., and Pandit, A. (2016). Effect of process intensifying parameters on the hydrodynamic cavitation based degradation of commercial pesticide (methomyl) in the aqueous solution. *Ultrasonics Sonochemistry* 28: 283–293.

77 Wang, X., Jia, J., and Wang, Y. (2017). Combination of photocatalysis with hydrodynamic cavitation for degradation of tetracycline. *Chemical Engineering Journal* 315: 274–282.

78 Gągol, M., Cako, E., Fedorov, K. et al. (2020). Hydrodynamic cavitation based advanced oxidation processes: studies on specific effects of inorganic acids on the degradation effectiveness of organic pollutants. *Journal of Molecular Liquids* 307: 113002.

79 Barik, A.J. and Gogate, P.R. (2018). Hybrid treatment strategies for 2,4,6-trichlorophenol degradation based on combination of hydrodynamic cavitation and AOPs. *Ultrasonics Sonochemistry* 40: 383–394.

80 Barik, A.J. and Gogate, P.R. (2016). Degradation of 4-chloro 2-aminophenol using a novel combined process based on hydrodynamic cavitation, UV photolysis and ozone. *Ultrasonics Sonochemistry* 30: 70–78.

81 Gogate, P.R. and Patil, P.N. (2015). Combined treatment technology based on synergism between hydrodynamic cavitation and advanced oxidation processes. *Ultrasonics Sonochemistry* 25: 60–69.

82 Niemczewski, B. (1980). A comparison of ultrasonic cavitation intensity in liquids. *Ultrasonics* 18 (3): 107–110.

83 Yi, C., Lu, Q., Wang, Y. et al. (2018). Degradation of organic wastewater by hydrodynamic cavitation combined with acoustic cavitation. *Ultrasonics Sonochemistry* 43: 156–165.

84 Jawale, R.H., Gogate, P.R., and Pandit, A.B. (2014). Treatment of cyanide containing wastewater using cavitation based approach. *Ultrasonics Sonochemistry* 21 (4): 1392–1399.

85 Bagal, M.V. and Gogate, P.R. (2013). Degradation of 2,4-dinitrophenol using a combination of hydrodynamic cavitation, chemical and advanced oxidation processes. *Ultrasonics Sonochemistry* 20 (5): 1226–1235.
86 Gogate, P.R. and Bhosale, G.S. (2013). Comparison of effectiveness of acoustic and hydrodynamic cavitation in combined treatment schemes for degradation of dye wastewaters. *Chemical Engineering and Processing: Process Intensification* 71: 59–69.
87 Braeutigam, P., Franke, M., Schneider, R.J. et al. (2012). Degradation of carbamazepine in environmentally relevant concentrations in water by Hydrodynamic-Acoustic-Cavitation (HAC). *Water Research* 46 (7): 2469–2477.
88 Wang, X. and Zhang, Y. (2009). Degradation of alachlor in aqueous solution by using hydrodynamic cavitation. *Journal of Hazardous Materials* 161 (1): 202–207.
89 Lu, G., Zhao, L., Zhu, M. et al. (2019). Effect of cavitation hydrodynamic parameters on bisphenol a removal. *Environmental Engineering Science* 36 (8): 873–882.
90 Rooze, J., Rebrov, E.V., Schouten, J.C., and Keurentjes, J.T.F. (2013). Dissolved gas and ultrasonic cavitation – a review. *Ultrasonics Sonochemistry* 20 (1): 1–11.
91 Mason, T.J. and Lorimer, J.P. (1988). *Sonochemistry*. Wiley.
92 Berlan, J., Trabelsi, F., Delmas, H. et al. (1994). Oxidative degradation of phenol in aqueous media using ultrasound. *Ultrasonics Sonochemistry* 1 (2): S97–S102.
93 Uddin, M.H. and Hayashi, S. (2009). Effects of dissolved gases and pH on sonolysis of 2,4-dichlorophenol. *Journal of Hazardous Materials* 170 (2–3): 1273–1276.
94 Zhang, H., Zhang, J., Zhang, C. et al. (2009). Degradation of CI Acid Orange 7 by the advanced Fenton process in combination with ultrasonic irradiation. *Ultrasonics Sonochemistry* 16 (3): 325–330.
95 Eren, Z. (2012). Ultrasound as a basic and auxiliary process for dye remediation: a review. *Journal of Environmental Management* 104: 127–141.
96 Harada, H. and Kumagai, H. (2003). Effect of dissolved oxygen in water on H_2O_2 production rate by sonication. *Japanese Journal of Applied Physics* 42(Part 1, No. 5B): 2958–2962.
97 Kidak, R. and Ince, N.H. (2006). Ultrasonic destruction of phenol and substituted phenols: a review of current research. *Ultrasonics Sonochemistry* 13 (3): 195–199.
98 Gireesan, S. and Pandit, A.B. (2017). Modeling the effect of carbon-dioxide gas on cavitation. *Ultrasonics Sonochemistry* 34: 721–728.
99 Wu, C.D., Zhang, Z.L., Wu, Y. et al. (2015). Effects of operating parameters and additives on degradation of phenol in water by the combination of H_2O_2 and hydrodynamic cavitation. *Desalination and Water Treatment* 53 (2): 462–468.
100 Suslick, K.S., Mdleleni, M.M., and Ries, J.T. (1997). Chemistry induced by hydrodynamic cavitation. *Journal of the American Chemical Society* 119 (39): 9303–9304.
101 Yang, G., Chen, H., Qin, H. et al. (2017). Effect of nitrogen doping on the catalytic activity of activated carbon and distribution of oxidation products in

catalytic wet oxidation of phenol. *The Canadian Journal of Chemical Engineering* 95 (8): 1518–1525.

102 Ansari, K.B., Loke, N.H., Pandit, A.B. et al. (2015). Process intensification of upgradation of crude oil and vacuum residue by hydrodynamic cavitation and microwave irradiation. *Indian Chemical Engineer* 57 (3–4): 256–281.

103 Pandur, Ž., Dogsa, I., Dular, M., and Stopar, D. (2020). Liposome destruction by hydrodynamic cavitation in comparison to chemical, physical and mechanical treatments. *Ultrasonics Sonochemistry* 61: 104826.

104 Chitsaz, H., Omidkhah, M., Ghobadian, B., and Ardjmand, M. (2018). Optimization of hydrodynamic cavitation process of biodiesel production by response surface methodology. *Journal of Environmental Chemical Engineering* 6 (2): 2262–2268.

105 Mansouri, L., Tizaoui, C., Geissen, S.-U., and Bousselmi, L. (2019). A comparative study on ozone, hydrogen peroxide and UV based advanced oxidation processes for efficient removal of diethyl phthalate in water. *Journal of Hazardous Materials* 363: 401–411.

106 Liu, Z., Hosseinzadeh, S., Wardenier, N. et al. (2019). Combining ozone with UV and H_2O_2 for the degradation of micropollutants from different origins: lab-scale analysis and optimization. *Environmental Technology* 40 (28): 3773–3782.

107 Bavasso, I., Poggi, C., and Petrucci, E. (2020). Enhanced degradation of paracetamol by combining UV with electrogenerated hydrogen peroxide and ozone. *Journal of Water Process Engineering* 34: 101102.

108 Xu, M., Deng, J., Cai, A. et al. (2020). Comparison of UVC and UVC/persulfate processes for tetracycline removal in water. *Chemical Engineering Journal* 384: 123320.

109 Čehovin, M., Medic, A., Scheideler, J. et al. (2017). Hydrodynamic cavitation in combination with the ozone, hydrogen peroxide and the UV-based advanced oxidation processes for the removal of natural organic matter from drinking water. *Ultrasonics Sonochemistry* 37: 394–404.

110 Thanekar, P., Lakshmi, N.J., Shah, M. et al. (2020). Degradation of dimethoate using combined approaches based on hydrodynamic cavitation and advanced oxidation processes. *Process Safety and Environmental Protection* 143: 222–230.

111 Jiang, B., Zheng, J., Qiu, S. et al. (2014). Review on electrical discharge plasma technology for wastewater remediation. *Chemical Engineering Journal* 236: 348–368.

112 Abramov, V.O., Abramova, A.V., Cravotto, G. et al. (2021). Flow-mode water treatment under simultaneous hydrodynamic cavitation and plasma. *Ultrasonics Sonochemistry* 70: 105323.

113 Maršálek, B., Maršálková, E., Odehnalová, K. et al. (2020). Removal of microcystis aeruginosa through the combined effect of plasma discharge and hydrodynamic cavitation. *Water* 12 (1): 8.

114 Patil, P.N. and Gogate, P.R. (2015). Degradation of dichlorvos using hybrid advanced oxidation processes based on ultrasound. *Journal of Water Process Engineering* 8: e58–e65.

115 Kumar, M.S., Sonawane, S.H., Bhanvase, B.A., and Bethi, B. (2018). Treatment of ternary dye wastewater by hydrodynamic cavitation combined with other advanced oxidation processes (AOP's). *Journal of Water Process Engineering* 23: 250–256.

116 Titchou, F.E., Zazou, H., Afanga, H. et al. (2021). Removal of organic pollutants from wastewater by advanced oxidation processes and its combination with membrane processes. *Chemical Engineering and Processing: Process Intensification* 169: 108631.

117 Choi, J., Cui, M., Lee, Y. et al. (2019). Hybrid reactor based on hydrodynamic cavitation, ozonation, and persulfate oxidation for oxalic acid decomposition during rare-earth extraction processes. *Ultrasonics Sonochemistry* 52: 326–335.

118 Neyens, E. and Baeyens, J. (2003). A review of classic Fenton's peroxidation as an advanced oxidation technique. *Journal of Hazardous Materials* 98 (1–3): 33–50.

119 Ghaly, M.Y., Jamil, T.S., El-Seesy, I.E. et al. (2011). Treatment of highly polluted paper mill wastewater by solar photocatalytic oxidation with synthesized nano TiO_2. *Chemical Engineering Journal* 168 (1): 446–454.

120 Long, Z., Li, Q., Wei, T. et al. (2020). Historical development and prospects of photocatalysts for pollutant removal in water. *Journal of Hazardous Materials* 395: 122599.

121 Loeb, S.K., Alvarez, P.J.J., Brame, J.A. et al. (2019). The technology horizon for photocatalytic water treatment: sunrise or sunset? *Environmental Science & Technology* 53 (6): 2937–2947.

122 Manassero, A., Satuf, M.L., and Alfano, O.M. (2017). Photocatalytic reactors with suspended and immobilized TiO_2: comparative efficiency evaluation. *Chemical Engineering Journal* 326: 29–36.

123 Çalışkan, Y., Yatmaz, H.C., and Bektaş, N. (2017). Photocatalytic oxidation of high concentrated dye solutions enhanced by hydrodynamic cavitation in a pilot reactor. *Process Safety and Environmental Protection* 111: 428–438.

8

Pre-treatment of Biomass for Enhancing Biofuel Yields

8.1 Introduction

The growing global population and the unbounded desire for material and life comforts have resulted in an increased energy demand. A significant proportion of this demand is due to requirements of electricity and transportation fuel. Currently, this demand is largely met by fossil fuels (>80%, as shown in Figure 8.1), and it is debated that fossil fuels will be depleted sooner rather than later. The undebatable and undeniable fact, however, is that the use of fossil fuels increases greenhouse gas emissions and contributes to global warming and climate change. To mitigate such adverse impacts, efforts are worldwide to shift the focus toward renewable energy that is sustainable. For instance, in 1990, renewable energy sources contributed 1.1 million ktoe (kilotonne of oil equivalent) to the global energy supply. However, the capacity increased by over 75% in 2018 [1]. Specifically, the contribution of biofuels increased by ~50% over the past three decades.

Biofuels are renewable energy sources produced from biomass feedstock. Biomass such as lignocellulosic biomass (LCB), wastewater sludge, waste vegetable oils, animal fats, and micro and macro algae can be utilized for the production of biofuels. The corresponding biofuels produced from these sources of feedstock are biogas, bioethanol, and biodiesel. Biogas and bioethanol are produced via biochemical (anaerobic fermentation) pathways, whereas biodiesel is produced via a thermocatalytic pathway. This chapter will specifically deal with the use of hydrodynamic cavitation as a biomass pre-treatment method for enhancing the yield of biogas and bioethanol. A brief discussion relating to enhancing biodiesel yields via hydrodynamic cavitation can be found in Chapter 9.

Biomass valorization to biofuels and value-added chemicals is currently of prime interest due to its potential to contribute towards net neutrality and the circular economy. Of particular interest among the available biomass feedstock is LCB such as agricultural and forest residues. The humongous quantities of this LCB range between 4 and 9 billion tons/year (equating to an energy potential of 17-82 EJ) [2]. Furthermore, when LCB is appropriately utilized to its maximum potential, fertilizers and high-value chemicals beyond biofuels could be produced. Biomass, in general, can be valorized using biochemical or thermocatalytic routes. The mode of valorization is, however, dependent on the type and availability of the

Hydrodynamic Cavitation: Devices, Design, and Applications, First Edition.
Vivek V. Ranade, Vinay M. Bhandari, Sanjay Nagarajan, Varaha P. Sarvothaman and Alister T. Simpson.

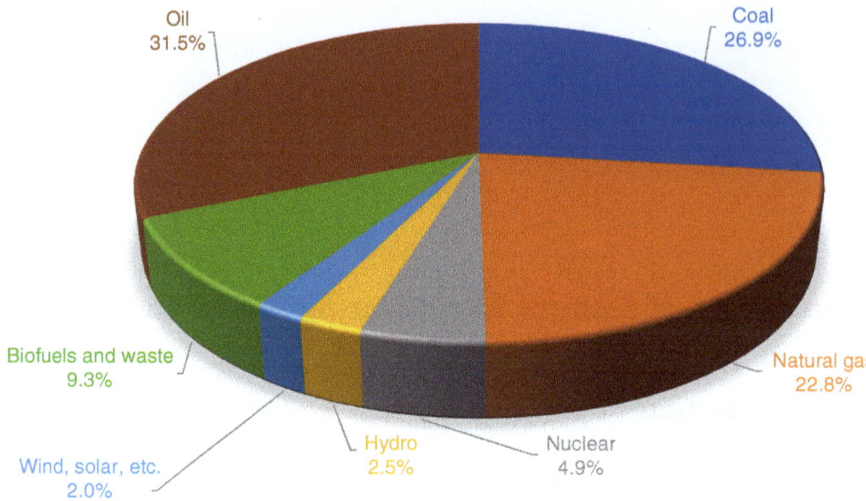

Figure 8.1 World total energy supply by source in 2018 (Source: Adapted from [1]).

biomass. In addition, the economic and social acceptance of biomass-based fuels also plays a role. Two biofuels of interest are discussed in this chapter – bioethanol and biogas – that are produced via anaerobic fermentation – sugar fermentation and anaerobic digestion, respectively. Bioethanol is a petrol/gasoline alternative and can be produced from classical pure-culture-based sugar fermentation. These fermentation processes require stringent process control with pure microbial starter cultures often used for ethanol generation. While there are microbial factors such as the growth, productivity, and ethanol tolerance that affect the bioethanol yield, there is also a techno-economic hindrance [3] that limits the fermentation process, namely the availability of highly pure sugars as feedstock for fermentation. The current practice of extensive fractionation of biomass (steam hydrolysis with acid) followed by enzymatic saccharification leads to the production of the required sugars [4, 5]. However, it may also generate toxins and growth inhibitors that adversely affect fermentation if not detoxified. Furthermore, the increase in the number of unit operations also increases the cost of bioethanol. For instance, Yang and Wyman [6] pointed out that up to 40% of the operating costs for bioethanol production is attributed to the pre-treatment and enzymatic saccharification steps which convert the cellulose in LCB to the monomeric sugar – glucose – required for bioethanol production. Cellulose in LCB is tightly bound to hemicellulose and lignin molecules, thus making it recalcitrant for bioprocessing without any pre-treatment [7]. Thus, development of an appropriate and effective pre-treatment process is important to realize a sustainable bioethanol production process.

While LCB is the primary feedstock for bioethanol production, biogas production can utilize any organic matter as feedstock. Biogas is primarily a mixture of methane and carbon dioxide, and when upgraded to a methane content of >95%, yields compressed biogas (CBG) which is similar to compressed natural gas (CNG). Biogas is produced in the absence of oxygen by the action of an anaerobic microbial consortia (bacteria and archaea) via a process known as anaerobic digestion (AD) – one of the world's oldest and most mature technologies. AD was conventionally used to treat wastewater sludge in the secondary wastewater treatment stage to reduce the sludge quantity and generate biogas. However, due to the microbial synergy that exists between various bacteria and archaea in AD, its potential to utilize more complex organic feedstock such as LCB was exploited. Furthermore, AD has the capability to produce energy throughout the year, unlike other seasonal renewable energy technologies based on solar and wind power. The main drawback of AD, however, is its high residence times ($\sim 10^1$ days) when digesting complex feedstock. Pre-treatment of biomass is needed to improve the rate of digestion, reduce the residence time, and maximize the biogas yields. When sludge is used as feedstock for biogas generation, a pre-treatment method such as thermal hydrolysis is often employed to disperse and solubilize the sludge to enhance the biogas generation [8–10]. In the case of LCB, opening the matrix to make the components more bioavailable is crucial for improving the biogas yield [11].

Pre-treatment is broadly classified under four categories, namely, physical, chemical, biological, and physicochemical [12, 13]. A generic overview of the types of pre-treatment is shown in Figure 8.2. Physical pre-treatments focus on reducing the particle size of the biomass and modifying its surface morphology, thereby improving its specific surface area and affecting the cellulose crystallinity. This is often achieved by milling the feedstock. The specific milling energy is dependent on the moisture content and the initial and final particle sizes of the biomass. The specific milling energy is proportional to the comminution ratio (ratio of the final particle size to the initial particle size) and the moisture content. Haque and Somerville reported that the energy required for drying is in the range of 1.4–2.9 MJ/kg total solids [14]. Whereas for milling, it has been reported that the energy requirement could be in the range of 0.1–2.9 MJ/kg total solids [15, 16]. Although the advantages of reduced crystallinity, higher specific surface area, and no inhibitor formation favor effective saccharification and biogas generation, the high energy requirements limit its use. Chemical pre-treatment, on the other hand, uses chemicals such as acid or alkali to partially hydrolyze the biomass. Inhibitors are often generated during chemical pre-treatment of biomass, hence necessitating a detoxification and sugar purification step. Biological pre-treatment exploits the potential of microbes or highly specific enzymes for pre-treating biomass. This includes lignolytic white or brown rot fungi or other cellulolytic microbes. High residence times are a typical requirement for biological pre-treatment and the cost of enzyme recovery and reuse adds to the downside further, making it expensive. Physicochemical methods, by contrast, combine the consequential effects of both physical and chemical pre-treatment and are comparatively less energy-intensive than milling. A commonly used physicochemical pre-treatment method is steam explosion, which consumes 70% less energy than comminution [12]. Hydrodynamic cavitation also falls under this category and is an emerging and promising method for biomass pre-treatment, with an energy consumption in the range of 0.5–2.4 MJ/kg total solids [17–19]. Hydrodynamic cavitation can be used either exclusively or in conjunction with chemicals where intensification of pre-treatment

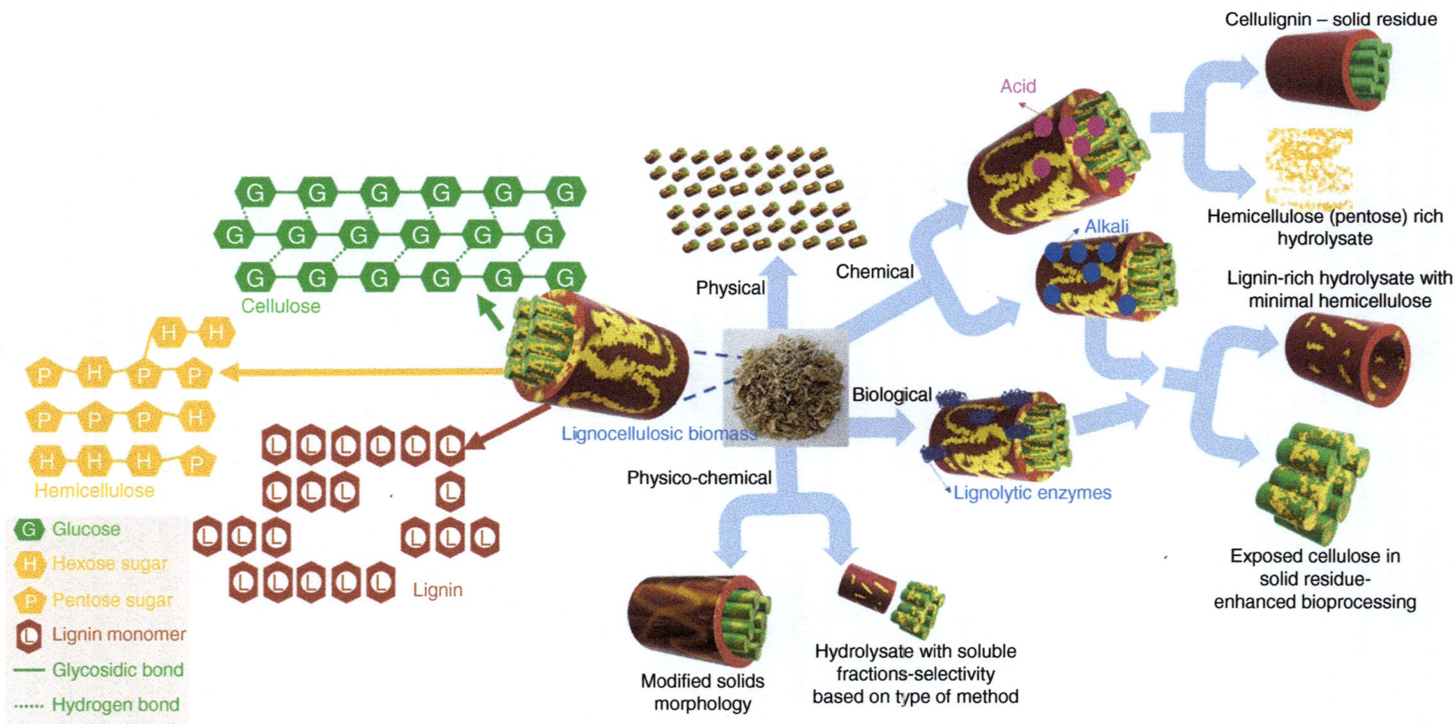

Figure 8.2 Lignocellulosic biomass pre-treatment methods Source: Reprinted from [13] with permission from The Royal Society of Chemistry.

at minimal chemical load (unlike exclusive chemical pre-treatment) is possible. The vaporous cavities generated during hydrodynamic cavitation, upon collapse, generate high-speed jets and intense shear that pre-treat the biomass physically by modifying the particle surface morphology, reducing particle size, and opening up the LCB matrix [19]. The reactive oxygen species, predominantly OH radicals that are generated in situ from water, can partially hydrolyze and solubilize a fraction of the biomass. These physicochemical modifications of the biomass lead to its enhanced utilization and maximized valorization.

Cellulose is the primary component of interest in LCB because of its ability to be bioprocessed to produce a variety of compounds, including fuels. In the case of bioethanol production, microbes ferment glucose rather than cellulose. The theoretical yield of ethanol from glucose corresponds to 0.51 kg ethanol/kg glucose from stoichiometry. Considering the consumption of glucose for microbial growth and cellular functions, the achievable yield could be ~95% of the theoretical yield [20]. Xylose, one of the predominant components in hemicellulose, has also been used for ethanol fermentation with a similar theoretical yield, but a lower achievable yield of 80%. Therefore, the goal of the hydrodynamic cavitation pre-treatment process in these cases would be to make the hemicellulose and cellulose fractions in the LCB more accessible for enzymatic hydrolysis and thereby improve the monomer yields – corresponding to higher alcohol yields. This could be achieved by focusing on improving delignification via hydrodynamic cavitation to retain the polysaccharides of interest in the solid mass [21].

In the case of biogas generation, unlike ethanol fermentation, consortia of microbes are involved in AD. Therefore, the bioavailability of the cellulose and hemicellulose fractions is crucial for improving the yield rather than producing its monomers. An improved surface area and a disrupted biomass morphology will be sufficient to enhance the bioavailability of the biomass. This will ensure that the hydrolytic bacteria in the anaerobic microbial consortia can act upon the biomass to produce subsequent intermediates required for faster biogas generation. For instance, sugarcane bagasse has a theoretical biomethane yield of 0.44 m^3/kg volatile solids [13]. However, this can never be achieved as all the organics in bagasse cannot be utilized for biogas generation because it contains recalcitrant lignin. The achievable biomethane yield would therefore be ~0.30 m^3/kg volatile solids, considering the cellulose and hemicellulose fractions. Therefore, the role of hydrodynamic cavitation in this case would be to improve the bioaccessibility of these polysaccharides.

This chapter specifically focuses on the application of hydrodynamic cavitation for biomass pre-treatment and on enhancing biofuel yields. Enhanced bioprocessing leading to higher yields of the two end products of interest – bioethanol and biogas – are discussed here. A review of the existing literature on the hydrodynamic cavitation pre-treatment of LCB for bioethanol production is first presented (Section 8.2), followed by a discussion on biogas generation (Section 8.3). Conventional wastewater sludge pre-treatment via hydrodynamic cavitation for biogas production is discussed in Section 8.3.1, in addition to LCB pre-treatment for enhanced biogas production (Section 8.3.2). In addition, the limited commercial data available on biogas generation on hydrodynamic cavitation and the net energy gains from biofuel production as a result of hydrodynamic cavitation pre-treatment are also presented.

8.2 Hydrodynamic Cavitation for Enhancing Bioethanol Yield

Bioethanol is one of the first liquid biofuels to have been commercialized. It is primarily used as a transportation fuel, either exclusively or blended with petrol in different ratios (e.g. E5, E10, or E85). Since bioethanol is an oxygenated fuel, it does not emit particulates on combustion [22]. Compared to petrol, bioethanol has a higher octane number, flash point, and auto ignition temperature, thus avoiding engine knocking and other ignition-related safety concerns [23, 24]. Conventionally, bioethanol was produced from starchy crops such as corn and cassava or sugarcane juice. This was termed as first-generation bioethanol production where a food crop was the source for the production of bioethanol. Ethical debate on the usage of food crops for fuel production and utilizing land suitable for the growth of food crops to cultivate energy crops led to second-generation bioethanol production. In this framework, the feedstock used for bioethanol production comprises waste organic materials such as agricultural residues, forest residues, municipal solid waste, etc. While the second-generation feedstock does not compete with food, is cheap, and readily available, its recalcitrance poses a significant challenge. Due to the interlinked lignin and holocellulose, cellulose is not readily available for enzymatic saccharification prior to fermentation. To enhance the saccharification yields and thereby the bioethanol yield, pre-treatment is essential.

A range of methods have been developed and used for pre-treating second-generation feedstock, and several studies have been published (see for example, [25, 26] and references cited therein). Among the existing methods, steam explosion – a physicochemical method – is currently the industrial state of art [27–29]. Steam explosion is favorable for solubilizing the hemicellulose fraction of the biomass effectively, leaving behind the cellulose exposed with some lignin. The increased porosity and surface area of the biomass residue upon pre-treatment favors enhanced enzymatic saccharification. However, compounds such as HMF and other furans are produced during pre-treatment and are inhibitory towards fermentation. Therefore, an additional detoxification step after the pre-treatment is required.

Hydrodynamic cavitation is classified under physicochemical pre-treatments which can be used either in a standalone mode or in conjunction with acid or alkali. Unlike steam explosion, hydrodynamic cavitation offers several advantages such as no inhibitor formation, lower energy requirement, and scalability. These factors have been harnessed to pre-treat biomass for enhanced sugar and bioethanol yields. A summary of the existing literature using hydrodynamic cavitation for enhancing bioethanol/sugar yields is presented in Table 8.1.

One of the earliest reports on the use of hydrodynamic cavitation for biomass pre-treatment to enhance bioethanol production was a patent by Sprague [36].

Table 8.1 A summary of studies reporting hydrodynamic cavitation pre-treatment of lignocellulosic biomass for improving saccharification and/or bioethanol yield.

Feedstock	Particle size, mm	Biomass loading, g/L (volume, L)	Hydrodynamic cavitation device	Optimum conditions	Additive	Calculated energy consumption, MJ/kg biomass	Saccharification or bioethanol yield	Lignin removal	Reference
Wood	<0.6	50 (5)	Venturi	5 bars, 144 passes	50 g/L NaOH	1.5	—	49%	[21]
Reed	~10	118 (0.15)	Multi-hole orifice	5 bars, 738 passes, 77 °C	30 g/L NaOH	3.7	5-fold increase in sugar yield, 90% theoretical ethanol yield	42%	[30]
Corn Stover	<0.25	40 (0.4)	Venturi (1.8 mm throat)	2.2 bars, 480 passes, 30 °C	42.4 g/L Na_2CO_3, 20.4 g/L H_2O_2	2.7	2.2-fold higher sugar yield	—	[31]
			Venturi (1.4 mm throat)	3.4 bars, 360 passes, 30 °C		3.2	2.8-fold higher sugar yield		
Sugarcane bagasse (SCB)	<4.7	50 (2.5)	Multi-hole orifice	3 bars, 1000 passes, 70 °C	12 g/L NaOH	6.1	6-fold increase in saccharification yield	51%	[32]
SCB	<1.2	20 (3)	Multi-hole orifice	3 bars, 278 passes, 60 °C	11.6 g/L NaOH, 11.3 g/L H_2O_2	6.1	6-fold increase in saccharification yield	—	[33]
SCB	<0.25	30 (0.8)	Venturi	6.9 bar, 454 passes, 65 °C	50 g/L NaOH	10.6	—	56%	[34]
SCB	2	10 (55)	Venturi	2.8 bar, 172 passes, 22 °C	1 g/L $Ca(OH)_2$	4.9	4.8-fold increase in saccharification yield	—	[35]

In his proposed bioethanol production process, hydrodynamic cavitation was used in as many as three instances: (i) after dry milling, (ii) upon cooking the first hydrodynamic cavitation-treated biomass (grain), and (iii) post treating the liquefied grain slurry. This sequence was reported to enhance the bioethanol production without burdening the overall production costs. Interestingly, in addition to the conventional hydrodynamic cavitation devices, a novel vortex-based device incorporating a cyclonette and a vortex finder was also used for pre-treatment. It was claimed that the process was effective when operated at inlet pressures of ~2 bar. Similar to Sprague, an orifice-based device was used in the same three instances in the bioethanol production process and patented by Kozyuk [37]. He proposed that the saccharification of the feedstock is dependent on the particle size, which is a function of the inlet pressure of the slurry and the number of passes through the device. A 2–5% increase in bioethanol as a result of hydrodynamic cavitation pre-treatment was observed by Kozyuk. In 2010, Kamfe reported the use of hydrodynamic cavitation as one of the pre-treatment methods to improve the bioethanol yield from wild algae and cultured *Spirulina* [38]. A rotor-stator device, commercially known as the shockwave power reactor (SPR) and developed by Hydro Dynamics Inc. was used for this purpose. The device was operated at 3600 rpm in the inlet pressure range of 1.3–6.2 bar. No significant increase in saccharification or bioethanol yields was observed upon SPR hydrodynamic cavitation. However, it was argued that the results obtained were due to the use of a non-optimized hydrodynamic cavitation system for pre-treatment.

Initially, LCB pre-treatment using hydrodynamic cavitation in combination with chemicals (predominantly alkali) was performed for delignification [21, 39] and improving the biodegradability index [40]. One of the first peer-reviewed publications on the influence of hydrodynamic cavitation on bioethanol production was published only in 2015, by Kim et al. [30]. They used a multi-hole orifice plate (27 holes), with the diameter of each hole being 1 mm for pre-treatment. A 150 g reed of ~10 mm was wrapped in a wire cloth and placed downstream of the orifice. This zone was termed the "cavitation zone" where the bubble collapse occurred. Then, 3% NaOH was circulated through the loop for 40 minutes at a pressure drop across the device of ~5 bar. A solids recovery of 76% was reported after hydrodynamic cavitation treatment with 42% delignification. The corresponding sugar yield after hydrodynamic cavitation was found to be 4.3-fold higher than the untreated sample. The bioethanol yield from the pre-treated reed reached 90% of the theoretical yield of 0.46 g ethanol/g glucose. The calculated energy consumption of 3.65 MJ/kg biomass was multiple folds lower than the corresponding acoustic cavitation pre-treatment (14.4 MJ/kg biomass) for similar enhanced sugar yields. The comparison reinstated that hydrodynamic cavitation is energy efficient and scalable compared to acoustic cavitation.

Nakashima et al. performed hydrodynamic cavitation of finely milled corn stover (<0.25 mm particle size) to determine its effect on saccharification yield [31]. They used venturis of two different throat diameters – 1.8 and 1.4 mm – to perform the pre-treatment. They targeted delignification of the feedstock as a means to enhance the enzymatic hydrolysis yield and therefore performed hydrodynamic cavitation in sodium percarbonate solution. Delignification was confirmed via Fourier transform infrared (FTIR) Spectroscopy with the sugar yield increasing by 2.2-fold with the venturi hosting a 1.8 mm throat and 2.8-fold for the venturi with a smaller throat (1.4 mm). A higher pressure gradient and the intense cavitation offered by the 1.4-mm throat venturi were ascribed to be the reasons for the higher saccharification yield.

Terán Hilares et al. performed extensive work on utilizing multi-hole orifices for the pre-treatment of sugarcane bagasse (SCB) for enhancing the saccharification and bioethanol yields [32, 33, 41–44]. They used a similar configuration as Kim et al. [30], where the biomass was placed in a wire cloth in the "cavitation zone" and alkali solution was recirculated through the orifice at batch-cavitating conditions in a recirculating loop. In a continuous pre-treatment mode, a dilute slurry of the biomass in aqueous alkali was used [44]. Multiparameter optimization was performed in their series of experiments that included the effect of solid loading, pressure drop across the hydrodynamic cavitation device, alkali concentration, peroxide concentration, and time required for pre-treatment (number of passes through cavitation device) [32, 33, 42, 44]. The sugar yield increased proportionally with the number of holes in the orifice. The other optimum conditions achieved with their configuration, with which they were able to achieve ~sixfold increase in saccharification yield, are shown in Table 8.1. Bimestre et al.'s recent work [34] on using venturi to pre-treat SCB also yielded results similar to Terán Hilares et al.'s work. Terán Hilares et al. also performed hydrodynamic cavitation of SCB using a range of alkalis such as Na_2CO_3, KOH, NaOH, and $Ca(OH)_2$ at different concentrations, and determined that NaOH was the most effective in delignifying SCB and improving the sugar yields [45].

Madison et al. investigated the use of $Ca(OH)_2$ as the choice of alkali for aiding delignification after hydrodynamic cavitation [35] in a sequential process. Lime is known to be an effective delignification agent [46]. However, its low aqueous solubility limits its use and therefore affects the efficiency and subsequent saccharification yield. Despite the limitation, following the sequential treatment, a comparable enhancement in saccharification yield was observed (4.8-fold increase, Table 8.1). However, the extent of delignification was not reported. From an industrial perspective, it would be cheaper to utilize lime (£0.6/kg) rather than NaOH (£1.5/kg) (www.mistralni.co.uk), and therefore it could be further investigated as a preferred choice of alkali. Though the same feedstock (SCB) and alkali were used to investigate the extent of saccharification in this work as in Terán Hilares et al. [45], the type of pre-treatment (alkali + hydrodynamic cavitation and hydrodynamic cavitation followed by alkali, respectively) and the hydrodynamic cavitation device may have influenced the difference in results obtained.

Lab-scale batch hydrodynamic cavitation experiments were often reported for enhancing the saccharification or bioethanol yield from biomass. Ramirez-Cadavid et al., however, reported on the use of continuous hydrodynamic cavitation processing at pilot and industrial scale for enhancing the bioethanol production [47, 48]. They performed pilot-scale studies with an orifice plate assembly as the hydrodynamic cavitation device and treated a total volume of 379 L at a milled corn (<2.74 mm) loading of 30.9% wt/wt. They performed hydrodynamic cavitation pre-treatment of the slurry at three different energy inputs of 313, 940, and 1567 J/kg slurry, corresponding to inflow rates of 26.5, 37.8, and 56.8 L/min with the operating pressures in the range of 3.4–20 bars [49]. The specific energy input greater than 940 J/kg slurry was found to enhance the bioethanol yield by 2.3% and was mainly attributed to the reduction in particle size of the corn meal on hydrodynamic cavitation. This also corresponded to a net energy gain as calculated by the authors [47]. In the industrial trial, with a plant production capacity of 379 million L bioethanol per annum, hydrodynamic cavitation pre-treatment of the slurry was performed at an inflow rate of 7192 L/min corresponding to a specific energy input of 1504 J/kg slurry. A similar 2% increase in bioethanol was also reported at an industrial scale. With the running costs of the industrial hydrodynamic cavitation unit contributing to ~2.6% OPEX, a net profit was forecasted. Further optimization and enhancement in bioethanol yield of up to 4% was achieved in their lab-scale optimization study, where, in addition to hydrodynamic cavitation, the cellulase loading was also optimized. Figure 8.3 shows the pilot- and industrial-scale configurations used by Ramirez-Cadavid et al. [48].

Most reports focused on using NaOH as the preferred additive during hydrodynamic cavitation pre-treatment of LCB as it is effective in removing lignin, thereby improving the bioavailability of cellulose. Correlating and comparing the existing reports on the basis of delignification, it can be seen that as the specific energy consumption increases, the lignin removal also increases gradually (Figure 8.4). The same figure also represents the enhancement in saccharification yields from hydrodynamic cavitation-treated LCB. As seen, similar to the improvement in the percentage of delignification, an increase in saccharification yield is also observed with increase in specific energy. This is only a vague correlation and one still needs to consider the effect of initial particle size, type of biomass (influencing the type and content of lignin), and type and concentration of alkali used, before developing appropriate correlations.

From an energy consumption perspective, hydrodynamic cavitation is a promising pre-treatment method. In all the reported methods, however, milled biomass is used. The energy requirement for drying and milling, as mentioned earlier, must be considered to ensure the feasibility of the process at scale. In general, hydrodynamic cavitation pre-treatment of milled LCB alone is not sufficient to enhance the saccharification yield. This is because delignification should occur for the enzymes to access the cellulose in LCB effectively. A longer pre-treatment time (higher number of passes) would be needed to achieve this without the addition of chemicals. For instance, hydrodynamic cavitation pre-treatment of milled SCB alone improved the sugar yields by 20%, whereas combination with alkali improved the yield by ~fivefold [35]. Therefore, it is necessary to investigate hybrid hydrodynamic cavitation (with an additive) as a choice of pre-treatment for enhancing saccharification or bioethanol yields. The effect of initial particle size on the efficiency of hydrodynamic cavitation pre-treatment should also be established along with the hydrodynamic cavitation operating parameters, as they will depend on the feedstock itself and the extent of delignification required.

Figure 8.3 (a) Pilot-scale hydrodynamic cavitation unit consisting of (A) the cavitation device assembly, (B) holding tank, (C and D) sample ports and (E) pump (Source: [47], reproduced with permission from Elsevier). (b) Industrial-scale hydrodynamic cavitation unit used for pre-treating corn meal for enhanced bioethanol production. Source: Reprinted from [48], with permission from Springer Nature).

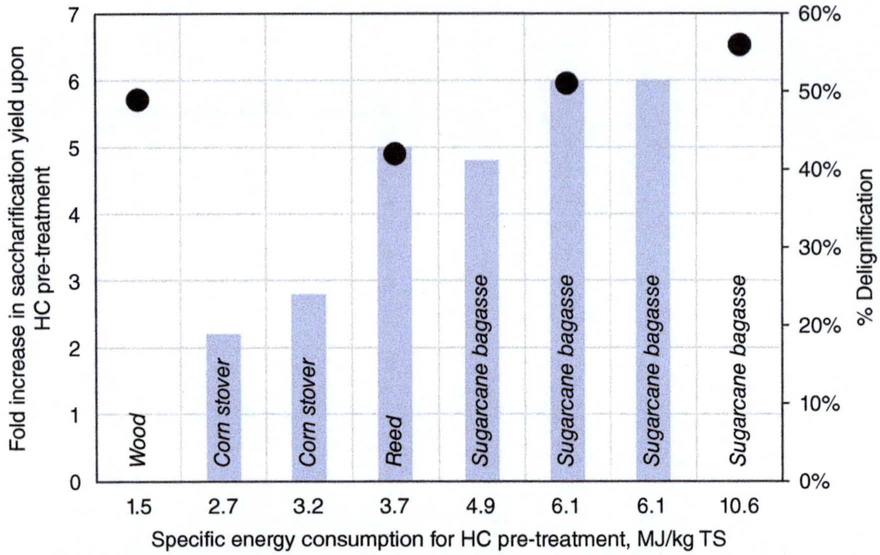

Figure 8.4 Effect of specific energy consumption on percentage of delignification (dots) and fold increase in saccharification yield (bars), where unavailable, data not included in the plot (Source: Data for figure taken from Table 8.1).

8.3 Hydrodynamic Cavitation for Enhancing Biogas Production

AD is one of the oldest known bioprocesses used to treat waste and as to produce renewable energy in the form of biogas. Upgradation of biogas yields CBG, which is an excellent fuel that can be used for heating, electricity generation, or as transportation fuel. CH_4 is a more potent greenhouse gas than CO_2. Although the combustion of CH_4 releases CO_2, it is still considered sustainable (biogenic CO_2) because the emissions are considered to be absorbed by natural carbon sinks such as forests, whose waste ends up as feedstock for biogas generation eventually. This circular framework of carbon recycling generates no net emissions. Hence, AD is considered a carbon-neutral process [50].

AD is a biochemically complex process where various classes of anaerobic bacteria and archaea work in synergy to breakdown organic matter to produce biogas. The four broad steps involved in biogas production via AD, namely:

(i) Hydrolysis – breakdown of biomass and solubilization of polymers to monomers.
(ii) Acidogenesis – volatile fatty acid, H_2, and CO_2 production from soluble organics.
(iii) Acetogenesis – production of acetic acid from volatile fatty acids.
(iv) Methanogenesis – Biogas production from acetic acid (acetoclastic methanogens) and H_2 and CO_2 (hydrogenotrophic methanogens).

Methanogenesis is typically the rate-limiting step, as the growth and metabolic rate of the methanogenic archaea are comparatively slower than the other bacteria in synergy. Hydrolysis may also become the rate-limiting step when complex LCB is used as feedstock. When complex biomass such as LCB is utilized for biogas production, the residence times required for digestion are often high, in the order of a few weeks. Therefore, from an industrial standpoint, it is crucial to lower the residence time without compromising on the biogas yields. Furthermore, a low productivity of typically 1 m^3 of biogas generated/m^3 of digester/day should be increased to maximize the offered decarbonization potential. It is therefore paramount to intensify the AD process to improve its performance and yield. A range of factors are known to influence AD productivity and yield [51, 52]. These include feedstock type, temperature, pH, mass transfer within the digester, inoculum characteristics, and digester start-up. Among these, the feedstock type is of primary importance as it critically affects all the other parameters. AD feedstock typically comprises recalcitrant LCB such as agricultural, farm, and forest residue. The characteristic feature of LCB is its recalcitrance due to the interlinked cellulose, hemicellulose, and lignin. Varying lignin content in LCB limits the bioavailability of the polysaccharides [53, 54]. If municipal solid waste or wastewater sludge is used as feedstock, the complexity of the chemical oxygen demand (COD) in the feedstock and the amount of suspended solids present in the feedstock limit its bioavailability. Irrespective of the type of feedstock, there exists a certain degree of complexity. This complexity of the feedstock affects the rate of biogas generation. Therefore, intensification strategies are required to enhance AD productivity. Several performance enhancement strategies to alleviate feedstock-induced recalcitrance have been reported in literature [55–76], namely:

(i) optimizing the organic loading rate and retention time,
(ii) varying the C/N ratio of the net substrate added to the digester,
(iii) co-digestion,
(iv) addition of micronutrients and trace elements,
(v) digestate recycling,
(vi) inoculum enrichment,
(vii) enhancing interspecies electron transfer,
(viii) multi-stage digestion,
(ix) pre-treatment,
(x) modeling to understand, and
(xi) digester scale-up.

Among the strategies, pre-treatment, especially hydrodynamic cavitation-based pre-treatment, is discussed in this section. In addition to improving overall rates of AD, pre-treatment may also be used to increase the overall biogas yield by opening up the LCB structures or increasing the soluble organic matter content. The commonly expected outcomes of pre-treatment on biogas generation are depicted in Figure 8.5.

The theoretical biogas yield of a feedstock can be determined using Buswell's formula (Reaction R8.1) [77]. It assumes a complete conversion of feedstock to

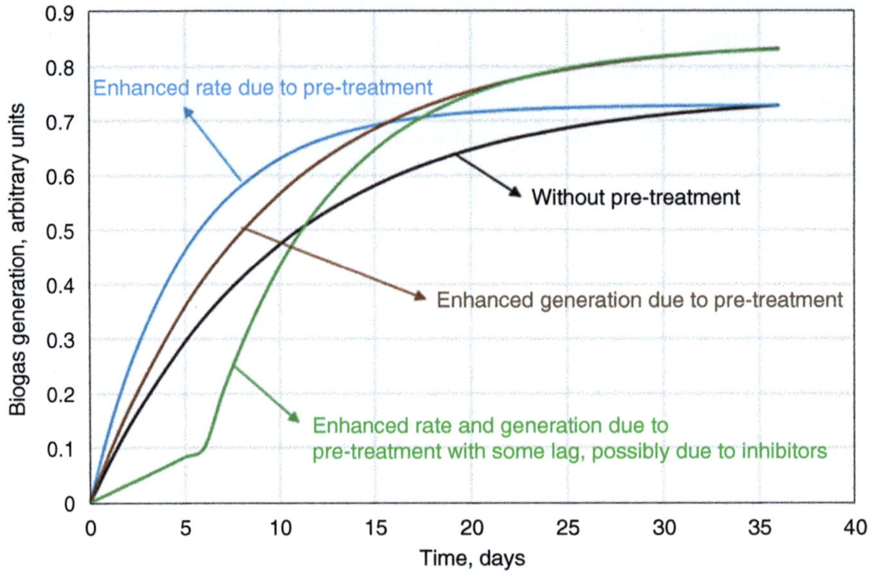

Figure 8.5 Effect of pre-treatment on biogas generation (Source: From [13]/with permission of Royal Society of Chemistry).

CH_4 and CO_2. However, this is not true, as a fraction of the biomass is recalcitrant and non-biodegradable and another fraction is utilized by the microbes for growth and metabolism [78]. It is therefore important to differentiate between the theoretical yield and the maximum achievable biogas yield. Complex biomass, when digested, ends up with sub-optimal biogas yields; pre-treatment can help in closing the gap between the sub-optimal yield and the maximum achievable yields (Figure 8.6):

$$C_xH_yO_z + \left(x - \frac{y}{4} - \frac{z}{2}\right) H_2O \rightarrow \left(\frac{x}{2} + \frac{y}{8} - \frac{z}{4}\right) CH_4 + \left(\frac{x}{2} - \frac{y}{8} + \frac{z}{4}\right) CO_2 \tag{8.1}$$

With the two categories of feedstock considered in this chapter, namely wastewater sludge and LCB, the commonly preferred pre-treatments are thermal hydrolysis [79] and physicochemical methods [78], respectively. While thermal hydrolysis of sludge uses high temperatures and pressures (~200 °C and up to 25 bars), physicochemical LCB pre-treatment methods use chemicals. These methods generate inhibitors that may induce a lag time and reduce the rate of methane production and the overall biomethane yield. Furthermore, the significant energy consumption and the requirement of heat integration and recovery options would induce an additional economic burden at industrial scale [9]. To overcome the issues posed by the state of art, hydrodynamic cavitation can be used as an alternative pre-treatment method as it is scalable and energy-efficient, and can be retrofitted. This section specifically presents a review of literature using hydrodynamic cavitation to enhance biogas generation from wastewater, sludge, and LCB.

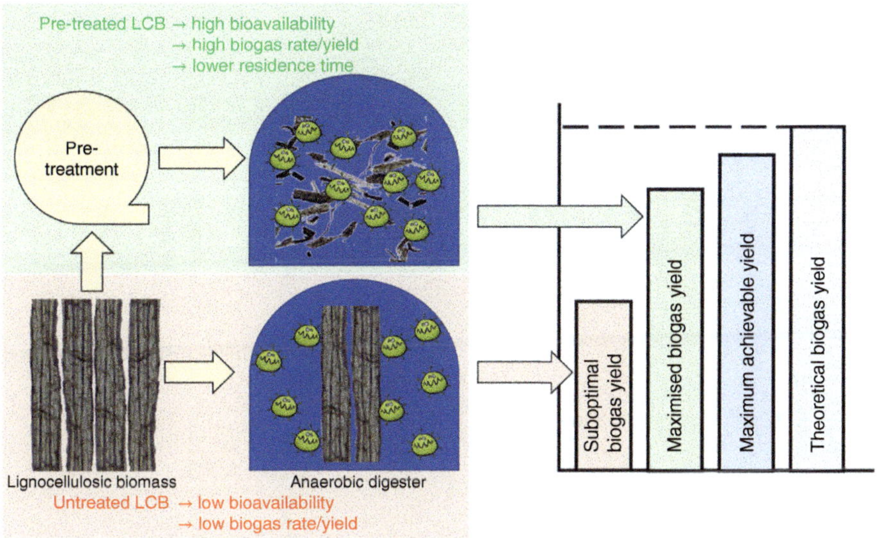

Figure 8.6 Role of pre-treatment in maximizing biogas yield and closing the gap between sub-optimal yields and the maximum achievable yields.

8.3.1 Wastewater and Sludge

First instances of sludge pre-treatment were performed using high-pressure homogenizers, where pressure gradients of up to 900 bars were reached [80, 81]. The specific energy input required for pre-treatment was high and reached almost 9 MJ/kg when these homogenizers were used. Hydrodynamic cavitation was found to consume energy an order of magnitude lower than these homogenizers and hence sparked interest [82]. A variety of studies reported the use of hydrodynamic cavitation for enhancing the biochemical methane potential (BMP) of sludge, as summarized in Table 8.2.

Preliminary studies by Nicholas and Vella [93] focused on improving the sludge solubilization by using a novel vortex-based cavitation device that generated rotational flow via two opposite flow-truncated nozzles (VRTX Technologies) [93]. Kim et al. revealed the dependence of sludge solubilization on the venturi design [94]. Grűbel and Machnicka investigated the use of a venturi as the hydrodynamic cavitation device for disintegrating activated sludge to enhance biogas generation [83]. With the specific energy consumption in the range of 5–12 MJ/kg, they were able to improve the degree of disintegration by fourfold. Increasing the degree of disintegration was supported by an increase in the amount of soluble organics (and chemical oxygen demand, COD) and thereby an increase in biogas generation. Maeng et al. then determined the effect of three venturis in series and either the suction or discharge pump mode of sludge solubilization [95]. Suction of sludge through the venturis in series was found to enhance the degree of disintegration for a similar number of passes and specific energy consumption (Figure 8.7). The lower total head loss in the suction mode was attributed to an intense cavitation in all the three venturis,

Table 8.2 A summary of literature reporting hydrodynamic cavitation for wastewater and sludge pre-treatment for enhancing biogas generation.

Feedstock	Total solids in sludge, g/L (Volume, L)	Hydrodynamic cavitation device	Additive (Concentration, g/L)	Optimum conditions	Calculated specific energy input, MJ/kg solids	AD conditions (mode, HRT, temperature)	Specific methane yield, L CH_4/kg VS	Increase in CH_4 yield	Reference
Activated sludge	(25)	Venturi	—	12 bars, 10 passes	5–12	Batch BMP, 22 days, 35 °C	-	Increased digester performance (m^3 CH_4/day/m^3 digester) by 94%	[83]
Activated sludge	9.9 (1.5)	Multi-hole orifice	—	6.3 bars, 356 passes	12	Batch BMP, 14 days, 35 °C	81.3	13%	[84]
			NaOH (200)				123.6	71%	
Activated sludge	12.5 (100)	Single-hole orifice	—	6.3 bars, 14 passes	3.5	Batch BMP, 35 days, 37 °C	162	2.5-fold	[85]
			80 V electric field applied to electrodes in tank				184	3.2-fold	
Anaerobic digestate	25 (260)	Venturi	—	2.5 barg back pressure, 100 passes	24	Batch BMP, 11 days, 36 °C	96	~60%	[86]
Thickened activated sludge	(196)	Rotor-stator	—	4.3 m^3/h, 0.8 mm clearance between rotor and stator, 20 passes	80.4[a)]	Batch BMP, 13 days, 40 °C	16.6[b)]	13% increase in biogas yield	[87]

Activated sludge	17.7 (25)	Venturi	NaOH (80)	12 bars, 10 passes	2	Batch BMP, 9 days, 35 °C for stage 1 and 15 days, 55 °C for stage 2	522[c)]	28%	[88]
Activated sludge	14 (10)	Single-hole orifice	—	6.3 bars, 125 passes	8.9	Batch BMP, 68 days, 35 °C	9.8	30%	[89]
			NaOH (0.8)				15.2	100%	
			$Ca(OH)_2$ (3.6)				11.4	50%	
			H_2O_2 (0.02)				13.6	80%	
Tannery effluent	19.7 (7)	Venturi	—	5 bars, 165 passes	6.3	Batch BMP, 60 days, 37 °C	36	2.1-fold increase in biogas	[90]
Thickened activated sludge	3.7 (10)	Rotor-stator	—	3.6 MJ/kg		Batch BMP, 27 days, 36 °C	126	2.5-fold	[91]
Distillery spent wash	162.5 (12)[d)]	Vortex-based device	—	3.9 barg, 2 passes	0.007	Batch BMP, 29 days, 41 °C	231[e)]	14%	[92]

a) MJ/m^3 sludge.
b) L biogas/L sludge.
c) L CH_4/kg VS removed, where VS is the volatile solids content.
d) industrial scale application for spent wash flow rates of 13–50 m^3/h showed an increase in biogas yield of up to 22% upon one-pass treatment.
e) indicates a net positive energy gain was obtained upon pre-treatment.

resulting in an enhanced degree of disintegration. While Maeng et al. reported the use of unconventionally large venturi throat diameters, Stoop et al. used a 5 mm venturi and performed hydrodynamic cavitation of anaerobically digested sludge [86]. The sludge was first filtered through a 4 mm mesh to avoid clogging of the venturi constriction. The effect of the number of passes (50, 75 and 100) at two different back pressures (2.9 and 8.5 barg) on BMP was investigated. No influence of back pressure on BMP could be established. However, the effect of the number of passes on BMP was distinct, as 50, 75, and 100 passes increased the methane yield by 17, 33, and 60%, respectively.

Lee and Han investigated the use of a multi-hole orifice (27 holes, 1 mm diameter) as the hydrodynamic cavitation device to pre-treat activated sludge to enhance its biogas yield [84]. Pressure drop across the device was established at 6.3 bars at a flow rate of 1.6 m^3/h. With an increase in hydrodynamic cavitation time (number of passes), the degree of disintegration was found to increase up to 23%, thereby enhancing the CH_4 yield. A 13% increase in CH_4 yield was observed after 356 passes corresponding to a specific energy input of 12 MJ/kg. They also compared the energy requirements with acoustic cavitation and determined that hydrodynamic cavitation utilized 3 times less energy for achieving the same degree of disintegration.

While linear flow devices were used for sludge disintegration and enhancing biogas generation, as mentioned earlier, rotational flow devices have also been used for this purpose. For instance, Petkovšek et al. modified a centrifugal pump to act as a rotor-stator-based hydrodynamic cavitation device for thickened activated sludge treatment [87]. The effect of clearance between the rotor and stator as well as the inflow rate and number of passes on degree of disintegration was investigated. The best case – with the clearance of 0.8 mm, flow rate of 4.3 m^3/h, and 20 passes – achieved up to 57% as the degree of disintegration and enhanced the biogas generation by 13%. Although an increase in biogas on pre-treatment was observed, a net positive gain could not be achieved. Similar to Petkovšek et al.'s study, a degree of disintegration of over 55% was observed in rotor-stator devices by other authors at much higher specific energy inputs of 167 MJ/kg TS [96, 97]. In contrast to the previous reports, Garlicka et al. achieved 4% degree of disintegration at their optimum hydrodynamic cavitation conditions with a rotor-stator device at a specific energy input of 3.6 MJ/kg TS [91]. With only a 4% degree of disintegration, they achieved a 2.5-fold increase in CH_4 yield, but without a net energy gain. At a higher energy input of 5.5 MJ/kg TS, the degree of disintegration increased to 48% but the increase in CH_4 yield fell from 2.5-fold to 53%. Re-agglomeration was ascribed as the reason for a lower enhancement in yield at a higher energy input. Another rotational flow device, the swirling jet cavitator, has also been reported for sludge disintegration [98–100]. For instance, Mancuso et al. reported the use of a swirling jet hydrodynamic cavitation device for sludge solubilization [99]. They observed that with a specific energy input of 1.6 MJ/kg TS, only 1.4% degree of disintegration could be achieved at 20 °C, whereas when the initial temperature was 50 °C at a specific energy input of 3.3 MJ/kg TS, 23.5% degree of disintegration was achieved. The higher temperature not only softens the sludge (and biomass) but also enhances extent of cavitation and thereby improves overall performance.

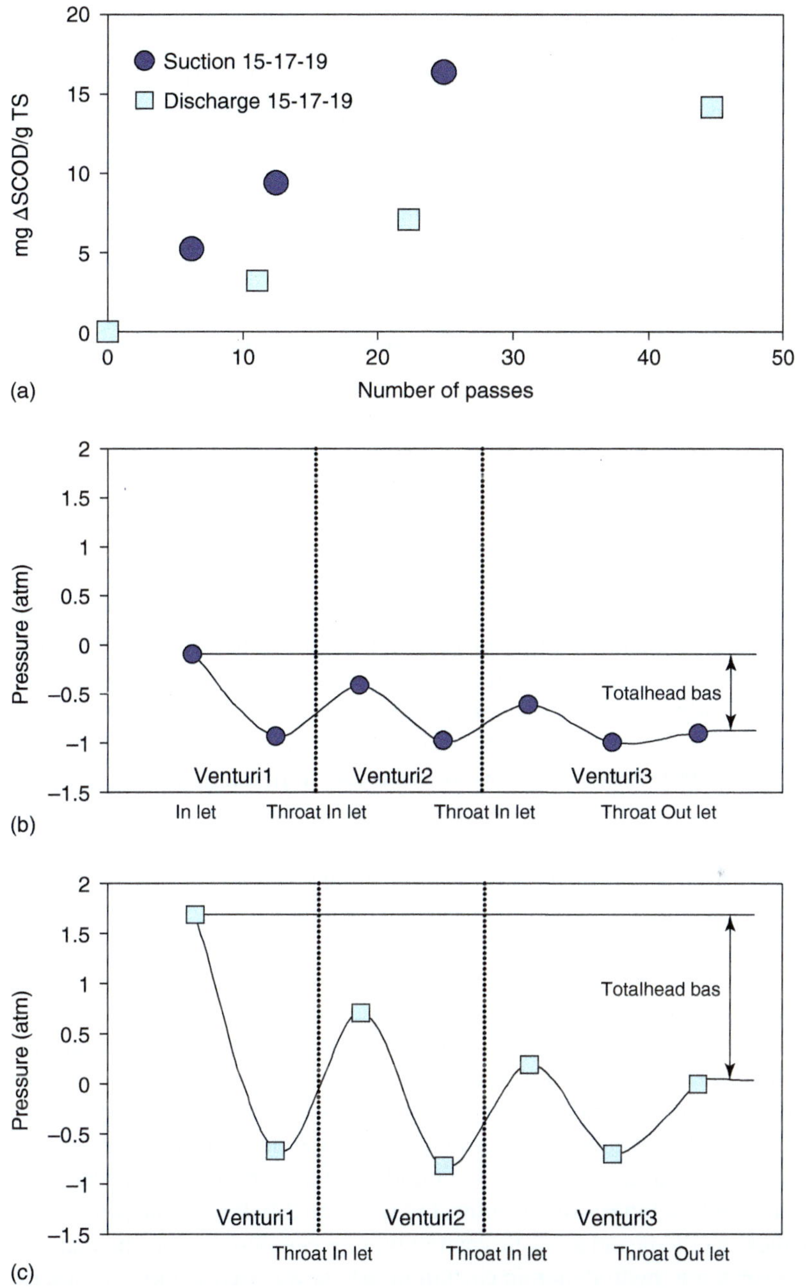

Figure 8.7 (a) Soluble COD (mg ΔSCOD) per unit weight of total solids (g TS) as a function of number of passes through the venturis in series in suction and discharge modes at a flow rate of 7.8 m^3/h, working volume of 650 L, and TS content of 28 g/L. 15-17-19 represent the throat diameters of the venturis in series as millimeters, as used by Maeng et al. [95], (b) total head loss for venturis in series operating in suction mode, and (c) total head loss for venturis in series operating in discharge mode (Source: From [95]/with permission of IWA Publishing).

Hybrid hydrodynamic cavitation (i.e. hydrodynamic cavitation in combination with other methods such as thermal, chemical, or electrical pre-treatment) has been investigated for solubilizing sludge [84, 88, 89, 99, 101, 102]. Among all the hybrid hydrodynamic cavitation methods investigated, hydrodynamic cavitation–alkali-mediated solubilization is the most frequently reported due to its ease of implementation. Lee and Han utilized 20% NaOH in combination with hydrodynamic cavitation to solubilize the sludge and increase CH_4 yield [84]. At a specific energy input of 12 MJ/kg TS, they were able to enhance the CH_4 yield by 71% compared to the untreated sludge. This was also higher than individual alkali and hydrodynamic cavitation pre-treatments by 28 and 53%, respectively. The initial cell disruption with alkali favored intracellular compound release from the biomass on hydrodynamic cavitation, thereby significantly increasing the BMP. Grübel and Suschka evaluated the effect of 8% NaOH and hydrodynamic cavitation on the BMP of sludge [88]. The pre-treated sludge was digested in series in two batch digesters – the first one operating at mesophilic conditions and the second digester at thermophilic conditions. Optimizing the residence time in each digester, an increase in biomethane yield of 28% was achieved compared to the control with pre-treated sludge. Mancuso et al. used hydrodynamic cavitation in conjunction with alkali (5 M NaOH) as well as with alkali and mild temperature (5 M NaOH and 50 °C) for sludge solubilization [99]. The degree of disintegration increased to 44.2 and 53%, respectively, when compared to hydrodynamic cavitation alone, at 1.4%. With the cost of sludge solubilization with only hydrodynamic cavitation calculated as €0.081/kg TS, alkali–hydrodynamic cavitation and alkali–thermal–hydrodynamic cavitation treatments cost twice and four times as much. In addition to synergistic pre-treatment, co-digestion along with hydrodynamic cavitation has also been utilized as a strategy to enhance the biogas generation [103–105].

Apart from wastewater sludge, effluent wastewater from industries has also been treated using hydrodynamic cavitation for producing biogas [90, 92]. For example, Saxena et al. utilized a venturi to pre-treat tannery effluent to enhance the BMP [90]. The effluent was pre-treated at a specific energy consumption of 6.3 MJ/kg and yielded a 2.1-fold increase in biogas yield. Although the yield was increased by over twofold, the absolute yield was only 36 L biogas/kg. Nagarajan and Ranade on the other hand used a novel vortex-based cavitation device to pre-treat distillery spent wash for enhancing its biogas yield with a net positive gain [92]. Spent wash is the wastewater stream produced from the distillery industry and is rich in complex COD and refractory compounds. AD is usually used to treat this wastewater to produce biogas and improve the water quality for discharge/recycle. While chemical methods and acoustic cavitation have been used to pre-treat spent wash at the lab scale, hydrodynamic cavitation was never used before Nagarajan and Ranade's work. Padoley et al. had, however, post-treated biomethanated spent wash using a venturi to improve its biodegradability index [106]. At a lab scale, Nagarajan and Ranade's study investigated pre-treatment of 12 L spent wash with a TS concentration of 162.5 g/L, at 3.9 barg inlet pressure, 2 passes through the novel vortex-based hydrodynamic cavitation device. They reported a CH_4 yield enhanced by 14% with a net positive energy gain (energy from excess methane – energy required for pre-treatment) of 0.98 MJ/kg VS. While the untreated spent wash showed diauxic digestion trends, it was not observed with any of the hydrodynamic cavitation pre-treated samples (2, 10, and 20 passes). The vortex-based cavitation device was also successfully scaled up. At an industrial scale, with the spent wash flow rates in the range of 13–50 m^3/h, enhancement in biogas yield in the range of 10–22% was achieved after a single pass through the vortex-based hydrodynamic cavitation device (Figure 8.8).

A range of devices have been used to disintegrate sludge and pre-treat effluents for enhancing biogas generation. While an increase in biogas yield is observed in all the hydrodynamic cavitation-treated cases, the process could not be effectively scaled up as no net energy gain could be obtained. While scale-up has its own difficulties, the primary determination of a positive net energy gain, at least at lab scale, if not pilot scale, is mandatory. Almost two and a half decades ago, when Weemaes and Verstraete compiled their review on the status of sludge pre-treatment, hydrodynamic cavitation was at its infancy; yet, they mentioned that it showed excellent promise as a sludge disintegration method and was cost-efficient [82]. With further advancement in device engineering and intensification strategies, this efficiency has increased (for instance, the lowest reported energy yet of 0.007 MJ/kg) [92].

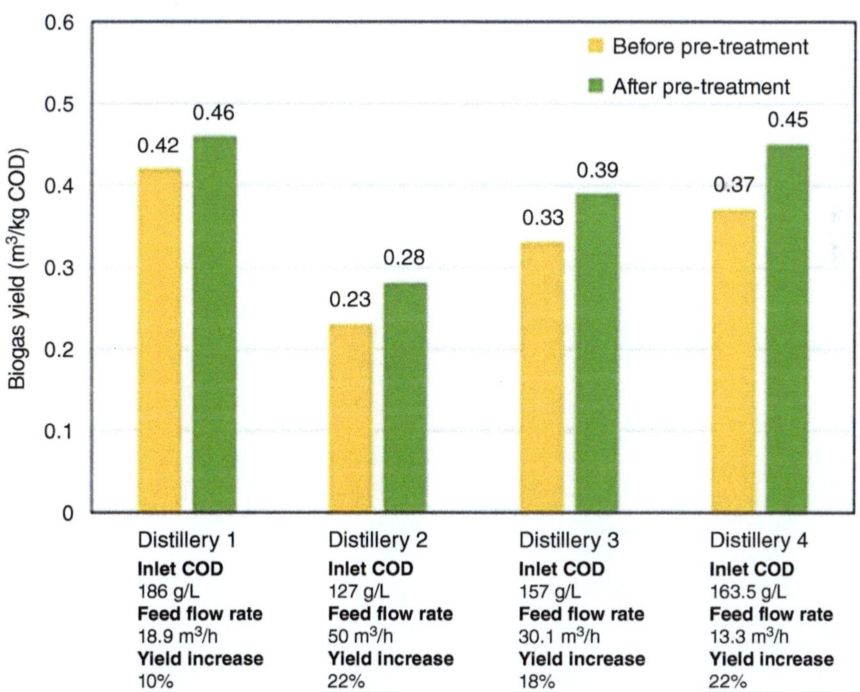

Figure 8.8 Increase in biogas yields from spent wash at industrial scale upon one-pass pre-treatment through the novel vortex-based hydrodynamic cavitation device (Source: From [92]/with permission of Elsevier).

8.3.2 Lignocellulosic Biomass (LCB)

LCB is typically composed of interlinked cellulose, hemicellulose, and lignin molecules, as shown in Figure 8.2. This interlinking makes LCB recalcitrant for bioprocessing, including AD. The fraction of each of these components however varies depending upon the type of biomass, time of harvest, stage of growth, and habitat [107]. Even within the same biomass, different parts of the plant can have different compositions. An example of variation in composition of grass is shown in Figure 8.9. For instance, sugarcane bagasse (SCB) has a cellulose, hemicellulose, and lignin content of 38, 31, and 22%, respectively, whereas sugarcane leaves – commonly known as sugarcane trash – have 33, 27, and 20%, respectively [109]. The differences between different kinds of biomass can however be more significant (Table 8.3). Although compositional differences are present, bioavailability is also dependent on other factors such as particle size, cellulose crystallinity, and type of lignin present. Specifically, for biogas production, extractives quantified as part of the LCB composition may also play a contributing role, as they are often composed of proteins, non-structural carbohydrates, and lipids. Therefore, it is important to determine the volatile solids (VS) content of LCB. It is the biochemical conversion of VS that yields biogas. Cellulose, hemicellulose, extractives, and a part of lignin contribute toward the VS content of LCB, although all of them cannot be readily converted (Figure 8.6). To improve the utilization of the VS component and to increase the biogas yield and production rate, pre-treatment is vital.

Unlike the pre-treatment required for bioethanol production, extensive fractionation of LCB followed by an enzymatic hydrolysis step is not required for biogas

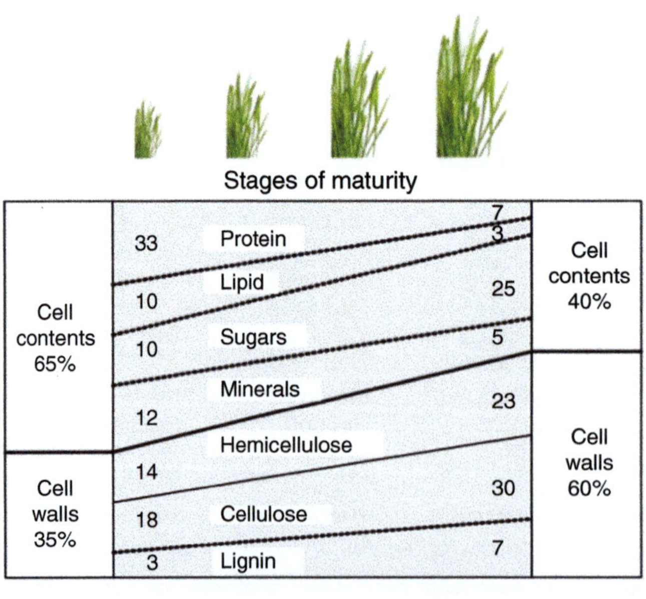

Figure 8.9 Changes in the composition of grass as a function of its age (Source: Reprinted from [108]/with permission from IEA Bioenergy).

Table 8.3 Composition of various types of lignocellulosic biomass.

Type of LCB	Cellulose (%)	Hemicellulose (%)	Lignin (%)
Switchgrass	30–50	10–40	5–20
Miscanthus	30–36	21–30	14–25
Wheat straw	33–40	20–25	15–20
Sorghum	31–35	23–24	17–21
Citrus fruit peels	12–14	5–6	1–2
Algae	0–20	20–43	0–14
Sugarcane bagasse	37–45	24–31	11–24
Grass silage	20–34	13–29	8–32
Palm	30–32	28–31	19–28
Aquatic plants	18–43	8–29	7–15
Softwood	35–40	25–30	27–30
Hardwood	45–50	20–25	20–25

Source: Data from [19, 109–112].

generation. Opening the LCB matrix and improving the surface area are sufficient to increase the bioavailability. While milling and chemical methods have been long considered [113, 114] as effective pre-treatment methods, the use of hydrodynamic cavitation as a pre-treatment method to enhance the productivity of AD is emerging [11, 18, 19]. The aspects of hydrodynamic cavitation pre-treatment, such as no use of chemicals, scalability, and energy efficiency have paved way for this increasing interest. Only a handful of reports are available at this point, and this section will specifically present a review of this literature.

In 2015, Abrahamsson, in her thesis, first reported the use of hydrodynamic cavitation for biomass pre-treatment with a view to enhance biogas generation [115]. She used a single-hole orifice to pre-treat a range of different types of biomass including food waste, seaweed, and grass. The biomass used in the study was blended to avoid clogging the orifice. Of the tested biomass, seaweed (*Ascophyllum nodosum*) showed the highest increase in the BMP by 44%. Food waste did not show a change in BMP upon hydrodynamic cavitation; however, a ~threefold increase in biomethane production rate was observed. Grass and the seaweed *Saccharina latissimi* showed a decrease in BMP by 29 and 18%, respectively. Release of respective inhibitors such as lignins and polyphenols were hypothesized to be the reasons for lower BMP compared to untreated feedstock.

Patil et al., in 2016, utilized a rotor-stator assembly to pre-treat finely powdered wheat straw for enhancing the BMP [116]. Experiments were performed to optimize the solid loading, treatment time, and rotational speed. However, important information on flow rates, inlet pressures, and pump selection was missing. It was also mentioned that, at a solid loading of greater than 1.5%, it was not possible to operate the hydrodynamic cavitation reactor due to issues with clogging. Within the investigated conditions – an increase in treatment time, rotational speed, and solids loading – the soluble COD content increased. However, this did not effectively reflect in the increase in BMP. All the pre-treated samples produced a higher BMP than the untreated wheat straw sample, but the trend was not similar to the trend observed with the soluble COD. When the feedstock was pre-soaked in KOH for 48 hours followed by hydrodynamic cavitation, the increase observed was similar to that of hydrodynamic cavitation alone (2.5-fold increase). The energy consumption required for pre-treatment was not mentioned, nor were the other operating parameters that were required for its calculation. While an increase in BMP was observed after hydrodynamic cavitation pre-treatment, it was significantly lower than the maximum achievable BMP of ~300 L CH_4/kg VS [117].

Zieliński et al. performed hydrodynamic cavitation pre-treatment of homogenized cattle manure and wheat straw (in the ratio 2 : 1 based on wet weight) using a hydrosonic pump [118]. The specific energy required for pre-treatment varied between 0 and 8 MJ/kg TS and its effect on solubilization and BMP were determined. Soluble COD increased by 30%, up to 3.3 MJ/kg TS, and plateaued with any further increase in specific energy. While organic carbon solubilization also plateaued after a similar specific energy input of 4 MJ/kg TS, total nitrogen solubilization increased up to 8 MJ/kg TS. An increase in BMP with increase in specific energy input of up to 4 MJ/kg TS was reported, beyond which no significant increase was observed. This corresponded to an increase in BMP by 40% compared to the untreated feedstock. It was also pointed out that a higher availability of soluble nitrogen as a result of hydrodynamic cavitation can induce ammonia formation during AD. However, no investigation along these lines was presented. It is important to take this into consideration, especially when working with high N-containing feedstock such as cattle manure. Langone et al. also performed hydrodynamic cavitation of cattle manure, but using a swirling jet device, and investigated the effect of inlet pressure (specific energy input) on a variety of parameters including BMP [17]. Similar to Zieliński et al.'s work, they also observed an increase in soluble COD as well as total ammoniacal nitrogen content, with increase in specific energy input. In addition, the particle size distribution revealed that a higher fraction of smaller particles (<0.1 mm) was produced upon hydrodynamic cavitation treatment, coupled with a reduction in viscosity with increase in specific energy. A meagre increase of ~3% in BMP was observed at an energy input of 2 MJ/kg TS at 7 bars with one hour of treatment. The net energy yield was calculated as the difference between the energy required for pre-treatment and total methane yield; however, this is not the correct method of calculating the net energy yield. The proper methods of calculating net energy gains are discussed later in Section 8.4.

Zieliński et al. also performed hydrodynamic cavitation of a mixture of *Sida hermaphrodita* silage and cattle manure using a rotor-stator device for enhancing the BMP [18]. At a fixed rotational speed of 2800 rpm of the rotor, the effect of treatment time on BMP was tested. The treatment time was varied from 0 to 20 minutes, and it was determined that with increase in treatment time, an increase in soluble carbohydrate content with a corresponding decrease of cellulose and hemicellulose in the solid fraction was observed. Unlike previously reported works, no change in total nitrogen content was observed on pre-treatment. An increase in BMP of ~20% was observed after five minutes of treatment and did not increase significantly beyond this pre-treatment time. Similar results with maize silage was also reported by Zubrowska-Sudol et al. [119] who achieved a 34% increase in BMP at a specific energy input of 0.3 MJ/kg TS, but a drop in BMP enhancement beyond this was observed [119].

While the most common devices used to pre-treat biomass for bioethanol production and wastewater sludge disintegration for biogas production were linear flow devices, the main problem posed by the conventional devices – clogging – was foreseen and therefore not used in the majority of the reported work utilizing LCB for biogas generation. As an exception, even though Abrahamsson [115] used an orifice, the feedstock was homogenized before hydrodynamic cavitation pre-treatment to mitigate the clogging-related risks. Therefore, the preferred device used for LCB pre-treatment for biogas generation has been the rotor-stator-based devices, although at an industrial scale, a mechanical shredder is still used before hydrodynamic cavitation [11]. The rotor-stator devices involve moving parts and have heavy demands on maintenance. The vortex-based device without any moving parts, reported for this purpose by Nagarajan and Ranade [19], appears to overcome these limitations.

Nagarajan and Ranade [19] utilized this novel vortex-based device to pre-treat two types of feedstock, namely grass silage and SCB. At a constant inlet pressure of 3.9 bars, the effect of number of passes through the device on BMP of the feedstock was investigated. Either the solid residue or the pre-treated slurry as a whole was utilized for BMP experiments. For the grass silage samples, nine passes through the device were found to be optimum to enhance the BMP of the solids by 12%. When the slurry samples were investigated for BMP upon hydrodynamic cavitation, nine passes were still found to be the optimum yielding a 15% increase in BMP. Any further increase in the number of passes yielded a relatively lower enhancement in BMP. This was ascribed to the possible release of inhibitors in solution that could have shifted the microbial community diversity and abundance as a result of pre-treatment. In contrast to the results obtained with grass silage, SCB solids did not show a difference in BMP upon hydrodynamic cavitation pre-treatment. The slurry samples however showed a significant increase in BMP of up to 24% on nine passes. At extremities (117 passes), a drop in BMP enhancement was observed and was hypothesized to be due to the release of inhibitors. This hypothesis was supported by the calculated rate of methane production, which was higher at nine passes by 9% and lower at 117 passes by 13% when compared to the untreated SCB. With both types of feedstock, nine passes were also found to yield a net energy gain. While this study utilized milled feedstock for hydrodynamic cavitation pre-treatment and BMP, it is possible to use as-received biomass with vortex-based devices without the risk of clogging. This offers a tremendous advantage over other devices for which milling becomes a mandatory step prior to hydrodynamic cavitation.

Overall, it is evident that hydrodynamic cavitation has the potential to increase the BMP of a range of LCBs, as shown in Figure 8.10. However, an increase in BMP alone is not sufficient to sustain the process economically. It is therefore important

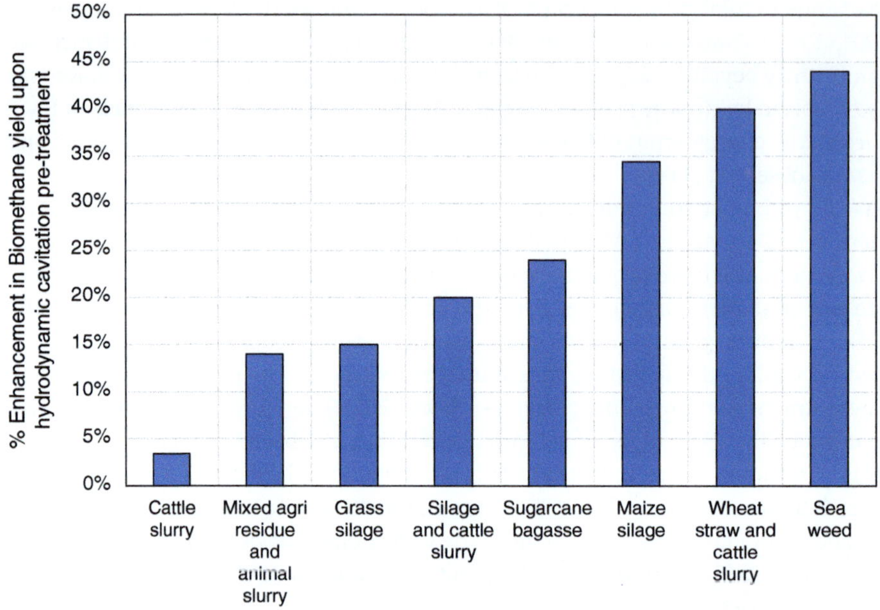

Figure 8.10 Percentage of enhancement in biomethane yield from a range of hydrodynamic cavitation-pre-treated lignocellulosic biomasses (Source: Data obtained from Table 8.4).

to determine whether this increase in biomethane yield is in conjunction with a net positive energy gain. The next section will discuss the net energy gain calculation and its importance in biofuel production processes.

8.4 Net Energy Gains

In general, as seen from previous sections, it is evident that hydrodynamic cavitation was successful in achieving the goal of biomass pre-treatment and solubilization. It would, however, only be meaningful when the achieved outcome of pre-treatment is supported by a net energy gain. While the initial studies tried to establish the cavitational effects on biomass and the consequences of treatment on end applications, the importance of net gains has only recently been highlighted. Application of hydrodynamic cavitation for LCB pre-treatment falls in TRL 4-6 [18, 19, 47] and in some cases even TRL 8-9 [11, 48, 92]. Therefore, as a primary indicator, the calculation of net energy gains is of utmost importance.

As an example, the energy requirements for LCB pre-treatment for biogas energy using a vortex-based cavitation device are presented here. Consider a continuous pre-treatment system, as shown in Figure 8.11, where the net flow rate to AD for biogas generation is q, m^3/s. Depending on the number of passes, n_p, the flow rate through the cavitation device can be set to Q, m^3/s and calculated as $n_p q$. In a batch system, n_p is calculated as Q/V, where V is the working volume, m^3.

Table 8.4 A summary of literature reporting hydrodynamic cavitation for lignocellulosic biomass pre-treatment for enhancing biogas generation.

Feedstock	Initial particle size (mm)	Total solids, g/L (Volume, L)	Hydrodynamic cavitation device	Additive (concentration, g/L)	Optimum conditions	Specific energy input, MJ/kg TS	AD conditions (mode, HRT, temperature)	Specific methane yield, L CH_4/kg VS	Increase in CH_4 yield	Reference
Food waste	1.3×10^{-6}	50 (2)	Single-hole orifice	—	10 L/min, 10 passes	~0.2	Batch BMP, 60 days, 37 °C	621	0%	[115]
Saccharina latissimi (seaweed)	1.2×10^{-6}							260	−18%[a)]	
Ascophyllum nodosum (seaweed)	1.1×10^{-6}							113	44%	
Grass	0.5×10^{-6}							323	−29%[a)]	
Wheat straw	0.125–1	5 (15)	Rotor-stator	—	2300 rpm, 2 min	—	Batch BMP, 16 days, 37 °C	77.9[b)]	2.4-fold	[116]
		5 (15)		Soaked in KOH (16.8)	2500 rpm, 4 min	—		~65[b)]	2.5-fold	
Homogenized Wheat straw + cattle manure	—	255 (5)	Hydrosonic pump	—	48.5 °C, 22.4 s	4	Batch BMP, 20 days, 35 °C	423	40%	[118]
Cattle manure	—	80 (50)	Swirling Jet	—	7 bars, 1 h	2	Batch BMP, 26 days, 38 °C	305	3.4%	[17]

(Continued)

Table 8.4 (Continued)

Feedstock	Initial particle size (mm)	Total solids, g/L (Volume, L)	Hydrodynamic cavitation device	Additive (concentration, g/L)	Optimum conditions	Specific energy input, MJ/kg TS	AD conditions (mode, HRT, temperature)	Specific methane yield, L CH_4/kg VS	Increase in CH_4 yield	Reference
Shredded digestate composed of pig slurry, corn meal, maize silage, triticale silage, and beet molasses	—	82 (1365 × 10^3)[c]	Rotor—stator	—	2 bars, 1 pass	0.7	Batch BMP, 27 days, 38 °C	159	14%	[11]
						0.5	Continuous large-scale AD for 6 months, 42–43 °C	332	6.4%	
Sida hermaphrodita silage + cattle manure	—	40 (25)	Rotor-stator	—	2800 rpm, 5 min	1	Batch BMP, 40 days, 37 °C	410[d]	20%	[18]
Soy grass	<2	10 (15)	Vortex based device	—	3.5–3.9 bars	—	Batch BMP, 21 days, 41 °C	264	60%	[120]
SCB	<2	10 (15)	Vortex-based device	—	3.5–3.9 bars, 9 passes	0.5	Batch BMP, 19 days, 41 °C	229[d]	24%	[19]
Grass silage							Batch BMP, 24 days, 41 °C	433[d]	15%	
Maize silage	—	50	Rotor-stator	—	3000 rpm,	0.3	Batch BMP, 37 °C	492[d]	34.4%	[119]

a) Decrease in methane yield.
b) L CH_4/kg TS.
c) Full-scale AD plant.
d) Indicates a net positive energy gain was achieved.

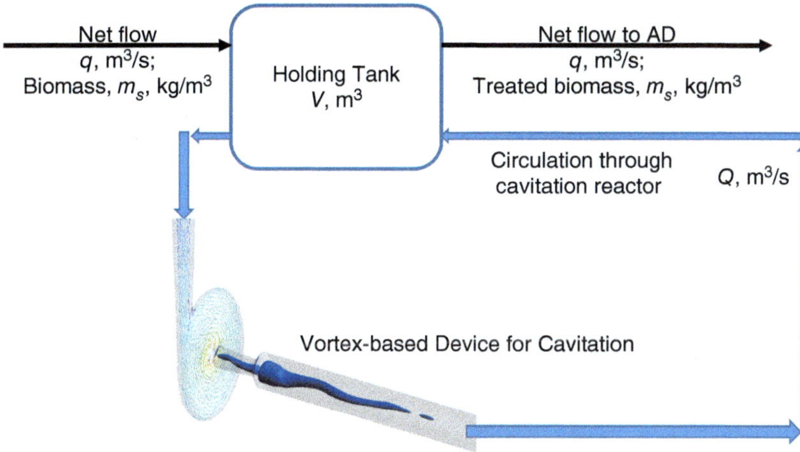

Figure 8.11 Proposed pre-treatment using vortex-based hydrodynamic cavitation (Source: From [19]/with permission of American Chemical Society).

If hydrodynamic cavitation pre-treatment does not involve any additives (typically in the case of biogas production), then the cost of pre-treatment is dependent on the specific energy requirement. However, in the case of bioethanol production, since alkali-assisted hydrodynamic cavitation is the common pre-treatment used, the price of alkali, P_{alk} (cost of alkali used to pre-treat unit weight of biomass, £/kg alkali/kg biomass), and the price of enzyme, P_{enz} (cost of enzyme used for hydrolysis of unit weight of biomass, £/kg enzyme/kg biomass), need to be included. In both the cases of biogas and bioethanol production, if the biomass has undergone a drying and comminution stage, then the specific energy requirements for drying (E_{dry}, MJ/kg TS) and milling (E_{mill}, MJ/kg TS) must be included.

The energy required for hydrodynamic cavitation pre-treatment, $E_{HC,}$ can be calculated using Eq. (8.1).

$$E_{HC} = \frac{\Delta P\, Q}{10^6\, q\, m_s\, \eta}\, \frac{\text{MJ}}{\text{kg TS}} \tag{8.2}$$

where ΔP is pressure drop across the cavitation device, Pa; m_s is the TS concentration in feed slurry, kg/m^3, and η is the pump efficiency.

Energy gained due to pre-treatment, $E_{gain,}$ can be calculated using Eq. (8.2).

$$E_{gain} = \Delta H_{cal}\,(F_{HC} - F_0)\,\frac{\text{MJ}}{\text{kg TS}} \tag{8.3}$$

where ΔH_{cal} is calorific value of methane (for biogas production) or bioethanol, MJ/m^3, and F is the fuel yield, m^3/ton; the subscripts "HC" and "0" denote the F_{max} upon hydrodynamic cavitation pre-treatment and that obtained with the untreated feedstock, respectively.

Net energy gain, $E_{Net\,gain}$ can therefore be calculated using Eq. (8.3).

$$E_{Net\,gain} = E_{gain} - E_{HC} - E_{dry} - E_{mill}\,\frac{\text{MJ}}{\text{kg TS}} \tag{8.4}$$

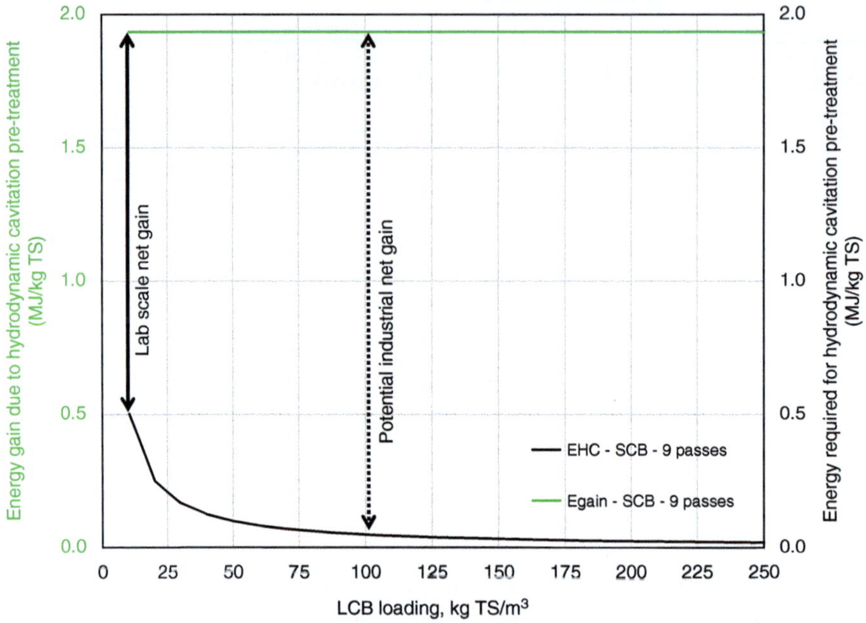

Figure 8.12 Net energy gain obtained in a lab-scale batch BMP and the potential gain possible at an industrial scale with a higher solid loading upon hydrodynamic cavitation pre-treatment of SCB for biogas generation (Source: Adapted from [19]/with permission of American Chemical Society).

A typical plot of the net energy gain with AD on hydrodynamic cavitation treatment and the influence of solid loading are shown in Figure 8.12. Data from nine-pass hydrodynamic cavitation-pre-treated SCB, reported by Nagarajan and Ranade [19], was used for this purpose. At an inlet pressure of 3.9 barg, they reported an E_{HC} for the pre-treatment of 1% SCB slurry (lab scale) as 0.5 MJ/kg TS. Even at such a low biomass loading of 1%, an $E_{Net\ gain}$ of 1.4 MJ/kg TS was obtained (solid line in Figure 8.12). At an industrial scale, the solid loading is expected to increase, and considering E_{gain} to be a constant across loading (although unlikely), at a higher loading of 10% TS, the gain looks significant at ~1.9 MJ/kg TS (dotted line in Figure 8.12). The net gain is a function of number of passes and the solid loading. Therefore, the pumpability of the slurry and the kinetics of digestion have to be considered when choosing a solid loading that is ideal for hydrodynamic cavitation pre-treatment and AD.

It is important to note that in the case shown in Figure 8.12, E_{mill} and E_{dry} were not considered. It is necessary to include these energy inputs when evaluating the energy requirements as well as net gain. As milling is dependent on the moisture content and the comminution ratio, it is important to account for these inputs when considering the overall picture. As an alternative, rotary cutter pumps can be utilized prior to hydrodynamic cavitation to chop the wet LCB and to pump the slurry, or a lobe pump or a progressive cavity pump can be used to pump the as-received fibrous LCB slurry, and the effect of hydrodynamic cavitation pre-treatment on unchopped LCB should be investigated. These strategies can help mitigate the dry-milling energy requirement.

As an example for the bioethanol production case, consider the results reported by Kim et al. [30]. While hydrodynamic cavitation pre-treatment was able to increase the saccharification yield to 85% compared to 16% with the untreated reed, leading to an ethanol yield of ~90% of theoretical yield, a net gain of 1.1 MJ/kg TS could be achieved, with E_{HC} reported to be 3.65 MJ/kg biomass (4.3 MJ/kg TS). However, it should also be considered that at an industrial scale, the biomass loading could be significantly higher than that reported at the lab scale, thereby leading to an increase in $E_{Net\,gain}$ (similar to enhanced biogas yields at an industrial scale, Figure 8.12). The possibilities of maximizing biomass valorization in biorefineries by producing high-value products in addition to biofuels should also be considered to improve net gains.

Overall, it is important to understand the parameters that influence the hydrodynamic cavitation pre-treatment of biomass leading to an enhanced biofuel yield. The most common parameters that affect the efficacy of hydrodynamic cavitation pre-treatment are the number of passes through the hydrodynamic cavitation device and the pressure drop (and mass inflow rates) across the device. These parameters predominantly control the specific energy required for pre-treatment. It is of utmost importance to understand the characteristics of hydrodynamic cavitation devices (as discussed in Chapters 2–5) before developing a biomass pre-treatment process. The choice of hydrodynamic cavitation device is crucial when using it for biomass pre-treatment. For instance, conventional devices such as the orifice or venturi possess small constrictions and always require the biomass to be milled prior to hydrodynamic cavitation [31]. This incurs additional costs and may therefore not be suitable. Similarly, rotor-stator devices have moving parts and require a high CAPEX as well as frequent maintenance. On the other hand, a vortex-based hydrodynamic cavitation device does not have any small constrictions and may be used for pre-treating unmilled biomass [19]. Furthermore, the advantages offered by the vortex-based hydrodynamic cavitation device – such as no device erosion due to cavitational collapse along the core, early inception, and improved mass transfer (interaction between cavities and biomass particles) – compared to conventional devices makes it suitable for biomass-based applications.

8.5 Summary and Path Forward

This chapter discussed the application of hydrodynamic cavitation as a potential LCB pre-treatment method as well as a sludge solubilization technique. The main end applications of this pre-treatment are in key renewable energy production areas such as bioethanol fermentation and biogas production via AD. The key differences in the pre-treatment of biomass for these applications were discussed. With respect to bioethanol production, hybrid hydrodynamic cavitation (with alkali) seems to be an effective strategy to delignify the biomass so that a cellulose-rich solid residue is obtained. Such a high cellulose-containing solid residue is beneficial for achieving high enzymatic saccharification yields and thereby higher bioethanol yields. In the case of biogas generation with wastewater or sludge, reducing the complexity of the COD or sludge disintegration is sufficient to enhance biogas generation. When LCB is used as feedstock for AD, hydrodynamic cavitation pre-treatment should be focused on opening up the matrix and increasing the available surface area to improve the bioavailability of the biomass fractions rather than extensive fractionation. Thus, chemical additives are not necessary when hydrodynamic cavitation is utilized for pre-treating biomass for AD.

In both the applications, there were multiple factors that were varied across studies. These include hydrodynamic cavitation device type, biomass type, initial particle size, solid loading, and specific energy input. The studies cannot be compared based on the type of hydrodynamic cavitation devices used, as there are no standards for the hydrodynamic cavitation device design. However, the operating parameters of the hydrodynamic cavitation device determine the extent of cavitation and the specific energy input achieved during pre-treatment that could be used as a metric to compare studies (without additives). Also, from an industrial perspective, it is important to focus on net energy gain as an indicator of technology acceptance. Compared to other available technologies such as acoustic cavitation and milling, hydrodynamic cavitation is significantly more energy-efficient (Figure 8.13). Modified rotor-stator devices [11, 119] and the novel vortex-based hydrodynamic cavitation devices [19, 92] have been able to achieve this net gain both at pilot and industrial scales for biogas generation. Vortex-based cavitation devices appear to be the most suitable for treating biomass slurries.

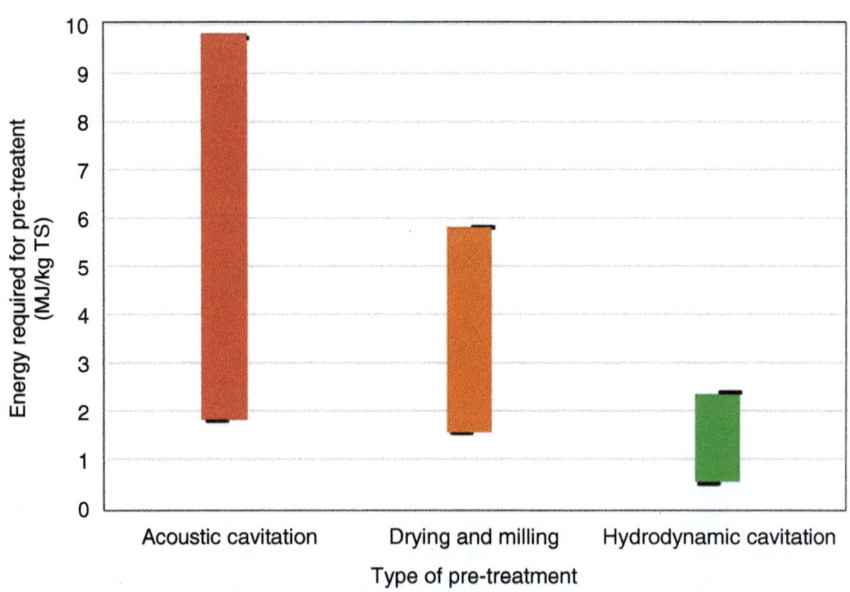

Figure 8.13 Energy required for lignocellulosic biomass pre-treatment (Source: Data based on [15–19]).

Future directions for the research and application of hydrodynamic cavitation for LCB pre-treatment lie in using optimized device designs to pre-treat biomass. Optimal device designs would lead to the generation of cavitating conditions at lower specific energy inputs. All available literature currently uses milled biomass for hydrodynamic cavitation treatment. It is important to consider the energy requirements of milling and therefore initiate research with the use of unchopped biomass. The current limitations to such investigations are the hydrodynamic cavitation device design and pumping ability. Modeling tools in predicting pre-treatment performance are also lacking currently. For instance, in the case of AD, the modeling of biogas generation and feedstock conversion is well established. However, multilayer models to connect pre-treatment performance to biogas generation are not yet available. Some efforts on using data-driven models (artificial neural network [ANN]-based models) to estimate performance of hydrodynamic cavitation-based pre-treatment on enhancement in biogas yield are being made [121]. Further research on developing high-fidelity computational models (physics-based as well as machine learning-based) is needed for making further progress in optimizing and scaling up the hydrodynamic cavitation pre-treatment processes.

On a broader perspective, considering the goal of net neutrality, sustainability, and a circular economy, hydrodynamic cavitation has more to offer. Beyond the production of biofuels from biomass, hydrodynamic cavitation has the potential to be used as an intensification tool to realize viable biorefineries. A couple of such biorefinery examples based on AD have been discussed by Konde et al. [13] and Nagarajan and Ranade [120]. With hydrodynamic cavitation pre-treatment of LCB expected to increase the biomethane yield with a net gain, the digestate could further be treated using hydrodynamic cavitation to enhance the residual biomethane potential. In this case, the hydrodynamic cavitation pre-treatment and primary AD (biological treatment) would have significantly disrupted the fibers and with a secondary hydrodynamic cavitation treatment post digestion, the net energy gain could be further enhanced. While hydrodynamic cavitation can improve the richness of cellulose in the solid residue on pre-treatment, the liquor rich in lignin and hemicellulose could be further valorized. Options such as biogas production from the liquor or selective C5 fermentation to xylitol could be further explored to generate net profits [122]. Overall, hydrodynamic cavitation shows immense promise for LCB pre-treatment and needs further investigation along the aforementioned areas to establish as the industrial state-of-art.

Notations

F_0	Maximum fuel yield from untreated biomass, m^3/ton
F_{HC}	Maximum fuel yield from hydrodynamic cavitation-treated biomass, m^3/ton
E_{dry}	Specific energy requirements for drying, MJ/kg TS
E_{gain}	Energy gained due to pre-treatment, MJ/kg TS
E_{HC}	Energy required for hydrodynamic cavitation pre-treatment, MJ/kg TS

E_{mill}	Specific energy requirements for milling, MJ/kg TS
$E_{Net\ gain}$	Net energy gain, MJ/kg TS
m_s	TS concentration in feed slurry, kg/m^3
n_p	Number of passes
P_{alk}	Cost of alkali used to pre-treat unit weight of biomass, £/kg alkali/kg biomass
P_{enz}	Cost of enzyme used for hydrolysis of unit weight of biomass, £/kg enzyme/kg biomass
q	Net flow rate to AD for biogas generation, m^3/s
Q	Flow rate through the cavitation device, m^3/s
V	Working volume, m^3

Symbols

ΔP	Pressure drop across the cavitation device, Pa
ΔH_{cal}	Calorific value of methane (for biogas production) or bioethanol, MJ/m^3
η	Pump efficiency

Abbreviations

AD	Anaerobic digestion
ANN	Artificial neural network
BMP	Biochemical methane potential
CAPEX	Capital expenditure
CBG	Compressed biogas
CNG	Compressed natural gas
COD	Chemical oxygen demand
EJ	Exa joule
FTIR	Fourier transform infrared
HC	Hydrodynamic cavitation
HMF	Hydroxymethylfurfural
ktoe	Kilo tons oil equivalent
LCB	Lignocellulosic biomass
MJ	Mega joule
OPEX	Operating expenditure
SCB	Sugarcane bagasse
SPR	Shockwave power reactor
TS	Total solids
VS	Volatile solids

References

1 International Energy Agency (2020). Data and statistics 2020. https://www.iea.org/data-and-statistics?country=EU28&fuel=CO2%20emissions&indicator=CO2%20emissions%20by%20sector (accessed 30 December 2020).

2 World Bioenergy Association (2019). Global bioenergy statistics 2019. http://www.worldbioenergy.org/uploads/191129%20WBA%20GBS%202019_LQ.pdf (accessed 17 March 2021).

3 Su, T., Zhao, D., Khodadadi, M., and Len, C. (2020). Lignocellulosic biomass for bioethanol: recent advances, technology trends, and barriers to industrial development. *Current Opinion in Green and Sustainable Chemistry* 24: 56–60.

4 Chen, H. and Fu, X. (2016). Industrial technologies for bioethanol production from lignocellulosic biomass. *Renewable and Sustainable Energy Reviews* 57: 468–478.

5 Klinke, H.B., Thomsen, A.B., and Ahring, B.K. (2004). Inhibition of ethanol-producing yeast and bacteria by degradation products produced during pre-treatment of biomass. *Applied Microbiology and Biotechnology* 66 (1): 10–26.

6 Yang, B. and Wyman, C.E. (2008). Pretreatment: the key to unlocking low-cost cellulosic ethanol. *Biofuels, Bioproducts & Biorefining* 2 (1): 26–40.

7 Himmel, M.E., Ding, S.-Y., Johnson, D.K. et al. (2007). Biomass recalcitrance: engineering plants and enzymes for biofuels production. *Science* 315 (5813): 804–807.

8 Toutian, V., Barjenbruch, M., Unger, T. et al. (2020). Effect of temperature on biogas yield increase and formation of refractory COD during thermal hydrolysis of waste activated sludge. *Water Research* 171: 115383.

9 Cano, R., Pérez-Elvira, S.I., and Fdz-Polanco, F. (2015). Energy feasibility study of sludge pretreatments: a review. *Applied Energy* 149: 176–185.

10 Barber, W.P.F. (2016). Thermal hydrolysis for sewage treatment: a critical review. *Water Research* 104: 53–71.

11 Garuti, M., Langone, M., Fabbri, C., and Piccinini, S. (2018). Monitoring of full-scale hydrodynamic cavitation pretreatment in agricultural biogas plant. *Bioresource Technology* 247: 599–609.

12 Sun, Y. and Cheng, J. (2002). Hydrolysis of lignocellulosic materials for ethanol production: a review. *Reviews Issue* 83 (1): 1–11.

13 Konde, K.S., Nagarajan, S., Kumar, V. et al. (2021). Sugarcane bagasse based biorefineries in India: potential and challenges. *Sustainable Energy & Fuels* 5: 52–78.

14 Haque, N. and Somerville, M. (2013). Techno-economic and environmental evaluation of biomass dryer. *5th BSME International Conference on Thermal Engineering*, vol. 56, pp. 650–655.

15 Miao, Z., Grift, T.E., Hansen, A.C., and Ting, K.C. (2011). Energy requirement for comminution of biomass in relation to particle physical properties. *Industrial Crops and Products* 33 (2): 504–513.

16 Priyanto, D.E., Ueno, S., Kasai, H., and Mae, K. (2018). Rethinking the inherent moisture content of biomass: its ability for milling and upgrading. *ACS Sustainable Chemistry & Engineering* 6 (3): 2905–2910.

17 Langone, M., Soldano, M., Fabbri, C. et al. (2018). Anaerobic digestion of cattle manure influenced by swirling jet induced hydrodynamic cavitation. *Applied Biochemistry and Biotechnology* 184 (4): 1200–1218.

18 Zieliński, M., Rusanowska, P., Krzywik, A. et al. (2019). Application of hydrodynamic cavitation for improving methane fermentation of sida hermaphrodita silage. *Energies* 12 (3): 525.

19 Nagarajan, S. and Ranade, V.V. (2019). Pretreatment of lignocellulosic biomass using vortex-based devices for cavitation: influence on biomethane potential. *Industrial and Engineering Chemistry Research* 58 (35): 15975–15988.

20 Krishnan, M.S., Ho, N.W.Y., and Tsao, G.T. (1999). Fermentation kinetics of ethanol production from glucose and xylose by recombinant Saccharomyces 1400(pLNH33). *Applied Biochemistry and Biotechnology* 78 (1): 373–388.

21 Baxi, P.B. and Pandit, A.B. (2012). Using cavitation for delignification of wood. *Bioresource Technology* 110 (Supplement C): 697–700.

22 Balat, M., Balat, H., and Öz, C. (2008). Progress in bioethanol processing. *Progress in Energy and Combustion Science* 34 (5): 551–573.

23 Alternative Fuels Data Center (2020). Fuel properties comparison 2020. https://afdc.energy.gov/fuels/properties (accessed 30 December 2020).

24 Bailey, B.K. (1996). Performance of ethanol as a transportation fuel. In: *Handbook on Bioethanol* (ed. C.E. Wyman), 37–60. Routledge.

25 Aditiya, H.B., Mahlia, T.M.I., Chong, W.T. et al. (2016). Second generation bioethanol production: a critical review. *Renewable and Sustainable Energy Reviews* 66: 631–653.

26 Alvira, P., Tomás-Pejó, E., Ballesteros, M., and Negro, M.J. (2010). Pretreatment technologies for an efficient bioethanol production process based on enzymatic hydrolysis: a review. *Special Issue on Lignocellulosic Bioethanol: Current Status and Perspectives* 101 (13): 4851–4861.

27 Pal, S., Joy, S., Kumbhar, P. et al. (2017). Pilot-scale pretreatments of sugarcane bagasse with steam explosion and mineral acid, organic acid, and mixed acids: synergies, enzymatic hydrolysis efficiencies, and structure-morphology correlations. *Biomass Conversion and Biorefinery* 7 (2): 179–189.

28 Pal, S., Satyendra, W.J., and Ravikumar, R.P., inventors (2014). Preparation of hydrolysate of lignocellulosic materials patent. WO2014203271A2.

29 Silveira, M.H.L., Chandel, A.K., Vanelli, B.A. et al. (2018). Production of hemicellulosic sugars from sugarcane bagasse via steam explosion employing industrially feasible conditions: pilot scale study. *Bioresource Technology Reports* 3: 138–146.

30 Kim, I., Lee, I., Jeon, S.H. et al. (2015). Hydrodynamic cavitation as a novel pretreatment approach for bioethanol production from reed. *Bioresource Technology* 192: 335–339.

31 Nakashima, K., Ebi, Y., Shibasaki-Kitakawa, N. et al. (2016). Hydrodynamic cavitation reactor for efficient pretreatment of lignocellulosic biomass. *Industrial and Engineering Chemistry Research* 55 (7): 1866–1871.

32 Terán Hilares, R., de Almeida, G.F., Ahmed, M.A. et al. (2017). Hydrodynamic cavitation as an efficient pretreatment method for lignocellulosic biomass: a parametric study. *Bioresource Technology* 235: 301–308.

33 Terán Hilares, R., Kamoei, D.V., Ahmed, M.A. et al. (2018). Santos JCd. A new approach for bioethanol production from sugarcane bagasse using hydrodynamic cavitation assisted-pretreatment and column reactors. *Ultrasonics Sonochemistry* 43: 219–226.

34 Bimestre, T.A., Júnior, J.A.M., Botura, C.A. et al. (2020). Theoretical modeling and experimental validation of hydrodynamic cavitation reactor with a Venturi tube for sugarcane bagasse pretreatment. *Bioresource Technology* 311: 123540.

35 Madison, M.J., Coward-Kelly, G., Liang, C. et al. (2017). Mechanical pretreatment of biomass – part I: acoustic and hydrodynamic cavitation. *Biomass and Bioenergy* 98: 135–141.

36 Sprague, A., inventor; Google Patents, assignee (2008). Alcohol production using hydraulic cavitation. US Patent US20080277264A1.

37 Kozyuk, O.V., inventor; Google Patents, assignee (2010). Apparatus and method for increasing alcohol yield from grain patent. US7667082B2.

38 Kampfe, S.K. (2010). *Processing and Conversion of Algae to Bioethanol*. Virginia: The College of William and Mary.

39 Badve, M.P., Gogate, P.R., Pandit, A.B., and Csoka, L. (2014). Hydrodynamic cavitation as a novel approach for delignification of wheat straw for paper manufacturing. *Ultrasonics Sonochemistry* 21 (1): 162–168.

40 Montusiewicz, A., Pasieczna-Patkowska, S., Lebiocka, M. et al. (2017). Hydrodynamic cavitation of brewery spent grain diluted by wastewater. *Chemical Engineering Journal* 313: 946–956.

41 Terán Hilares, R., Dionízio, R.M., Prado, C.A. et al. (2019). Pretreatment of sugarcane bagasse using hydrodynamic cavitation technology: semi-continuous and continuous process. *Bioresource Technology* 290: 121777.

42 Terán Hilares, R., dos Santos, J.C., Ahmed, M.A. et al. (2016). Hydrodynamic cavitation-assisted alkaline pretreatment as a new approach for sugarcane bagasse biorefineries. *Bioresource Technology* 214: 609–614.

43 Terán Hilares, R., Ramos, L., da Silva, S.S. et al. (2018). Hydrodynamic cavitation as a strategy to enhance the efficiency of lignocellulosic biomass pretreatment. *Critical Reviews in Biotechnology* 38 (4): 483–493.

44 Terán Hilares, R., Dionízio, R.M., Sánchez Muñoz, S. et al. (2020). Hydrodynamic cavitation-assisted continuous pre-treatment of sugarcane bagasse for ethanol production: effects of geometric parameters of the cavitation device. *Ultrasonics Sonochemistry* 63: 104931.

45 Terán Hilares, R., Ienny, J.V., Marcelino, P.F. et al. (2017). Ethanol production in a simultaneous saccharification and fermentation process with interconnected reactors employing hydrodynamic cavitation-pretreated sugarcane bagasse as raw material. *Bioresource Technology* 243: 652–659.

46 Chang, V.S., Nagwani, M., and Holtzapple, M.T. (1998). Lime pretreatment of crop residues bagasse and wheat straw. *Applied Biochemistry and Biotechnology* 74 (3): 135–159.

47 Ramirez-Cadavid, D.A., Kozyuk, O., Lyle, P., and Michel, F.C. (2016). Effects of hydrodynamic cavitation on dry mill corn ethanol production. *Process Biochemistry* 51 (4): 500–508.

48 Ramirez-Cadavid, D.A., Kozyuk, O., and Michel, F.C. (2014). Improvement in commercial scale dry mill corn ethanol production using controlled flow cavitation and cellulose hydrolysis. *Biomass Conversion and Biorefinery* 4 (3): 211–224.

49 Ramirez, D.A. (2012). *Improvement of Ethanol Production on Dry-Mill Process Using Hydrodynamic Cavitation Pretreatment*. The Ohio State University.

50 Furtado Amaral, A., Previtali, D., Bassani, A. et al. (2020). Biogas beyond CHP: the HPC (heat, power & chemicals) process. *Energy* 203: 117820.

51 Bajpai, P. (2017). Process parameters affecting anaerobic digestion. In: *Anaerobic Technology in Pulp and Paper Industry* (ed. P. Bajpai), 13–27. Singapore: Springer Singapore.

52 Uddin, M.N., Siddiki, S.Y.A., Mofijur, M. et al. (2021). Prospects of bioenergy production from organic waste using anaerobic digestion technology: a mini review. *Frontiers in Energy Research* 9: 33.

53 Siqueira, G., Arantes, V., Saddler, J.N. et al. (2017). Limitation of cellulose accessibility and unproductive binding of cellulases by pretreated sugarcane bagasse lignin. *Biotechnology for Biofuels* 10 (1): 176.

54 Mulat, D.G., Dibdiakova, J., and Horn, S.J. (2018). Microbial biogas production from hydrolysis lignin: insight into lignin structural changes. *Biotechnology for Biofuels* 11 (1): 61.

55 Zealand, A.M., Roskilly, A.P., and Graham, D.W. (2017). Effect of feeding frequency and organic loading rate on biomethane production in the anaerobic digestion of rice straw. *Applied Energy* 207: 156–165.

56 Ferguson, R.M.W., Coulon, F., and Villa, R. (2016). Organic loading rate: a promising microbial management tool in anaerobic digestion. *Water Research* 100: 348–356.

57 Bi, S., Qiao, W., Xiong, L. et al. (2019). Effects of organic loading rate on anaerobic digestion of chicken manure under mesophilic and thermophilic conditions. *Renewable Energy* 139: 242–250.

58 Wang, X., Yang, G., Feng, Y. et al. (2012). Optimizing feeding composition and carbon–nitrogen ratios for improved methane yield during anaerobic co-digestion of dairy, chicken manure and wheat straw. *Bioresource Technology* 120: 78–83.

59 Marañón, E., Castrillón, L., Quiroga, G. et al. (2012). Co-digestion of cattle manure with food waste and sludge to increase biogas production. *Waste Management* 32 (10): 1821–1825.

60 Xue, S., Wang, Y., Lyu, X. et al. (2020). Interactive effects of carbohydrate, lipid, protein composition and carbon/nitrogen ratio on biogas production of different food wastes. *Bioresource Technology* 312: 123566.

61 Yan, L., Liu, C., Zhang, Y. et al. (2021). Effects of C/N ratio variation in swine biogas slurry on soil dissolved organic matter: content and fluorescence characteristics. *Ecotoxicology and Environmental Safety* 209: 111804.

62 Rabii, A., Aldin, S., Dahman, Y., and Elbeshbishy, E. (2019). A review on anaerobic Co-digestion with a focus on the microbial populations and the effect of multi-stage digester configuration. *Energies* 12 (6): 1106.

63 Siddique, M.N.I. and Wahid, Z.A. (2018). Achievements and perspectives of anaerobic co-digestion: a review. *Journal of Cleaner Production* 194: 359–371.

64 Hagos, K., Zong, J., Li, D. et al. (2017). Anaerobic co-digestion process for biogas production: progress, challenges and perspectives. *Renewable and Sustainable Energy Reviews* 76: 1485–1496.

65 Choong, Y.Y., Norli, I., Abdullah, A.Z., and Yhaya, M.F. (2016). Impacts of trace element supplementation on the performance of anaerobic digestion process: a critical review. *Bioresource Technology* 209: 369–379.

66 Chen, R., Li, Z., Feng, J. et al. (2020). Effects of digestate recirculation ratios on biogas production and methane yield of continuous dry anaerobic digestion. *Bioresource Technology* 316: 123963.

67 Hu, Y., Shen, F., Yuan, H. et al. (2014). Influence of recirculation of liquid fraction of the digestate (LFD) on maize stover anaerobic digestion. *Biosystems Engineering* 127: 189–196.

68 Vats, N., Khan, A.A., and Ahmad, K. (2020). Options for Enhanced Anaerobic Digestion of Waste and Biomass – a Review. *Journal of Biosystems Engineering* 45 (1): 1–15.

69 Monlau, F., Sambusiti, C., Ficara, E. et al. (2015). New opportunities for agricultural digestate valorization: current situation and perspectives. *Energy & Environmental Science* 8 (9): 2600–2621.

70 Li, Y., Zhang, Y., Sun, Y. et al. (2017). The performance efficiency of bioaugmentation to prevent anaerobic digestion failure from ammonia and propionate inhibition. *Bioresource Technology* 231: 94–100.

71 Wang, Z., Wang, T., Si, B. et al. (2021). Accelerating anaerobic digestion for methane production: potential role of direct interspecies electron transfer. *Renewable and Sustainable Energy Reviews* 145: 111069.

72 Schievano, A., Tenca, A., Lonati, S. et al. (2014). Can two-stage instead of one-stage anaerobic digestion really increase energy recovery from biomass? *Applied Energy* 124: 335–342.

73 Li, Y., Liu, C., Wachemo, A.C. et al. (2017). Serial completely stirred tank reactors for improving biogas production and substance degradation during anaerobic digestion of corn stover. *Bioresource Technology* 235: 380–388.

74 Atelge, M.R., Atabani, A.E., Banu, J.R. et al. (2020). A critical review of pre-treatment technologies to enhance anaerobic digestion and energy recovery. *Fuel* 270: 117494.

75 Batstone, D.J., Keller, J., Angelidaki, I. et al. (2002). The IWA Anaerobic Digestion Model No 1 (ADM1). *Water Science and Technology* 45 (10): 65–73.

76 Ersahin, M.E. (2018). Modeling the dynamic performance of full-scale anaerobic primary sludge digester using Anaerobic Digestion Model No. 1 (ADM1). *Bioprocess and Biosystems Engineering* 41 (10): 1539–1545.

77 Buswell, A.M. and Mueller, H.F. (1952). Mechanism of methane fermentation. *Industrial and Engineering Chemistry* 44 (3): 550–552.

78 Carlsson, M., Lagerkvist, A., and Morgan-Sagastume, F. (2012). The effects of substrate pre-treatment on anaerobic digestion systems: a review. *Waste Management* 32 (9): 1634–1650.

79 Gurieff, N., Bruus, J., Nielsen, B. et al. (2012). Moving towards an energy neutral WWTP – the positive impact of ExelysTM continuous thermal hydrolysis in achieving this goal. *Water Practice Technology* 7 (2): wpt2012034.

80 Kopp, J., Miiller, J., Dichtl, N., and Schwedes, J. (1997). Anaerobic digestion and dewatering characteristics of mechanically disintegrated excess sludge. *Water Science and Technology* 36 (11): 129–136.

81 Müller, J., Lehne, G., Schwedes, J. et al. (1998). Disintegration of sewage sludges and influence on anaerobic digestion. *Water Science and Technology* 38 (8): 425–433.

82 Weemaes, M.P.J. and Verstraete, W.H. (1998). Evaluation of current wet sludge disintegration techniques. *Journal of Chemical Technology and Biotechnology* 73 (2): 83–92.

83 Grűbel, K. and Machnicka, A. (2009). Use of hydrodynamic disintegration to accelerate anaerobic digestion of surplus activated sludge. *Water Environment Research* 81 (12): 2420–2426.

84 Lee, I. and Han, J.-I. (2013). The effects of waste-activated sludge pretreatment using hydrodynamic cavitation for methane production. *Ultrasonics Sonochemistry* 20 (6): 1450–1455.

85 Jung, K.-W., Hwang, M.-J., Yun, Y.-M. et al. (2014). Development of a novel electric field-assisted modified hydrodynamic cavitation system for disintegration of waste activated sludge. *Ultrasonics Sonochemistry* 21 (5): 1635–1640.

86 Stoop, A.H., Bakker, T.W., and Kramer, H.J.M. (2015). Cavitation for improved sludge conversion into biogas. *Journal of Physics: Conference Series* 656: 012118.

87 Petkovšek, M., Mlakar, M., Levstek, M. et al. (2015). A novel rotation generator of hydrodynamic cavitation for waste-activated sludge disintegration. *Ultrasonics Sonochemistry* 26: 408–414.

88 Grübel, K. and Suschka, J. (2015). Hybrid alkali-hydrodynamic disintegration of waste-activated sludge before two-stage anaerobic digestion process. *Environmental Science and Pollution Research* 22 (10): 7258–7270.

89 Şağban, F.O.T., Dindar, E., Cirakoglu, C., and Keskinler, B. (2017). Hydrodynamic cavitation of waste-activated sludge. *Environmental Engineering Science* 35 (8): 775–784.

90 Saxena, S., Saharan, V.K., and George, S. (2019). Modeling & simulation studies on batch anaerobic digestion of hydrodynamically cavitated tannery waste effluent for higher biogas yield. *Ultrasonics Sonochemistry* 58: 104692.

91 Garlicka, A., Zubrowska-Sudol, M., Umiejewska, K. et al. (2020). Effects of thickened excess sludge pre-treatment using hydrodynamic cavitation for anaerobic digestion. *Energies* 13 (10): 2483.

92 Nagarajan, S. and Ranade, V.V. (2020). Pre-treatment of distillery spent wash (vinasse) with vortex based cavitation and its influence on biogas generation. *Bioresource Technology Reports* 11: 100480.

93 Nicholas, D. and Vella, P. (2008). Wastewater digester enhancement and sludge volume reduction through cell lysis with controlled hydrodynamic cavitation. *Proceedings of the Water Environment Federation Technical Exhibition and Conference, WEFTEC*. 15: 2019–2029.

94 Kim, H.J., Nguyen, D.X., and Bae, J.H. (2008). The performance of the sludge pretreatment system with venturi tubes. *Water Science and Technology* 57 (1): 131–137.

95 Maeng, J.W., Lee, E.Y., and Bae, J.H. (2010). Optimization of the hydrodynamic sludge pre-treatment system with venturi tubes. *Water Practice Technology* 5 (2): wpt2010034.

96 Kim, H., Sun, X., Koo, B., and Yoon, J.Y. (2019). Experimental investigation of sludge treatment using a rotor-stator type hydrodynamic cavitation reactor and an ultrasonic bath. *Processes* 7 (11): 790.

97 Kim, H., Koo, B., Sun, X., and Yoon, J.Y. (2020). Investigation of sludge disintegration using rotor-stator type hydrodynamic cavitation reactor. *Separation and Purification Technology* 240: 116636.

98 Ferrentino, R. and Andreottola, G. (2020). Investigation of sludge solubilization and phosphorous release in anaerobic side-stream reactor with a low pressure swirling jet hydrodynamic cavitation treatment. *Journal of Environmental Chemical Engineering* 8 (5): 104389.

99 Mancuso, G., Langone, M., Andreottola, G., and Bruni, L. (2019). Effects of hydrodynamic cavitation, low-level thermal and low-level alkaline pre-treatments on sludge solubilisation. *Ultrasonics Sonochemistry* 59: 104750.

100 Mancuso, G., Langone, M., and Andreottola, G. (2017). A swirling jet-induced cavitation to increase activated sludge solubilisation and aerobic sludge biodegradability. *Ultrasonics Sonochemistry* 35 (Pt A): 489–501.

101 Chanda, S.K. (2012). *Disintegration of Sludge Using Ozone-Hydrodynamic Cavitation*. University of British Columbia.

102 Cai, M., Hu, J., Lian, G. et al. (2018). Synergetic pretreatment of waste activated sludge by hydrodynamic cavitation combined with Fenton reaction for enhanced dewatering. *Ultrasonics Sonochemistry* 42: 609–618.

103 Habashi, N., Mehrdadi, N., Mennerich, A. et al. (2016). Hydrodynamic cavitation as a novel approach for pretreatment of oily wastewater for anaerobic co-digestion with waste activated sludge. *Ultrasonics Sonochemistry* 31: 362–370.

104 Habashi, N., Alighardashi, A., Mennerich, A. et al. (2018). Improving biogas production from continuous co-digestion of oily wastewater and waste-activated sludge by hydrodynamic cavitation pre-treatment. *Environmental Technology* 39 (8): 1017–1024.

105 Vilarroig, J., Martínez, R., Zuriaga-Agustí, E. et al. (2020). Design and optimization of a semi-industrial cavitation device for a pretreatment of an anaerobic digestion treatment of excess sludge and pig slurry. *Water Environment Research* 92 (12): 2060–2071.

106 Padoley, K.V., Saharan, V.K., Mudliar, S.N. et al. (2012). Cavitationally induced biodegradability enhancement of a distillery wastewater. *Journal of Hazardous Materials* 219–220: 69–74.

107 Jeffries, T.W. (1994). Biodegradation of lignin and hemicelluloses. In: *Biochemistry of Microbial Degradation* (ed. C. Ratledge), 233–277. Dordrecht: Springer Netherlands.

108 Murphy, J., Braun, R., Weiland, P., and Wellinger, A. (2011). *Biogas from Crop Digestion, Task 37 Energy from Biogas Brochure, IEA Bioenergy*. International Energy Agency.

109 Rabemanolontsoa, H. and Saka, S. (2013). Comparative study on chemical composition of various biomass species. *RSC Advances* 3 (12): 3946–3956.

110 McKendry, P. (2002). Energy production from biomass (part 1): overview of biomass. *Bioresource Technology* 83 (1): 37–46.

111 Jung, S.-J., Kim, S.-H., and Chung, I.-M. (2015). Comparison of lignin, cellulose, and hemicellulose contents for biofuels utilization among 4 types of lignocellulosic crops. *Biomass and Bioenergy* 83: 322–327.

112 Ververis, C., Georghiou, K., Danielidis, D. et al. (2007). Cellulose, hemicelluloses, lignin and ash content of some organic materials and their suitability for use as paper pulp supplements. *Bioresource Technology* 98 (2): 296–301.

113 Taherzadeh, J.M. and Karimi, K. (2008). Pretreatment of lignocellulosic wastes to improve ethanol and biogas production: a review. *International Journal of Molecular Sciences* 9 (9): 1621–1651.

114 Kainthola, J., Kalamdhad, A.S., and Goud, V.V. (2019). A review on enhanced biogas production from anaerobic digestion of lignocellulosic biomass by different enhancement techniques. *Process Biochemistry* 84: 81–90.

115 Abrahamsson, L. (2015). *Improving Methane Production Using Hydrodynamic Cavitation as Pre-Treatment*. Sweden: Linköping University.

116 Patil, P.N., Gogate, P.R., Csoka, L. et al. (2016). Intensification of biogas production using pretreatment based on hydrodynamic cavitation. *Ultrasonics Sonochemistry* 30: 79–86.

117 Victorin, M., Davidsson, Å., and Wallberg, O. (2020). Characterization of mechanically pretreated wheat straw for biogas production. *Bioenergy Research* 1–12.

118 Zieliński, M., Dębowski, M., Kisielewska, M. et al. (2017). Comparison of ultrasonic and hydrothermal cavitation pretreatments of cattle manure mixed with straw wheat on fermentative biogas production. *Waste and Biomass Valorization* 10: 747.

119 Zubrowska-Sudol, M., Dzido, A., Garlicka, A. et al. (2020). Innovative hydrodynamic disintegrator adjusted to agricultural substrates pre-treatment aimed at methane production intensification—CFD modelling and batch tests. *Energies* 13 (16): 4256.

120 Nagarajan, S. and Ranade, V.V. (2021). Valorizing waste biomass via hydrodynamic cavitation and anaerobic digestion. *Industrial and Engineering Chemistry Research* 60 (46): 16577–16598.

121 Ranade, N.V., Nagarajan, S., Sarvothaman, V., and Ranade, V.V. (2021). ANN based modelling of hydrodynamic cavitation processes: biomass pre-treatment and wastewater treatment. *Ultrasonics Sonochemistry* 72: 105428.

122 Prabhu, A.A., Bosakornranut, E., Amraoui, Y. et al. (2020). Enhanced xylitol production using non-detoxified xylose rich pre-hydrolysate from sugarcane bagasse by newly isolated Pichia fermentans. *Biotechnology for Biofuels* 13 (1): 209.

9

Other Applications of Hydrodynamic Cavitation

9.1 Introduction

Hydrodynamic cavitation has been extensively investigated for intensifying a variety of physical, chemical, and biological processes and applications. The key applications relevant to water and energy – relating to sustainability and a circular economy – were discussed in Chapters 6–8. Considering the numerous applications of hydrodynamic cavitation (see recent reviews: Mancuso et al. [1]; Panda et al. [2]; Holkar et al. [3]; Asaithambi et al. [4]), it is not possible to review all the key applications in a single chapter. Instead, we have attempted to present a brief overview of a few other important applications; we have organized the discussion based on the classification of multiple phases present in the system (Figure 9.1).

Hydrodynamic cavitation is inherently a multiphase (gas–liquid) system since cavities are generated via local degassing and evaporation of liquid. The classification of multiphase systems mentioned in Figure 9.1 is based on phases excluding these in-situ-generated cavities. In reality, the systems will contain cavities in addition to the other phases mentioned therein. The applications shown in blue in Figure 9.1 are briefly discussed in the following sections. An attempt is made in this chapter to present a generic approach toward applying hydrodynamic cavitation to intensify such applications. The goal is to convey a general approach so that the material presented in this book can be used to further develop these applications or to develop new applications.

9.2 Gas–Liquid Applications

Hydrodynamic cavitation generates intense localized shear and high-velocity jets. The generated shear and jets can be harnessed to cause breakage of gas bubbles and to generate a very high gas–liquid interfacial area. Hydrodynamic cavitation can produce bubble sizes of the order of 1 to 10 μm and therefore generate very high interfacial area, which can be used to develop its numerous applications in intensifying gas–liquid systems. This is illustrated by discussing the application to flotation in the following section. Some comments on similar application of hydrodynamic cavitation for other gas–liquid applications are included as well.

Hydrodynamic Cavitation: Devices, Design, and Applications, First Edition.
Vivek V. Ranade, Vinay M. Bhandari, Sanjay Nagarajan, Varaha P. Sarvothaman and Alister T. Simpson.

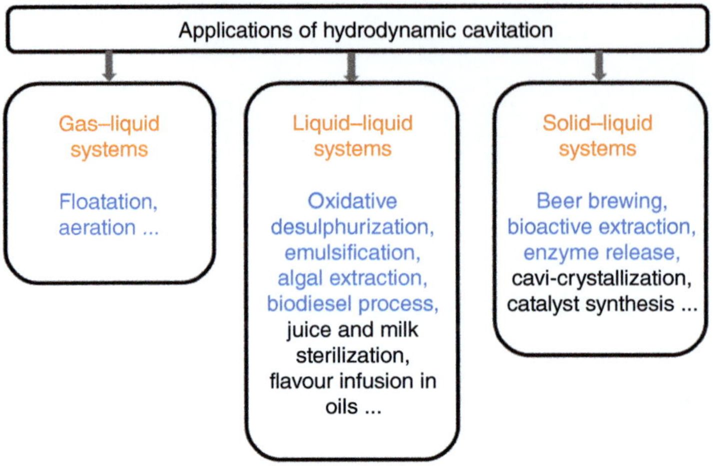

Figure 9.1 Other applications of hydrodynamic cavitation.

The process of flotation for mineral recovery has been described from as early as 1900. Since the mid-twentieth century, mechanical flotation cells and flotation columns have been extensively used for mineral recovery due to the simplicity in their operation. Flotation is a physical separation process based on surface property, used for recovering minerals. It depends on using chemical reagents to alter the interactions with minerals and to enable collection of the valuable content in the minerals by upward transport, using gas bubbles [5, 6]. The flotation process can be divided into two stages – the conditioning stage and the separation stage. In the conditioning stage, minerals and reagents are sufficiently dispersed to form the so-called "pulp."

Three important chemicals, namely collector, frother, and modifier are used in this stage to facilitate the recovery of minerals. The role of these chemical agents is to alter the interactions between the minerals of interest and the bulk media. Collectors are a class of surfactants whose role is to increase the hydrophobicity. The role of the frother is to stabilize the bubbles (froth) and entrap the hydrophobic particles, to ensure that they rise by buoyancy to the surface. The modifier acts as an agent to improve the minerals' interaction with the collector or to improve selectivity of a certain mineral over the others. The collective interaction of these chemical agents, sparged gas bubbles, and mineral particles facilitates separation. Hydrophobic particles adhere to gas bubbles, and because of buoyancy, these particle– gas aggregates rise to the suspension surface and form the froth layer. The hydrophilic particles settle to the bottom of the flotation cell as tailings. The typical equipment used for flotation is shown in Figure 9.2. Both these cells have found applicability on an industrial scale for a number of minerals, namely iron, phosphate, and coal [7, 8].

In both these types of flotation equipment, the bubble size (distribution), which controls the effective gas–liquid interfacial area and the residence time of bubbles,

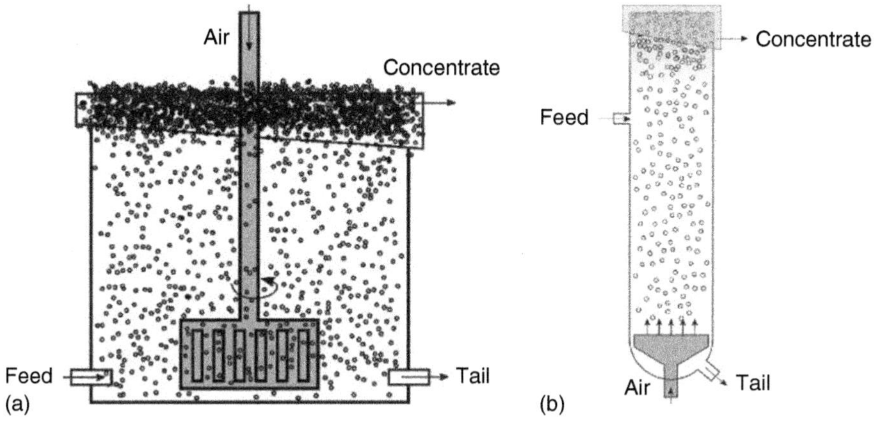

Figure 9.2 Schematic of flotation equipment. (a) mechanical floatation cell. (b) pneumatic flotation cell (Source: From [7]/with permission of Elsevier).

is a crucial parameter influencing overall effectiveness of the separation process. Bubble size in the pulp influences particle adhesion in two ways [9]. The first is via probability and the efficiency of particle–bubble collisions. The probability of collision is a function of the ratio of particle size to bubble size. The second is via attachment. The first step of particle–bubble collision must result in successful attachment of the particles to bubbles for particle recovery to take place. The attachment is governed by various processes like thinning of film, rupture, and formation of three-phase contact. These surface processes again are strongly influenced by bubble sizes and their motion. In general, it has been shown that smaller bubbles improve particle–bubble collisions and enhance attachment due to enhanced residence time. For example, Ahmed and Jameson [10] have reported an increase in mineral recovery of up to 100-fold when the bubble size was reduced from 655 to 75 μm. It is therefore desirable to create smaller-sized bubbles. Several techniques such as ultrasound and electrochemical means, which are rather expensive, have been explored for generating smaller bubbles [11]. Hydrodynamic cavitation can be harnessed for inexpensively generating smaller bubbles. Hydrodynamic cavitation acts as an energy concentrator and generates intense localized shear and energy dissipation rates (see for example, Pandit et al. [12]). Such localized shears and high energy dissipation zones have been shown to generate very small bubbles (of the order of 1 μm or lower). See for example, Pourkarimi et al. [11], who used hydrodynamic cavitation in the absence or at a lower dosage of flocculent to achieve sub-micron-sized bubbles, leading to better mineral recovery.

One of the earliest setups employing hydrodynamic cavitation in improving the performance of flotation is shown in Figure 9.3a, for a study that focused on recovering 1% w/w silica. The conventional flotation setup was altered in a manner to suit the incorporation of an in situ aeration mechanism. The venturi was used as the hydrodynamic cavitation device. The venturi, operated at cavitating conditions, generated smaller bubbles and improved recovery (see Figure 9.3b). Illustrative references for the flotation process assisted by hydrodynamic cavitation are tabulated

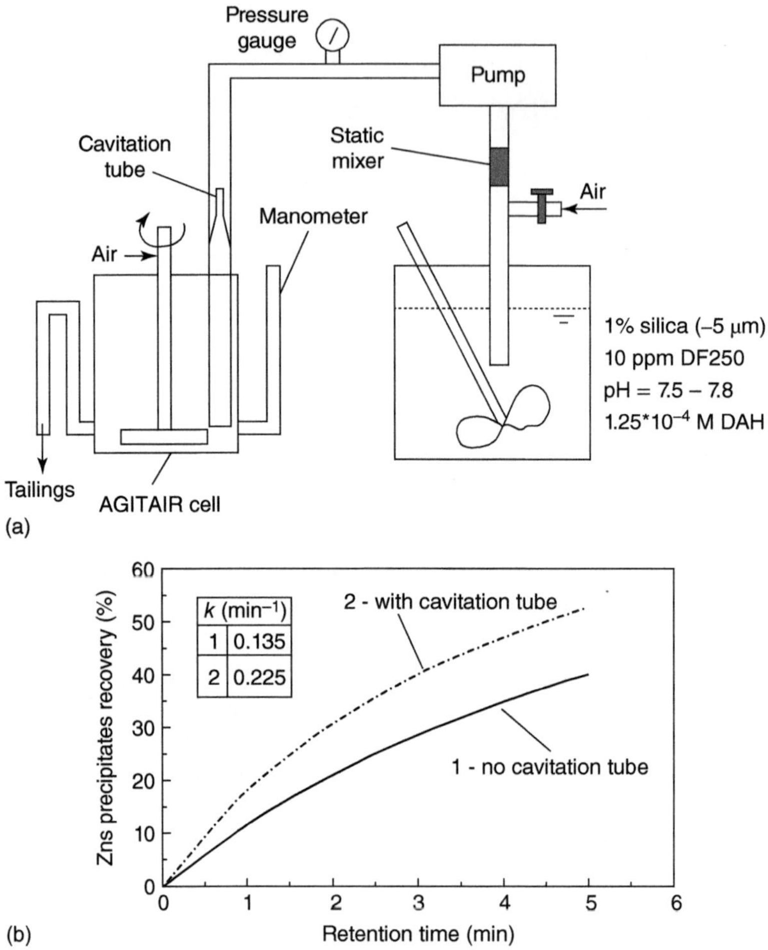

Figure 9.3 Use of hydrodynamic cavitation for flotation. (a) Schematic of set-up for laboratory based flotation test. (b) Comparison of ZnS precipitate flotation with and without cavitation tube (Source: From [15]/with permission of Elsevier).

in Table 9.1. As seen in the table, most reported studies use venturi-type cavitation devices for flotation applications. Li et al. [13] have simulated the outlet angle and the throat diameter ratio for optimizing the venturi device design required for flotation. Pourkarimi et al. [14] varied the venturi's internal diameter as 2.2 and 1.5 mm, and found that smaller bubble diameters were generated by the venturi with smaller internal diameter for the same inflow conditions.

In addition to the design of the cavitation devices, another important parameter of interest is the operating pressure drop across the cavitation device. In a study which varied the throat velocity (and as a result, the pressure drop) from 7.1 to 12.7 m/s, there was a marked increase in recovery of fine coal up to 8 m/s (see Figure 9.4) [8]. Upon further increase in throat velocity to 12.7 m/s, the recovery increased by nearly seven times. In another study which varied throat velocity for the recovery

Table 9.1 An overview of Gas–Liquid applications of hydrodynamic cavitation.

Hydrodynamic cavitation device	Hydrodynamic cavitation parameters	Other process parameters	Significant outcome due to hydrodynamic cavitation	Reference
Application: Flotation				
Key issue: Generation and stability of microbubbles/high dosage of additives				
Role of hydrodynamic cavitation: Acceleration of particle–bubble attachment/ maximizing scale of flotation/enhanced mass transfer				
Venturi "Cavitation tube"	Time – 10 min, Coal (10 g/L), working volume – 1.6 L	Excess air of 2.5 mL/min	Increase in fine coal recovery by 11-fold	[8]
Venturi	Time – 5 min, Silica (1 g/L), working volume – 3 L	Collector – 1.25×10^{-4} M Dodecylamine, Frother – 10 ppm Dowfroth 250, Air – 0.05 vvm	77% increase in mineral recovery	[15]
Venturi	25% solids	Collector – Fatty acid (FA-18G): fuel oil (1 : 1 wt ratio), Frother – Glycol (F-507)	Collector dosage decreased by 50–66% Frother dosage decreased by 50%	[18]
Venturi	Pressure – 1.5 to 3.5 bars	Collector-Frother mix – Methyl Isobutyl Carbinol, Pine oil, PEB 70, Mixed fatty acid and Apirole	Increasing frother dosage and pressure drop decreased bubble size from microns to nanometers	[14]
Venturi	Flow rate – 16.6 to 84 LPM	Frother – Methyl Isobutyl Carbinol, Sodium Oleate and *n*-pentanol	At 30 mg/L frother and pH 6, the size of bubbles varied between 150 and 650 nm	[20]

of silica, there was a maxima observed with respect to throat velocity (~15, 18, and 25 m/s), suggesting that excessive bubble production hinders the recovery [15]. While Figure 9.4 presents a monotonic increase in the recovery of coal particles, a study with 1% w/w silica showed a throat velocity (pressure drop) for the recovery in the absence of aeration (Figure 9.5), this relative increase was 1.2 times that of the conventional flotation process. With an aeration-assisted process, hydrodynamic cavitation was able to enhance the conventional flotation to about 1.7 times at the

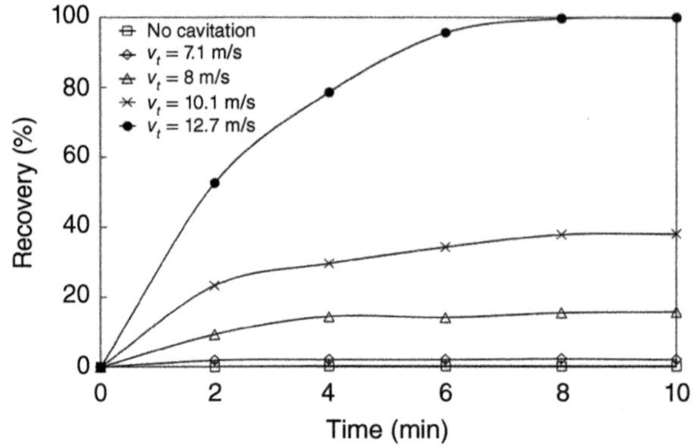

Figure 9.4 Recovery of coal particles as a function of flotation time (Source: Data from [8]/with permission of Elsevier). Hydrodynamic cavitation device: venturi. Coal added = 3%. Maximum recovery = 2.7% (value used to normalize the recovery % shown here).

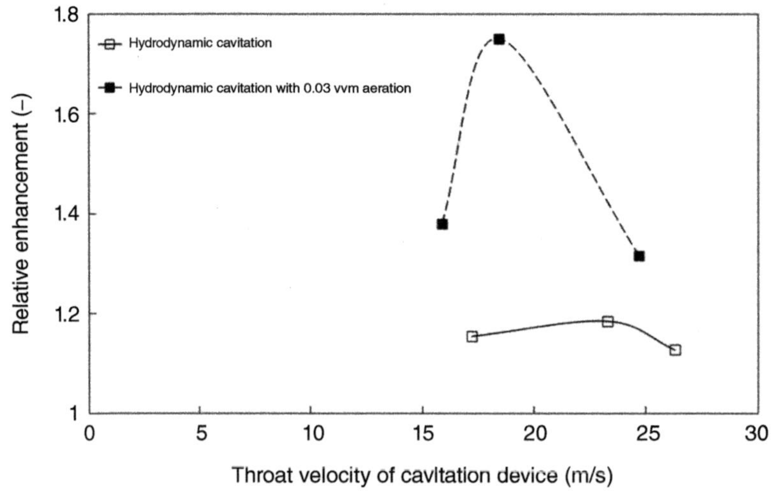

Figure 9.5 Recovery of silica by hydrodynamic cavitation and hydrodynamic cavitation assisted by aeration, as a function of throat velocity (Source: Data from [15]).

optimal throat velocity. In another study comparing the floatation process with and without hydrodynamic cavitation, in the case of three different surfactants, the incorporation of the cavitation process uniformly increased the recovery of coarse particles [16]. Similar results have also been reported by Zhou et al. [17] and Fan et al. [18]. This indicates that hydrodynamic cavitation devices are influential in enhancing the recovery of minerals by the action of in situ bubble generation.

The relevant parameters to flotation-based hydrodynamic cavitation can be grouped as device parameters (device type, pressure drop), operating parameters (pH, temperature, dissolved air flow rate), and additive parameters (frother type and dosage) [14]. Till the end of the 1990s, studies on hydrodynamic cavitation-based flotation focused on evaluating and extending the applicability to different minerals [15, 19]. The more recent studies have shifted their focus to investigation of different parameters and the application of hydrodynamic cavitation towards more difficult-to-recover fine minerals. For a process significantly impacted by aeration, and given the fact that cavitation performance is impacted by the role of dissolved gas media, it is surprising that the role of different dissolved gases on the particle recovery is limited [17]. The role of different gas media can possibly help clarify if hydrodynamic cavitation can be incorporated in the next generation of flotation cells [17].

The generation of micron or submicron bubbles is of interest to multiple domains of application such as wastewater treatment, microalgal harvesting, flotation separations, or bioremediation of contaminants. For example, Raut et al. [21] demonstrated the use of a venturi-type hydrodynamic cavitation device to produce bubbles smaller than 10 μm. It is otherwise extremely difficult to generate such smaller bubbles by mechanical agitation. Zimmerman et al. [22] have highlighted the use of hydrodynamic cavitation for microbubble generation, and its importance in areas of cell culture and microalgal growth. The application of compact hydrodynamic cavitation devices like those investigated by Simpson and Ranade [23] for generating smaller bubbles is expected to beneficially impact various gas–liquid reactors and separators.

9.3 Liquid–Liquid Applications

The performance of many liquid–liquid applications is controlled by interphase mass transfer. Similar to gas–liquid applications discussed in Section 9.2, hydrodynamic cavitation can also be used to intensify liquid–liquid mass transfer. Appropriate cavitation devices and operating conditions may be designed so that one or more liquid media cavitate. The generation of cavitating conditions leads to intense mixing, droplet breakage, and the generation of highly reactive radicals that can intensify liquid–liquid mass transfer and reactions. Illustrative applications such as oxidative desulfurization, emulsification, microalgal extraction, and the bio-diesel process using hydrodynamic cavitation are discussed in this section. A brief overview of published studies on these applications is provided in Table 9.2.

9.3.1 Oxidative Desulfurization

The consumption of petroleum and other liquid fossil fuels worldwide is beyond 5 billion tons per year [24]. The raw material for such fuels – crude oil – contains sulfur in the elemental form, hydrogen disulfide, or as organo-sulfur (thiols, sulfide, thiophene, sulfoxide, sulfone, sulfonate, sulfate, etc.) compounds in the range of 1000–60 000 ppm [25]. Sulfur compounds in automotive fuels such as diesel or gasoline emit SO_x on combustion, which can lead to human respiratory problems, reduced crop yields, and acid rains [26]. In view of these harmful effects, various countries in the world have stipulated stringent restrictions on the sulfur content in liquid fuels, with most developed countries having adopted a limit of 10 ppm of sulfur in fuels since 2018. Gasoline or any such fuel for use in fuel cells may require sulfur levels below 1 ppm to prevent catalyst poisoning.

Table 9.2 An overview of Liquid–Liquid applications of hydrodynamic cavitation.

Hydrodynamic cavitation device	Hydrodynamic cavitation parameters	Other process parameters	Significant outcome due to hydrodynamic cavitation	Reference
Application: Desulfurization and viscosity reduction of heavy oils				
Key issue: Capital-intensive reactors for hydrodesulfurization, inferior mass transfer in suggested alternative reactors				
Role of hydrodynamic cavitation: Intensified mass transfer, catalyst-free removal of certain "S" species, and reduced oxidant load for other "S" species				
Orifice	Time – 30 min, Liquid 1 – gasoline, Liquid 2 – heavy oil, Working volume – 6–8 L	2% v/v of hydrogen donor (gasoline) in heavy oil	Gasoline addition during hydrodynamic cavitation decreased the heavy oil viscosity up to 35%	[37]
Orifice, Vortex-based cavitation device	Pressure – 2 bars (Orifice) and 0.5 bars (Vortex device), Time – 120 min, Liquid 1 – Water, Liquid 2 – Diesel, Initial S content – 100–300 ppm, Working volume – 12–20 L	2.5–10% v/v of organic liquids and diesel as model fuel in water	~55% thiophene removal without catalyst and oxidant, for 10% v/v of diesel case, energy expense for sulfur removal was 176 kWh/kg (~100% removal for low organic loadings of 2.5% v/V) For 10% v/v of diesel for sulfur removal, energy expense was 41 kWh/kg	[33]
Vortex-based cavitation device	Pressure – 5 bars, Time – 60 min, Liquid 1 – Pyrolysis tire oil, Liquid 2 – Water, Initial S content – 10 500 ppm, Working volume – 15 L	$40 \pm 5\,°C$	Increasing v/v from 2.5 to 7.5%, S removal increased by over fourfold: from 7 to 30%	[36]
Orifice	Pressure – 3–6 bars, Time – 30 min, Liquid 1 – Diesel, Liquid 2 – Formic acid and hydrogen peroxide (<1% v/v), Initial S content – 1750 ppm, Working volume – 5 L	<1% v/v of aqueous phase ($HCOOH + H_2O_2$) in hydrotreated diesel. $50 \pm 5\,°C$. Oxidation succeeded by acetonitrile extraction.	Acid to oxidant ratio of 3.2 and oxidant to sulfur ratio of 40, effective in >90% desulfurization	[34]

Application: Emulsification				
Key issue: Conventional homogenizers use high-pressure and high-speed rotors				
Role of hydrodynamic cavitation: Production of stable nano emulsions				
Circular and slit venturi	Pressure – 5 to 20 bars, Time – 120 minutes, Working volume – 5 L	Coconut oil in water. Tween 80 and Span 80 as surfactants	Increase in pressure led to a decrease in droplet size up to 233 nm	[38]
Multiple devices – 5 × orifices and 2 × venturis	Pressure – 5–15 bars, Time – 120 min	Mustard oil as dispersed phase (in water). Tween 80 and Span 80 as surfactants	Stable droplets (<100 nm); Hydrodynamic cavitation – 11 times more energy efficient than acoustic cavitation	[39]
Vortex-based cavitation devices	Pressure – 50–300 kPa, up to 100 passes	Rapeseed oil as dispersed phase in water. Tween 20 as surfactant	Hydrodynamic cavitation device was found to approach energy efficiency of 1% which is significantly higher compared to conventional devices	[40]
Application: Microalgal oil extraction/biodiesel production/transesterification				
Key issue: Mass transfer limitations and high catalyst/solvent to oil molar ratio				
Role of hydrodynamic cavitation: intensify mixing, enhance mass transfer, effective alcohol utilization				
Multi-hole orifice	Pressure – 10 bars, Passes – 39, Working volume – 40 L, Algal species – *Tetraselmis suecica*, Algal concentration – 22.5 g dry algae/L, Temperature – 35 °C, Specific energy consumption – 3 MJ/kg	Hydrodynamic cavitation – lipid release *n*-hexane treatment – lipid extraction	2.5-fold increase in lipid yield vs. control	[41]
Multi-hole orifice	Pressure – 3.6 bars, Passes – 270, Algal species – *Nannochloropsis salina*, Algal concentration – 20 g dry algae/L, Specific energy consumption – 3 MJ/kg	Simultaneous lipid release and extraction (1:0.8 v/v, slurry:n-hexane)	Two-fold increase in lipid yield vs hydrodynamic cavitation only	[42]

(Continued)

Table 9.2 (Continued)

Hydrodynamic cavitation device	Hydrodynamic cavitation parameters Other process parameters	Significant outcome due to hydrodynamic cavitation	Reference
Two multi-hole orifice plates in series	Pressure – 17 bars, Time – 5 min, Oil – Pig fat, Methanol:oil molar ratio: 6:1, NaOH – 1 wt%, Temperature – 60 °C, Specific energy consumption – 0.1 MJ/kg	90% conversion to biodiesel, 2.5-fold higher yield vs. control	[43]
Multi-hole orifice	Pressure – 7 bars, Time – 5 min, Oil – Waste cooking oil, Methanol:oil molar ratio: 6.8 : 1, NaOH – 1 wt%, Temperature – 35 °C, Specific energy consumption – 0.14 MJ/kg	99% conversion and cavitational yield of 5.6 kg/MJ	[44]
Orifice	Pressure – 7 bars, Time – 30 min, Oil – Soyabean oil, Methanol:oil molar ratio: 6 : 1, KOH – 1 wt% Temperature – 45 °C, Specific energy consumption – 0.66 MJ/kg	100% conversion, 2.7-fold less energy consumption vs. mixing	[45]
Multi-hole orifice	Pressure – 3 bars, Time – 10 min, Oil – Frying oil Methanol:oil molar ratio: 6 : 1, KOH – 1 wt% Temperature – 60 °C, Specific energy consumption – 0.74 MJ/kg	95% conversion, 60% increase in conversion compared to non-cavitating flow	[46]
Multi-hole orifice	Pressure – 2 bars, Time – 15 min, Oil – Waste cooking oil, Methanol:oil molar ratio: 6 : 1, KOH – 1 wt%, Temperature – 60 °C, Specific energy consumption – 0.91 MJ/kg	98% conversion, eightfold more energy efficient vs. mixing	[47]
Orifice	Time – 90 min, Oil – C8-C10 fatty acids, Methanol:oil molar ratio: 10 : 1, H_2SO_4 – 2 wt%, Temperature – 28 °C, Specific energy consumption – 5.9 MJ/kg	95% conversion	[48]
Slit venturi	Pressure – 3 bars, Time – 30 min, Oil – Waste cooking oil, Methyl acetate:oil molar ratio: 12 : 1, NaOH – 1 wt%	89% conversion, 27% increase in biodiesel yield, ~twofold increase in transesterification rate vs control	[49]
Rotor-stator	Flow rate – 390 L/h, Time – 30 min Oil – Palm oil, Methanol:oil molar ratio: 6 : 1, NaOH – 0.4 wt%, Temperature – 55 °C	99% conversion, 8.15 kg/MJ cavitational yield	[50]

To achieve the required limits, the widely adopted practice or industry standard has been catalytic hydrodesulfurization. In this process, the liquid fuel containing sulfur comes into contact with hydrogen at 300–350 °C, 20 bars over a $CoMo/Al_2O_3$ or $NiMo/Al_2O_3$ catalyst [26]. The process involves a high capital cost, and the removal is less favorable for sulfur moieties with higher carbon number as a consequence of resonance stability and steric hindrance. As an alternative to hydrodesulfurization, oxidative desulfurization has been investigated for nearly three decades with a potential to operate at relatively benign temperatures (50–60 °C). Significant work has been undertaken in this field to develop different routes for removing or converting sulfur [26, 27]. The oxidative desulfurization process consists of the use of an oxidant with or without a catalyst. It occurs in two stages: the first is the oxidation stage, where the oxidant and fuel come into contact and the sulfur-containing species are oxidized to form sulfones or sulfoxides. In the second stage, the sulfones are extracted in suitable immiscible aqueous phase to reduce the sulfur content in the fuel. The oxidative desulfurization process needs to overcome key challenges such as reliance on catalyst (activity, recyclability), oxidant (handling, safety), and solvents (reusability, cost). Several laboratory studies have used ultrasound cavitation to intensify the process and provide a means to lower the usage of oxidant and catalyst by enhancing the mass transfer [28–31]. However, the potential of such processes is limited because of inherent difficulties in scaling up ultrasound cavitation.

Hydrodynamic cavitation can potentially be used instead of ultrasound cavitation. It can generate strongly oxidizing hydroxyl radicals in situ with water, which can also be used for preferentially extracting polar sulfones formed via oxidation of sulfur-containing species. Suryawanshi et al. [32, 33] have investigated the use of hydrodynamic cavitation for removing a model sulfur compound, thiophene, from three organic solvents (*n*-octane, toluene, *n*-octanol). The influence of the type of cavitation device and operating parameters on performance was studied, and in some cases, a very high removal – to the extent of 95% – was demonstrated. The results were also verified using commercial diesel (see Figure 9.6). It was found that cavitation yield is significantly higher for the vortex diode compared to the orifice.

In the works of Suryawanshi et al. [32, 33], no external catalyst or oxidant was used. There are several other studies where an external catalyst or oxidant has been used to enhance overall performance. For example, Baradaran and Sadeghi [34, 35] used hydrogen peroxide and formic acid (as catalyst) to remove sulfur from non-hydrotreated kerosene and a non-hydrotreated diesel. A sample of their results is shown in Figure 9.7. Their results show an enhancement due to hydrodynamic cavitation (for inlet pressure >400 kPa) when compared to non-cavitating flows (inlet pressure <400 kPa). The improvement in performance was attributed to the strong mixing conditions offered by hydrodynamic cavitation to enhance the mass transfer between the oxidant and catalyst with the fuel. The influence of device inlet pressure on oxidative desulfurization of fuel is illustrated in Figure 9.8. Though hydrodynamic cavitation looks very promising, further work is needed to fully realize its potential for intensifying oxidative desulfurisation processes.

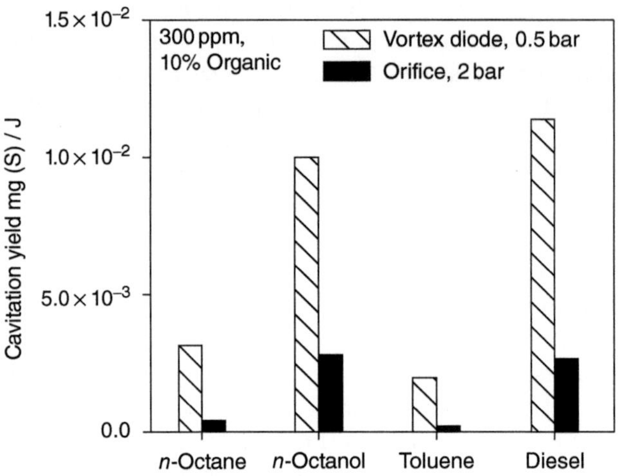

Figure 9.6 Comparison of cavitational yield for vortex diode and orifice (Source: From [33]/with permission of Elsevier).

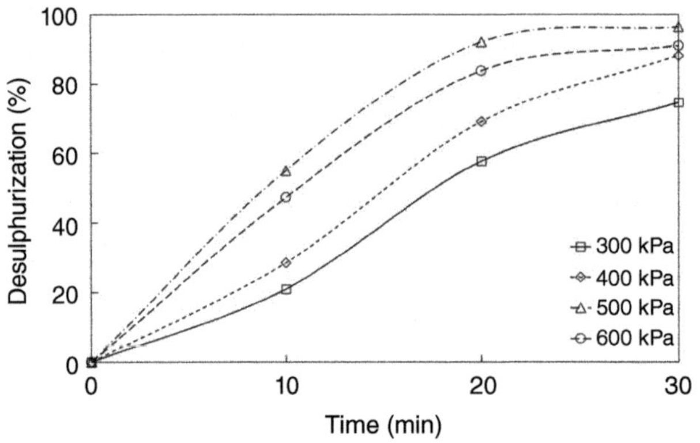

Figure 9.7 Effect of inlet pressure on kerosene desulfurization at $n_{oxidant}/n_{sulfur} = 30$, $n_{acid}/n_{sulfur} = 90$ and $T = 50\,°C$) (Source: Data from [35]/with permission of Elsevier).

9.3.2 Emulsification

Emulsification is a process of dispersing an immiscible liquid into a continuous phase. Such liquid–liquid emulsions are widely used in various industry sectors such as healthcare (drugs, active pharmaceutical ingredients), personal care (cosmetics, fragrances, beauty care), home care (paints, room fresheners), food and nutrition (milk products, ice creams, salad dressings), and agrochemicals (insecticides, fungicides, pesticides). In all of these sectors, there is an increasing demand for new ways of producing emulsions with desired droplet size distributions. Various types of equipment such as stirred tanks, high-pressure homogenizers, colloid mills, membranes, and microchannels have been used for producing emulsions. Hydrodynamic cavitation offers an attractive alternative for producing liquid–liquid emulsions at

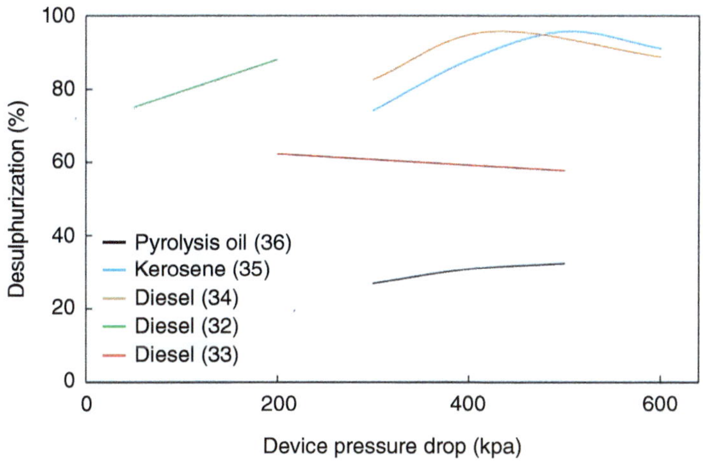

Figure 9.8 Influence of device pressure drop on oxidative desulfurization of fuel (Source: Data from: vortex – diode [36]; orifice [35]; orifice [34]; vortex – diode [32] and orifice [33]).

both the laboratory scale as well as at the industrial scale [40, 51]. It has been shown that stable emulsions can be produced with less than 10 passes through hydrodynamic cavitation devices [40, 52]. The type of device and operating pressure drop are the key factors influencing the quality of emulsions.

Ramisetty et al. [38] reported the use of two types of venturi devices: slit and circular. A minimum droplet size of 170 nm was achieved with the aid of an emulsifying agent of volume fraction of 0.11 (Tween 80 and Span 80). The inlet pressure was varied between 500 and 2000 kPa. The increase in inlet pressure from 500 kPa (after 13 passes) to 1000 kPa (after 18 passes) was found to be beneficial in lowering the droplet size from 711 to 332 nm. Carpenter et al. [39] used two venturis and five different orifice plates (different geometries) to compare their emulsification performance. The performance was evaluated at 500, 1000, and 1500 kPa inlet pressure conditions. The results from the study showed that increasing inlet pressure from 500 to 1000 kPa was beneficial in reducing the droplet size. However, further increase to a higher inlet pressure (1500 kPa in this case) did not induce a significant change. A similar observation was reported by Ramisetty et al. [38].

In addition to the droplet size, it is essential to arrive at an objective comparison in terms of energy consumption. Such an attempt is made in the study by Carpenter et al. [39]. Their results are presented in Figure 9.9. In order to compare the performance under similar energy consumption (90 ± 10 J/s), the results for different devices were compared at different values of operating pressure drop: 1000 kPa for the slit venturi, circular venturi, and the orifice plate; 500 kPa for the rest of the devices. It can be seen that droplet size is similar for the similar energy-consumption conditions. Zhang et al. reported the use of an orifice plate to generate droplet size of less than 100 nm (27 nm) for soybean oil in water [52]. Thaker and Ranade [40, 53] used rapeseed oil in water for emulsification and reported one order of magnitude reduction in droplet size in a single pass through

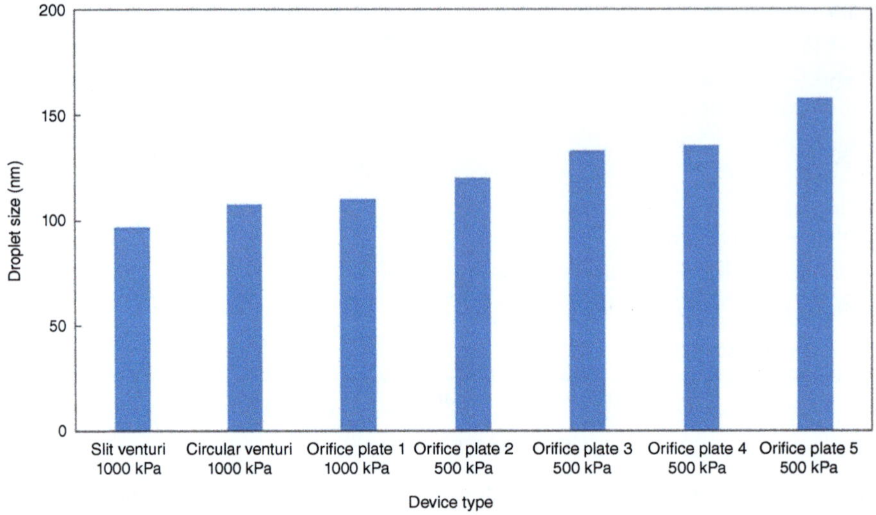

Figure 9.9 Influence of device type on droplet size of mustard oil-in-water emulsion for comparable power dissipation (Source: Data from [39]).

the hydrodynamic cavitation device. The studies used pressure drop conditions within 250 kPa, which are significantly lower than the commonly used pressure ratios for orifice and venturi devices for this application.

Thaker and Ranade [40] report the energy efficiency of the vortex-based device as 1% for oil volume fraction of 20%, which is significantly higher than conventional devices. The power consumption per unit mass of emulsion with drop size ~20 μm was found to be 260 J/kg (at 20% oil volume fraction) which is significantly lower than conventional devices. Hydrodynamic cavitation thus has a significant potential for lowering operational costs, especially with the vortex-based devices which offer a means to emulsify at low operating pressure drop.

9.3.3 Microalgal Oil Extraction

Microalgae are microscopic unicellular phytoplanktons that thrive in fresh water as well as the marine environment. Various metabolites are synthesized and stored in the microalgal cell during the growth process. These include proteins, carbohydrates, and lipids, and are present in varying quantities depending on the type of microalgae. For example, *Porphyridium cruentum* has a carbohydrate content as high as up to 57% [54], whereas *Neochloris oleoabundans* can have a high lipid content of up to 54% dry cell weight [55]. This lipid content is typically used for microalgae-based biodiesel process. The microalgal cell diameters are in the range of nm–μm. Because of their significant lipid content, the algal cells may be treated as liquid droplets rather than solid particles. It is essential to break the walls of algal cells to release the intra-cellular lipids. The disruption of cell walls typically makes the lipid extraction an energy-intensive process [41].

A range of methods – physical, chemical, and biological – have been developed to disrupt microalgae to release the oil for transesterification (biodiesel production) [56]. Recently, hydrodynamic cavitation was also used to disrupt microalgal cells for lipid extraction. The localized intense shear and energy dissipation is useful for cell disruption and subsequent release of cytoplasmic fluid (containing lipids and other value-added metabolites). Lee et al. probably used hydrodynamic cavitation for the first time to disrupt microalgae (*Tetraselmis suecica*) for lipid extraction [41]. They used a multi-hole orifice plate at 10 bars inlet pressure corresponding to a flow rate of $4\,m^3/h$ to disrupt the microalgal cells (15 kg dry algae/m^3 slurry) prepared in sea water. Upon 26 passes (calculated) with hydrodynamic cavitation treatment, corresponding to an energy requirement of 3 MJ/kg, followed by organic solvent extraction, ~0.5 g lipids/g dry weight was obtained. They proposed that the periplasmic breakdown of the microalgal cell by hydrodynamic cavitation was responsible for lipid extraction. Similar results were also obtained by Lee and Han who used a multi-hole orifice plate to disrupt *Nannochloropsis salina* cell wall [42]. After ~530 passes, at a pressure drop of 3.6 bars, a specific energy consumption of 5 MJ/kg, and a separate lipid extraction step, 70% lipid yield was obtained. When the oil extraction step was combined with cell disruption (algal slurry:*n*-hexane in 1 : 0.8 v/v), a lipid yield of 80% was obtained after 270 passes (calculated specific energy input of 3 MJ/kg). The immediate separation of the lipids from the cell debris was attributed to the increased lipid yield at a lower number of passes. Similar to other performance enhancement strategies discussed elsewhere in this book (Chapter 7), improved device designs [57] and augmented hydrodynamic cavitation with chemicals and enzymes [58, 59] can be used to improve the efficiency of cell disruption and lipid extraction. Some of the published results of lipid yield as a function of specific energy consumption are shown in Figure 9.10.

It can be seen that with an increase in specific energy consumption, the lipid yield (solid circles) increases up to a point, beyond which it decreases. These results were based on a process where hydrodynamic cavitation was used for disrupting microalgal cells, followed by a separate solvent-based lipid extraction step. It is also possible to carry out the disruption of cells and extraction in one step with hydrodynamic cavitation. The results from such a simultaneous process are indicated by the solid triangle in Figure 9.10. It can be seen that simultaneous disruption and extraction intensified by hydrodynamic cavitation further reduces energy consumption. This shows excellent promise for intensifying biodiesel processes based on microalgae.

9.3.4 Transesterification of Oils to Produce Biodiesel

Biodiesel is a renewable alternative to fossil diesel. It is found to contribute to a similar or better engine performance when compared to fossil-based diesel [61]. A lower overall emission from biodiesel and its blends has also been reported [61]. It is typically produced from fats and oils, and depending on the source of the oils, it can be classified under four generations [62]. Current trends lie in the utilization of second- and third-generation oil feedstock, namely nonedible oils (example, Jatropha oil)

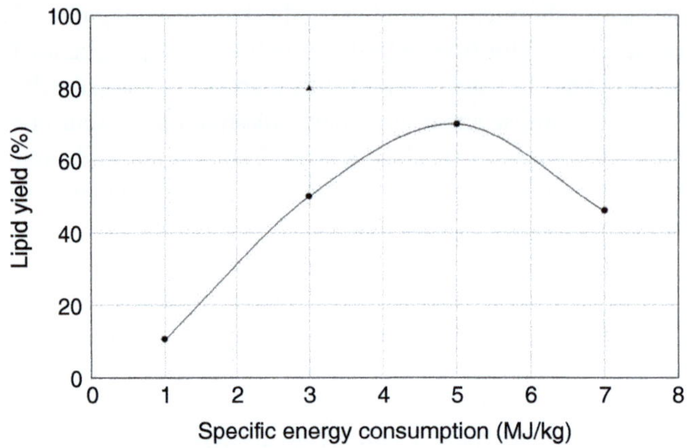

Figure 9.10 Effect of specific energy consumption for hydrodynamic cavitation on lipid extraction yield from microalgae. Solid circles represent studies performing hydrodynamic cavitation-based cell disruption followed by solvent-based lipid extraction. The solid triangle represents simultaneous cell disruption and oil extraction (Source: Data obtained from [41, 42, 57, 60]).

and microalgal lipids, respectively. In the second-generation biodiesel production process, the hindrance to cost-effective fuel production is the transesterification process, as larger than stoichiometric quantities of solvent (typically methanol) and catalyst (alkali) are required, thereby increasing the processing costs. In the case of third-generation biodiesel production, lower microalgal productivity leading to higher lipid extraction costs makes biodiesel expensive. To overcome these limitations, hydrodynamic cavitation can be used as a process intensification tool to enhance mass transfer and increase transesterification rates as well as lipid extraction yields for a higher throughput relevant for commercial applications.

The physcio-chemical effects of cavitation are known to promote chemical reactions by increasing the rate of the reactions and enhancing mass transfer [63]. Traditional transesterification reactions are alcoholysis reactions that convert triglycerides sequentially to their alkyl esters (biodiesel) (Scheme 9.1). As the rate of alcoholysis is generally slow, it is usually promoted by the presence of a catalyst [63]. This catalyst could be an enzyme (e.g. lipase), acid, or an alkali. Enzyme-mediated alcoholysis reactions are not practiced industrially as they require a longer reaction

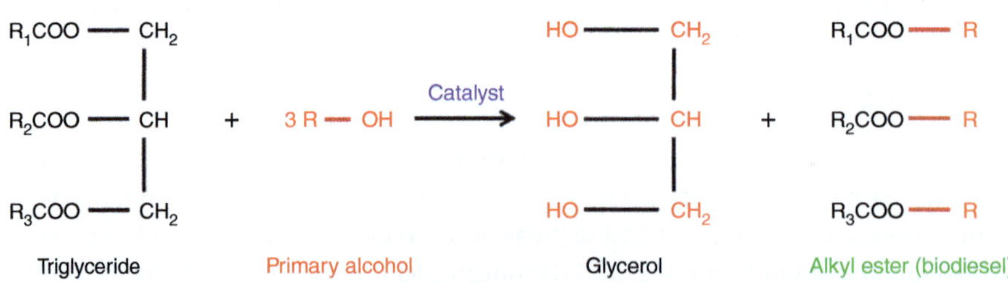

Scheme 9.1 Catalyst-mediated alcoholysis (transesterification) for the production of biodiesel.

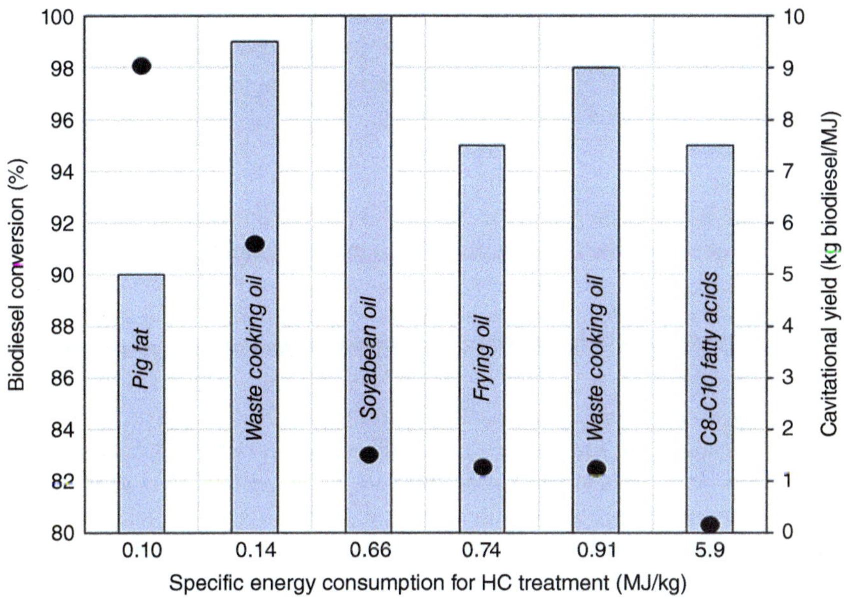

Figure 9.11 Hydrodynamic cavitation for biodiesel: conversion and cavitational yield (Source: Data obtained from Table 9.2 [Bars – Biodiesel conversion; Dots – cavitational yield]).

time to complete transesterification. Alkali is the most commonly used catalyst as it offers higher reactivity and shorter reaction times. The commercial production of biodiesel utilizes catalytic alcoholysis in batch-stirred tank reactors. The process, however, has two important hurdles.

- Mass transfer limitations – contact between catalyst/solvent and triglycerides
- High catalyst/solvent to oil molar ratio – prohibitive costs

Hydrodynamic cavitation can be harnessed to overcome these limitations and has been proposed as a potential alternative. Typical values of specific energy consumption, biodiesel conversion, and cavitational yield from published studies are shown in Figure 9.11.

In the early 1990s, Pandit and Joshi used a throttled valve as hydrodynamic cavitation device for the hydrolysis of fats/oils [64]. This initial promise later led to the use of hydrodynamic cavitation as a process intensification methodology for catalyst-mediated alcoholysis. For example, Ji et al. used hydrodynamic cavitation generated via an orifice plate for the alkali-mediated transesterification of soyabean oil [45]. They reported that hydrodynamic cavitation required lower energy consumption compared to stirring and sonication for complete transesterification of unit weight of soya bean oil. Kelkar et al. reported the use of orifice-generated hydrodynamic cavitation on the transesterification of vegetable oils in the presence of a homogeneous acid catalyst [48]. They have reported beneficial reduction of the methanol:oil molar ratio from 10 : 1 to 2 : 1. They also reported an improved cavitational yield (by over an order of magnitude) with hydrodynamic cavitation compared to the acoustic cavitation. Pal et al. investigated the use of a range of multi-hole orifice plates on the alkali-mediated alcoholysis of *Citrullus colocyntis* (Thumba) oil and observed that the initial rate of transesterification was higher with the increased number of holes in the orifice [65].

Unlike Pal et al. who investigated the use of orifice with 1, 3, 5, and 7 holes (of constant diameter of 3 mm), Ghayal et al. [46] investigated the effect of orifice with 1, 16, 20, and 25 holes of varying diameters and inlet pressures on transesterification. They concluded that the orifice with the smallest hole diameter had the highest biodiesel yield and attributed this to the formation of better emulsions and higher mass transfer with such devices. Similar to Ghayal et al., Adnađević et al. reported an increased conversion of triglyceride with an increase in inlet pressure for their venturi device [66]. Adnađević et al. reported that the triglyceride conversion increases with an increase in methanol:oil ratio, alkali loading, and number of passes. Other researchers have also used multi-hole orifice plates [47, 67, 68], venturi tubes [69], rotor-stator [70], or devices such as high-pressure homogenizers [71] to intensify the biodiesel production process. Orifice plates with much smaller holes (0.4 mm diameter) and orifices in series have also been investigated for biodiesel production but using animal fats as feedstock [43]. All these reports yielded a similar conclusion that with increase in specific energy (function of inlet pressure, flow rate, and number of passes), alkali loading, and alcohol:oil molar ratio, an improved biodiesel conversion was obtained. Temperature was found to have a less significant effect on the intensification process.

Similar to the microalgal lipid extraction process, transesterification is primarily influenced by the physical effects of cavitation – mainly intense mixing and generation of smaller droplets to overcome the mass transfer limitations. Hydrodynamic cavitation can therefore be used for reducing the chemical requirements and energy consumption while increasing the biodiesel yield.

9.3.5 Food (Juice and Milk) Sterilization

The demand for ready-to-eat foods and beverages is continuously increasing. Industrial food manufacturing typically uses thermal-based processes to reduce microbial and enzymatic activity in processed food and beverages. These processes use elevated temperatures of up to 120 °C to reduce the activity of microbe and enzyme [72] and therefore are energy-intensive. Furthermore, higher temperatures used in such processes can affect the quality of the product, especially when dealing with foods containing heat-sensitive compounds. Hydrodynamic cavitation offers an attractive method for disinfection, as discussed in Chapter 6 for water. The applications of hydrodynamic cavitation for sterilization can be extended beyond water for a variety of beverages and food products with the advantage of lower energy consumption over thermal and electric processing methods. The local hotspots, intense shear, and highly oxidizing radicals can realize effective sterilization by killing the microbes present in the food/beverages. Some of the examples reported in literature have used hydrodynamic cavitation to sterilize products such as skimmed milk and apple and tomato juices [4, 72]. In the coming years, hydrodynamic cavitation, either alone or in combination with other complementary techniques like UV, is expected to be increasingly used for developing improved sterilization processes with significantly reduced energy consumption.

9.4 Solid–Liquid Applications

The phrase "solid–liquid" is used here to bring under one umbrella any process that involves a solid phase suspended in a liquid phase (slurries). These can be, but are not limited to biomass pre-treatment, catalyst synthesis, crystallization, and bio-product extraction. Biomass pre-treatment for enhancing the yield of biogas and bioethanol have been discussed previously in Chapter 8. In most of the solid–liquid applications, localized intense shear and high energy dissipation rates are used for reducing particle size and enhancing solid–liquid mass transfer. Some of the published studies on using hydrodynamic cavitation for solid–liquid applications are summarized in Table 9.3.

The food and beverage industry often finds itself with the usage of slurries for extracting flavoring agents. Hydrodynamic cavitation is therefore identified as an intensification tool which will play a significant role within the industry. While the traditional use of hydrodynamic cavitation in the food industry started with sterilization of liquid streams of temperature-sensitive beverages, hydrodynamic cavitation is being increasingly explored for solid–liquid applications within the industry. Two illustrative applications, beer brewing and flavonoid extraction, are briefly discussed here.

9.4.1 Beer Brewing

Hydrodynamic cavitation has been effectively used to improve the quality of beer and the energy efficiency of the process, while being scalable. The specific applications in the brewing industry span wort sterilization, avoiding dry milling of malts, intensifying malt saccharification, heating the wort, and foam stabilization. Combined together, these applications within the industry make the hydrodynamic cavitation-assisted brewing process safe, reliable, and energy efficient [73]. For instance, Albanese et al. [73] were able to improve the saccharification efficiency of the malt by 30% compared to the traditional process while consuming significantly less energy with a hydrodynamic cavitation-based process (Figure 9.12). They were able to do so while completely avoiding the energy-intensive dry milling of grains and boiling. In addition, they were also able to demonstrate that "gluten-free" and "very-low gluten-containing" beer production would be possible with hydrodynamic cavitation-assisted brewing [74]. Furthermore, hydrodynamic cavitation was also found to be useful in enriching the polyphenols, flavonoids, and antioxidant concentration in beer during the "hopping" stage of brewing [80].

Table 9.3 An overview of Solid–Liquid applications of hydrodynamic cavitation.

Hydrodynamic cavitation device	Application	Feedstock	Significant outcome	Reference
Venturi	Beer brewing	Malt grain	Avoid malt milling, wort sterilization, lower water and energy foot print, improved saccharification	[73]
Venturi	Beer brewing	Malt grain	Completely remove gluten, potential to replace current state-of-the-art	[74]
Orifice	Invertase enzyme release	Yeast cells	Order of magnitude increase in selective invertase release vs. high-pressure homogenization	[75]
Venturi	Bioactive compound extraction	Waste orange peels	Pectin (with long-term stability), polyphenolic compounds and terpene (limonene) extraction at a significantly higher rate than state-of-the art (hydro-distillation)	[76]
Venturi	Bioactive compound extraction	Waste lemon peels	Extraction of pectin in water (compared to conventional mild acid extraction), improved anti-microbial and anti-oxidant activity of pectin	[77, 78]
Vortex diode	Anti-solvent crystallization	Paracetamol – methanol – water	In-line combination of crystallizer and hydrodynamic cavitation was found to reduce mean and variance of crystals	[79]

9.4.2 Bioactive Compound Extraction

The extraction of bioactive compounds and the simultaneous infusion of the flavor into beer are specific to brewing. However, biomass cell disruption for the release of bioactive high-value compounds has also been performed using hydrodynamic cavitation. The traditional use of high-pressure homogenization for cell disruption [81] can be replaced with hydrodynamic cavitation. For instance, Mevada et al. compared high-pressure homogenization and hydrodynamic cavitation for the release of the enzyme periplasmic invertase from yeast cells [75]. Invertase is an important enzyme with application in the food and beverage industry, used for the hydrolysis of sucrose. Conventional production uses high-pressure homogenization.

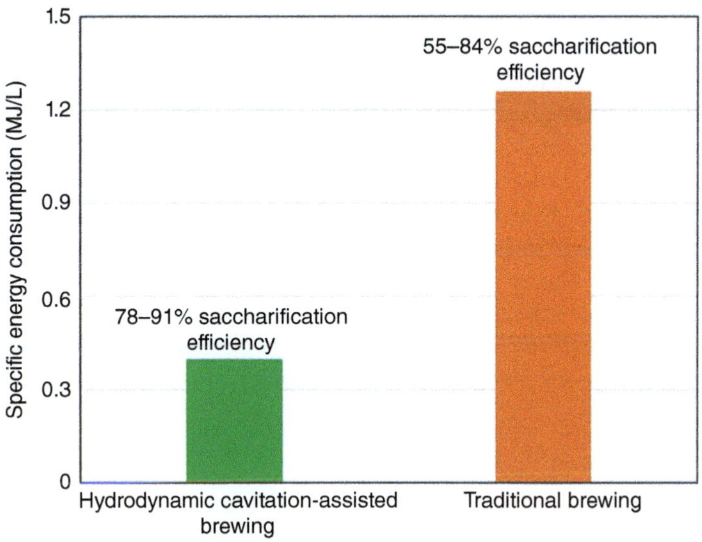

Figure 9.12 A comparison of saccharification efficiency between traditional brewing and hydrodynamic cavitation-assisted brewing across a range of specific energy consumptions (Source: Data based on [73]).

Mevada et al. [75] used a 1% yeast suspension in 5 L of 0.12 M acetate buffer (pH 5) and investigated the total protein and specific invertase release using hydrodynamic cavitation (single-hole orifice, 2 mm diameter) as a function of inlet pressure (4–8 bar) and number of passes (0–60 passes). Invertase release and selectivity were compared against high-pressure homogenization (~200 bar for 15 passes). At 6-bar inlet pressure and 10 passes, hydrodynamic cavitation was able to facilitate an invertase release of 12 U/mL (U is unit enzyme) with a selectivity of 301 U/mg. High-pressure homogenization, on the other hand, was able to release a higher concentration of invertase into solution (98 U/mL) but with lower selectivity (103 U/mg). In terms of energy efficiency for invertase release, high-pressure homogenization demonstrated 0.113 U/J whereas hydrodynamic cavitation was significantly more efficient (1.6 U/J).

In addition to revisiting conventional cell disruption applications, modern applications in the food and beverage industry have also been proposed. For instance, Meneguzzo et al. proposed the use of hydrodynamic cavitation generated using a venturi tube for pectin, polyphenols, and terpene (limonene) extraction from waste orange peels [76]. Water-soluble bioactive polyphenolic compounds were also extracted from waste orange peels via the mediated process. They reported the rate to be significantly higher than that of the state-of-the-art hydro-distillation process. Pectin extracted from the orange peels via hydrodynamic cavitation was found to be stable even after a storage period of 18 months. Similarly, they also demonstrated the extraction of pectin from waste lemon peels [77]. It was observed that unlike the conventional pectin extraction process involving mild acid, hydrodynamic cavitation-assisted extraction could be made possible in water. They further reported that the novel IntegroPectin, produced via hydrodynamic cavitation, is superior in antibacterial activity compared to the commercially available citrus pectins [78]. A similar solvent-free hydrodynamic cavitation-based process was also able to produce an antioxidant-rich aqueous solution of silver fir needles, whose antioxidant activities were better than common standards and beverages [82]. Interestingly, alongside other flavonoids reported earlier, it was possible to extract hesperidin and naringin from citrus fruit peels in solvent-free aqueous-based slurries [83]. These compounds have significant therapeutic properties and were especially found to alleviate the respiratory tract infection symptoms caused by COVID-19. With newer device designs with an ability to tailor number density and intensity of cavitation events, hydrodynamic cavitation is expected to find more applications in the extraction of various high-value bioactives.

9.4.3 Particle Size Reduction

Hydrodynamic cavitation-mediated particle size reduction is a broad term that covers a variety of applications. Catalyst synthesis with controlled particle size [84–87] and biochar production with an increased BET surface [88] area are the two main applications reported in this category. Particle size reduction is dependent on shear generated by the collapsing cavities. The extent of cavitation dependent on the specific energy input is related to the shear intensity, and therefore would influence the particle size distribution. For instance, Albanese et al. [88] investigated the use of a venturi-based hydrodynamic cavitation device to improve the surface area (reduce particle size) of biochar. Biochar finds its application as a soil additive due to its excellent water and nutrient adsorption capacities that can enable slow release of these components to the plants. A higher porosity with a higher surface area is therefore critical to achieve. Biochar is typically produced from slow pyrolysis, fast pyrolysis, or hydrothermal carbonization. The quantity and quality of biochar are dependent on the feedstock used. The effect of specific energy consumption on the biochar properties (mainly adsorption/desorption and surface area) was investigated by Albanese et al. [88]. They reported that with an increase in specific energy consumption up to 3.5 MJ/kg, the BET surface area increased from ~60 to ~131 m^2/g. Any further increase in specific energy consumption did not lead to a significant increase in the surface area. The increase in microporus and mesoporous surface areas (and pore volume) also followed a similar trend. Further implications such as the possibility of avoiding high temperatures and pressures for the pyrolytic production of biochars, as well as excessive alkalization could be avoided when a hydrodynamic cavitation-based processing step was used. Similar effects on particle size distribution of Pt/C catalysts by hydrodynamic cavitation were reported by Kuroki et al. [85]. They were able to produce sub-micron-sized catalyst particles by using hydrodynamic cavitation. The resulting size distribution was found to be similar to ball milling (with a narrow particle size distribution) as opposed to ultrasonication (with a broad particle size distribution). Furthermore, the performance of the fuel cells was comparable to that operated with ball-milled catalysts, thus showing immense promise. Madane and Ranade [79] reported the use of vortex-based hydrodynamic cavitation for reducing the mean and width of crystal size distributions obtained via anti-solvent crystallization. Considering the ability to realize localized intense shear and energy dissipation rates, combined with the wide choice of cavitation devices and scalability, the use of hydrodynamic cavitation for a wide range of applications requiring particle breakage is likely to expand in the coming years.

9.5 Summary, Outlook, and Conclusions

Hydrodynamic cavitation has been found to be an effective method for intensifying a wide range of processes relevant to various industry sectors ranging from petroleum to pharmaceutical and food industries. Illustrative applications involving gas–liquid, liquid–liquid, and solid–liquid systems are briefly discussed in this chapter. These applications are based on creatively using key physico-chemical effects of hydrodynamic cavitation, namely, generation of intense shear, high pressure and temperature zones, and hydroxyl radicals for process intensification. Intense shear, high-velocity jet, and high energy dissipation rates generated by collapsing cavities are used for breaking up gas, liquid, or solid particles to produce particles of ~1 μm size, thereby enhancing interfacial area by an order of magnitude more than conventional devices. Smaller bubbles and enhanced interfacial area are found beneficial for flotation as well as for many other gas–liquid applications where mass transfer has been found to be limiting. Similarly, for several other liquid–liquid and solid–liquid systems, the energy consumption required for generation of additional interfacial area is far less for hydrodynamic cavitation-based processes as compared to other conventional methods. Apart from enhanced interfacial area and transport rates, the hydroxyl radicals generated by hydrodynamic cavitation also contribute to intensification of reactions like oxidation or sterilization. Besides the applications discussed here, hydrodynamic cavitation has been used for a wide range of different applications such as food and beverage sterilization [4], viscosity reduction [89], internal surface finishing of metals [90], liposome degradation [91], particle size reduction [88], heat transfer [92], and many more.

The key to expanding application horizons of hydrodynamic cavitation or translating laboratory-scale results to commercial practice is clearly in establishing quantitative relationships among design of cavitation devices, operating parameters, and critical process parameters. It is essential to identify the performance-controlling step and improve its rate by leveraging the physico-chemical effects of hydrodynamic cavitation. It may be noted that in many applications, hydrodynamic cavitation may have to be used along with other complementary intensification strategies. The unique physico-chemical effects of hydrodynamic cavitation can be synergistically harnessed for a wide range of applications in process intensification as well as for new product development. The material presented in this book will be useful in achieving this. The illustrative applications discussed in Chapters 6–8 and in this chapter will also provide an insight into the overall methodology for developing various applications based on hydrodynamic cavitation. These will serve as general guidelines for developing any new applications.

References

1 Mancuso, G., Langone, M., and Andreottola, G. (2020). A critical review of the current technologies in wastewater treatment plants by using hydrodynamic cavitation process: principles and applications. *Journal of Environmental Health Science and Engineering* 18 (1): 311–333.

2 Panda, D., Saharan, V.K., and Manickam, S. (2020). Controlled hydrodynamic cavitation: a review of recent advances and perspectives for greener processing. *Processes* 8 (2): 220.

3 Holkar, C.R., Jadhav, A.J., Pinjari, D.V., and Pandit, A.B. (2019). Cavitationally driven transformations: a technique of process intensification. *Industrial & Engineering Chemistry Research* 58 (15): 5797–5819.

4 Asaithambi, N., Singha, P., Dwivedi, M., and Singh, S.K. (2019). Hydrodynamic cavitation and its application in food and beverage industry: a review. *Journal of Food Process Engineering* 42 (5): e13144.

5 Rao, S.R. (2013). *Surface Chemistry of Froth Flotation*, Fundamentals, vol. 1. Springer Science & Business Media.

6 Kumar, D. and Kumar, D. (2018). *Sustainable Management of Coal Preparation*. Woodhead Publishing.

7 Mesa, D. and Brito-Parada, P.R. (2019). Scale-up in froth flotation: a state-of-the-art review. *Separation and Purification Technology* 210: 950–962.

8 Li, H., Afacan, A., Liu, Q., and Xu, Z. (2015). Study interactions between fine particles and micron size bubbles generated by hydrodynamic cavitation. *Minerals Engineering* 84: 106–115.

9 Yoon, R.H. (2000). The role of hydrodynamic and surface forces in bubble–particle interaction. *International Journal of Mineral Processing* 58 (1–4): 129–143.

10 Ahmed, N. and Jameson, G.J. (1985). The effect of bubble size on the rate of flotation of fine particles. *International Journal of Mineral Processing* 14 (3): 195–215.

11 Pourkarimi, Z., Rezai, B., Noaparast, M. et al. (2021). Proving the existence of nanobubbles produced by hydrodynamic cavitation and their significant effects in powder flotation. *Advanced Powder Technology* 32 (5): 1810–1818.

12 Pandit, A.V., Sarvothaman, V.P., and Ranade, V.V. (2021). Estimation of chemical and physical effects of cavitation by analysis of cavitating single bubble dynamics. *Ultrasonics Sonochemistry* 77: 105677.

13 Li, M., Bussonnière, A., Bronson, M. et al. (2019). Study of Venturi tube geometry on the hydrodynamic cavitation for the generation of microbubbles. *Minerals Engineering* 132: 268–274.

14 Pourkarimi, Z., Rezai, B., and Noaparast, N. (2017). Effective parameters on generation of nanobubbles by cavitation method for froth flotation applications. *Physicochemical Problems of Mineral Processing* 53: 920–942.

15 Zhou, Z.A., Xu, Z., Finch, J.A. et al. (1997). Role of hydrodynamic cavitation in fine particle flotation. *International Journal of Mineral Processing* 51 (1–4): 139–149.

16 Nazari, S., Chelgani, S.C., Shafaei, S.Z. et al. (2019). Flotation of coarse particles by hydrodynamic cavitation generated in the presence of conventional reagents. *Separation and Purification Technology* 220: 61–68.

17 Zhou, Z.A., Xu, Z., Finch, J.A. et al. (2009). On the role of cavitation in particle collection in flotation – a critical review II. *Minerals Engineering* 22 (5): 419–433.

18 Fan, M. and Tao, D. (2008). A study on picobubble enhanced coarse phosphate froth flotation. *Separation Science and Technology* 43 (1): 1–10.

19 Rao, S.R., Finch, J.A., Zhou, Z., and Xu, Z. (1998). Relative flotation response of zinc sulfide: mineral and precipitate. *Separation Science and Technology* 33 (6): 819–833.

20 Zhang, X.-y., Wang, Q.-s., Wu, Z.-x., and Tao, D.-p. (2020). An experimental study on size distribution and zeta potential of bulk cavitation nanobubbles. *International Journal of Minerals, Metallurgy, and Materials* 27 (2): 152–161.

21 Raut, J.S., Stoyanov, S.D., Duggal, C. et al. (2012). Hydrodynamic cavitation: a bottom-up approach to liquid aeration. *Soft Matter* 8 (17): 4562–4566.

22 Zimmerman, W.B., Tesar, V., Butler, S., and Bandulasena, H.C.H. (2008). Microbubble generation. *Recent Patents on Engineering* 2 (1): 1–8.

23 Simpson, A. and Ranade, V.V. (2019). 110th Anniversary: comparison of cavitation devices based on linear and swirling flows: hydrodynamic characteristics. *Industrial & Engineering Chemistry Research* 58 (31): 14488–14509.

24 Isa, Y.M. and Ganda, E.T. (2018). Bio-oil as a potential source of petroleum range fuels. *Renewable and Sustainable Energy Reviews* 81: 69–75.

25 Shi, Q. and Wu, J. (2021). Review on sulfur compounds in petroleum and its products: state-of-the-art and perspectives. *Energy & Fuels* 35 (18): 14445–14461.

26 Rajendran, A., Cui, T.-y., Fan, H.-x. et al. (2020). A comprehensive review on oxidative desulfurization catalysts targeting clean energy and environment. *Journal of Materials Chemistry A* 8 (5): 2246–2285.

27 Avvaru, B., Venkateswaran, N., Uppara, P. et al. (2018). Current knowledge and potential applications of cavitation technologies for the petroleum industry. *Ultrasonics Sonochemistry* 42: 493–507.

28 Dana, M., Sobati, M.A., Shahhosseini, S., and Ansari, A. (2020). Optimization of a continuous ultrasound assisted oxidative desulfurization (UAOD) process of diesel using response surface methodology (RSM) considering operating cost. *Chinese Journal of Chemical Engineering* 28 (5): 1384–1396.

29 Ghahremani, H., Nasri, Z., and Eikani, M.H. (2021). Ultrasound-assisted oxidative desulfurization (UAOD) of Iranian heavy crude oil: investigation of process variables. *Journal of Petroleum Science and Engineering* 204: 108709.

30 Wu, Z. and Ondruschka, B. (2010). Ultrasound-assisted oxidative desulfurization of liquid fuels and its industrial application. *Ultrasonics Sonochemistry* 17 (6): 1027–1032.

31 Wan, M.W., Kan, C.C., and Yen, T.F. (eds.) (2010). Ultrasound-assisted oxidative desulfurization (UAOD) process on model sulfur compounds in polyoxometalate/H2O2 system and its kinetic study. *International Advanced Oxidation Processes Conference.*

32 Suryawanshi, N.B., Bhandari, V.M., Sorokhaibam, L.G., and Ranade, V.V. (2016). A non-catalytic deep desulphurization process using hydrodynamic cavitation. *Scientific Reports* 6 (1): 1–8.

33 Suryawanshi, N.B., Bhandari, V.M., Sorokhaibam, L.G., and Ranade, V.V. (2017). Developing techno-economically sustainable methodologies for deep desulfurization using hydrodynamic cavitation. *Fuel* 210: 482–490.

34 Baradaran, S. and Sadeghi, M.T. (2019). Intensification of diesel oxidative desulfurization via hydrodynamic cavitation. *Ultrasonics Sonochemistry* 58: 104698.

35 Baradaran, S. and Sadeghi, M.T. (2020). Desulfurization of non-hydrotreated kerosene using hydrodynamic cavitation assisted oxidative desulfurization (HCAOD) process. *Journal of Environmental Chemical Engineering* 8 (4): 103832.

36 Uebe, J., Kryževičius, Ž., Janutėnienė, J. et al. (2021). Desulfurizing of pyrolysis oil of used tires using a 3D-printed vortex diode and modeling of process. *Journal of Marine Science and Engineering* 9 (8): 876.

37 Askarian, M., Vatani, A., and Edalat, M. (2017). Heavy oil upgrading via hydrodynamic cavitation in the presence of an appropriate hydrogen donor. *Journal of Petroleum Science and Engineering* 151: 55–61.

38 Ramisetty, K.A., Pandit, A.B., and Gogate, P.R. (2014). Novel approach of producing oil in water emulsion using hydrodynamic cavitation reactor. *Industrial & Engineering Chemistry Research* 53 (42): 16508–16515.

39 Carpenter, J., George, S., and Saharan, V.K. (2017). Low pressure hydrodynamic cavitating device for producing highly stable oil in water emulsion: effect of geometry and cavitation number. *Chemical Engineering and Processing: Process Intensification* 116: 97–104.

40 Thaker, A.H. and Ranade, V.V. (2021). Drop breakage in a single pass through vortex based cavitation device: experiments and modelling. *AIChE Journal* e17512.

41 Lee, A.K., Lewis, D.M., and Ashman, P.J. (2015). Microalgal cell disruption by hydrodynamic cavitation for the production of biofuels. *Journal of Applied Phycology* 27 (5): 1881–1889.

42 Lee, I. and Han, J.-I. (2015). Simultaneous treatment (cell disruption and lipid extraction) of wet microalgae using hydrodynamic cavitation for enhancing the lipid yield. *Bioresource Technology* 186: 246–251.

43 Díez, D., Urueña, A., Barrios, C., and Antolín, G. (2019). Valorization of rendering fats to produce biodiesel by single and multi orifice plate cavitation reactor. *Waste and Biomass Valorization* 10 (12): 3773–3780.

44 Bargole, S., George, S., and Kumar, S.V. (2019). Improved rate of transesterification reaction in biodiesel synthesis using hydrodynamic cavitating devices of high throat perimeter to flow area ratios. *Chemical Engineering and Processing: Process Intensification* 139: 1–13.

45 Ji, J., Wang, J., Li, Y. et al. (2006). Preparation of biodiesel with the help of ultrasonic and hydrodynamic cavitation. *Ultrasonics* 44: e411–e414.

46 Ghayal, D., Pandit, A.B., and Rathod, V.K. (2013). Optimization of biodiesel production in a hydrodynamic cavitation reactor using used frying oil. *Ultrasonics Sonochemistry* 20 (1): 322–328.

47 Chuah, L.F., Yusup, S., Abd Aziz, A.R. et al. (2016). Cleaner production of methyl ester using waste cooking oil derived from palm olein using a hydrodynamic cavitation reactor. *Journal of Cleaner Production* 112: 4505–4514.

48 Kelkar, M.A., Gogate, P.R., and Pandit, A.B. (2008). Intensification of esterification of acids for synthesis of biodiesel using acoustic and hydrodynamic cavitation. *Ultrasonics Sonochemistry* 15 (3): 188–194.

49 Maddikeri, G.L., Gogate, P.R., and Pandit, A.B. (2014). Intensified synthesis of biodiesel using hydrodynamic cavitation reactors based on the interesterification of waste cooking oil. *Fuel* 137: 285–292.

50 Crudo, D., Bosco, V., Cavaglià, G. et al. (2016). Biodiesel production process intensification using a rotor-stator type generator of hydrodynamic cavitation. *Ultrasonics Sonochemistry* 33: 220–225.

51 Ranade, V.V. and Simpson, A. (2019). *Cavitation for Emulsions*. United Kingdom: Droplets Durham.

52 Zhang, Z., Wang, G., Nie, Y., and Ji, J. (2016). Hydrodynamic cavitation as an efficient method for the formation of sub-100 nm O/W emulsions with high stability. *Chinese Journal of Chemical Engineering* 24 (10): 1477–1480.

53 Thaker, A.H. and Ranade, V.V. (2020). Towards 'factory in a box' for personalised emulsions on demand: harnessing vortex based hydrodynamic cavitation. *Formulation* 41.

54 Markou, G., Angelidaki, I., and Georgakakis, D. (2012). Microalgal carbohydrates: an overview of the factors influencing carbohydrates production, and of main bioconversion technologies for production of biofuels. *Applied Microbiology and Biotechnology* 96 (3): 631–645.

55 Goh, B.H.H., Ong, H.C., Cheah, M.Y. et al. (2019). Sustainability of direct biodiesel synthesis from microalgae biomass: a critical review. *Renewable and Sustainable Energy Reviews* 107: 59–74.

56 Lee, A.K., Lewis, D.M., and Ashman, P.J. (2012). Disruption of microalgal cells for the extraction of lipids for biofuels: processes and specific energy requirements. *Biomass and Bioenergy* 46: 89–101.

57 Lee, I., Oh, Y.-K., and Han, J.-I. (2019). Design optimization of hydrodynamic cavitation for effectual lipid extraction from wet microalgae. *Journal of Environmental Chemical Engineering* 7 (2): 102942.

58 Waghmare, A., Nagula, K., Pandit, A., and Arya, S. (2019). Hydrodynamic cavitation for energy efficient and scalable process of microalgae cell disruption. *Algal Research* 40: 101496.

59 Lee, I. and Han, J.-I. (2021). Development of a pretreatment method based on Fenton-like reaction combined with hydrodynamic cavitation for lipid extraction from wet microalgae. *Renewable Energy* 175: 415–421.

60 Setyawan, M., Budiman, A., Mulyono, P., and Sutijan, S. (2018). Optimum extraction of algae-oil from microalgae using hydrodynamic cavitation. *International Journal of Renewable Energy Research* 8 (1): 451–458.

61 Tamilselvan, P., Nallusamy, N., and Rajkumar, S. (2017). A comprehensive review on performance, combustion and emission characteristics of biodiesel fuelled diesel engines. *Renewable and Sustainable Energy Reviews* 79: 1134–1159.

62 Singh, D., Sharma, D., Soni, S.L. et al. (2020). A review on feedstocks, production processes, and yield for different generations of biodiesel. *Fuel* 262: 116553.

63 Patil, A., Baral, S., and Dhanke, P. (2021). Hydrodynamic cavitation for process intensification of biodiesel synthesis – a review. *Current Research in Green and Sustainable Chemistry* 4: 100144.

64 Pandit, A.B. and Joshi, J.B. (1993). hydrolysis of fatty oils: effect of cavitation. *Chemical Engineering Science* 48 (19): 3440–3442.

65 Pal, A., Verma, A., Kachhwaha, S.S., and Maji, S. (2010). Biodiesel production through hydrodynamic cavitation and performance testing. *Renewable Energy* 35 (3): 619–624.

66 Adnađević, B.K., Nikolić, I.R., Milenković, S.A., and Jovanović, J.D. (2021). The effect of operating parameters of hydrodynamic cavitation – assisted alkaline catalyzed transesterification of sunflower oil with methanol on the degree of triglyceride conversion. *International Journal of Chemical Reactor Engineering* https://doi.org/10.1515/ijcre-2021-0131.

67 Chuah, L.F., Klemeš, J.J., Yusup, S. et al. (2017). Kinetic studies on waste cooking oil into biodiesel via hydrodynamic cavitation. *Journal of Cleaner Production* 146: 47–56.

68 Chuah, L.F., Aziz, A.R.A., Yusup, S. et al. (2015). Performance and emission of diesel engine fuelled by waste cooking oil methyl ester derived from palm olein using hydrodynamic cavitation. *Clean Technologies and Environmental Policy* 17 (8): 2229–2241.

69 Chitsaz, H., Omidkhah, M., Ghobadian, B., and Ardjmand, M. (2018). Optimization of hydrodynamic cavitation process of biodiesel production by response surface methodology. *Journal of Environmental Chemical Engineering* 6 (2): 2262–2268.

70 Chipurici, P., Vlaicu, A., Calinescu, I. et al. (2019). Ultrasonic, hydrodynamic and microwave biodiesel synthesis – a comparative study for continuous process. *Ultrasonics Sonochemistry* 57: 38–47.

71 Mohod, A.V., Gogate, P.R., Viel, G. et al. (2017). Intensification of biodiesel production using hydrodynamic cavitation based on high speed homogenizer. *Chemical Engineering Journal* 316: 751–757.

72 Arya, S.S., Sawant, O., Sonawane, S.K. et al. (2020). Novel, nonthermal, energy efficient, industrially scalable hydrodynamic cavitation – applications in food processing. *Food Reviews International* 36 (7): 668–691.

73 Albanese, L., Ciriminna, R., Meneguzzo, F., and Pagliaro, M. (2017). Beer-brewing powered by controlled hydrodynamic cavitation: theory and real-scale experiments. *Journal of Cleaner Production* 142: 1457–1470.

74 Albanese, L., Ciriminna, R., Meneguzzo, F., and Pagliaro, M. (2017). Gluten reduction in beer by hydrodynamic cavitation assisted brewing of barley malts. *LWT - Food Science and Technology* 82: 342–353.

75 Mevada, J., Devi, S., and Pandit, A. (2019). Large scale microbial cell disruption using hydrodynamic cavitation: energy saving options. *Biochemical Engineering Journal* 143: 151–160.

76 Meneguzzo, F., Brunetti, C., Fidalgo, A. et al. (2019). Real-scale integral valorization of waste orange peel via hydrodynamic cavitation. *Processes* 7 (9): 581–603.

77 Nuzzo, D., Cristaldi, L., Sciortino, M. et al. (2020). Exceptional antioxidant, non-cytotoxic activity of integral lemon pectin from hydrodynamic cavitation. *ChemistrySelect* 5 (17): 5066–5071.

78 Presentato, A., Scurria, A., Albanese, L. et al. (2020). Superior antibacterial activity of integral lemon pectin extracted via hydrodynamic cavitation. *ChemistryOpen* 9 (5): 628–630.

79 Madane, K. and Ranade, V.V. (2022). Using vortex based cavitation device for tailoring particle size distribution in anti-solvent crystallisation. SSPC Report: University of Limerick.

80 Ciriminna, R., Albanese, L., Di Stefano, V. et al. (2018). Beer produced via hydrodynamic cavitation retains higher amounts of xanthohumol and other hops prenylflavonoids. *LWT* 91: 160–167.

81 Navarro-López, E., Cerón-García, M.C., López-Rodríguez, M. et al. (2020). Biostimulants obtained after pilot-scale high-pressure homogenization of *Scenedesmus* sp. grown in pig manure. *Algal Research* 52: 102123.

82 Albanese, L., Bonetti, A., D'Acqui, L.P. et al. (2019). Affordable production of antioxidant aqueous solutions by hydrodynamic cavitation processing of silver fir (Abies alba Mill.) needles. *Food* 8 (2): 65–85.

83 Meneguzzo, F., Ciriminna, R., Zabini, F., and Pagliaro, M. (2020). Review of evidence available on hesperidin-rich products as potential tools against COVID-19 and hydrodynamic cavitation-based extraction as a method of increasing their production. *Processes* 8 (5): 549–566.

84 Ciocănea, A., Vasile, E., Budea, S. et al., eds. (2021). A hydrodynamic cavitation – high voltage discharge in water procedure for recycling platinum group metals contained in three-way catalyst converters. *2021 10th International Conference on ENERGY and ENVIRONMENT (CIEM)* (14–15 October 2021).

85 Kuroki, H., Onishi, K., Asami, K., and Yamaguchi, T. (2019). Catalyst slurry preparation using a hydrodynamic cavitation dispersion method for polymer electrolyte fuel cells. *Industrial and Engineering Chemistry Research* 58 (42): 19545–19550.

86 Ciocanea, A., Vasile, E., Ionescu, V. et al. (2020). Second life application of automotive catalysts: hydrodynamic cavitation recovery and photo water splitting. *Metals* 10 (10): 1307–1320.

87 Liu, L., Shen, Z., Yi, M. et al. (2014). A green, rapid and size-controlled production of high-quality graphene sheets by hydrodynamic forces. *RSC Advances* 4 (69): 36464–36470.

88 Albanese, L., Baronti, S., Liguori, F. et al. (2019). Hydrodynamic cavitation as an energy efficient process to increase biochar surface area and porosity: a case study. *Journal of Cleaner Production* 210: 159–169.

89 Gregersen, S.B., Wiking, L., Bertelsen, K.B. et al. (2019). Viscosity reduction in concentrated protein solutions by hydrodynamic cavitation. *International Dairy Journal* 97: 1–4.

90 Nagalingam, A.P., Thiruchelvam, V.C., and Yeo, S.H. (2019). A novel hydrodynamic cavitation abrasive technique for internal surface finishing. *Journal of Manufacturing Processes* 46: 44–58.

91 Pandur, Ž., Dogsa, I., Dular, M., and Stopar, D. (2020). Liposome destruction by hydrodynamic cavitation in comparison to chemical, physical and mechanical treatments. *Ultrasonics Sonochemistry* 61: 104826.

92 Mann, G.W., Madamadakala, G.R., and Eckels, S.J. (2016). Heat transfer characteristics of R-134a in a converging-Diverging nozzle. *International Journal of Heat and Fluid Flow* 62: 464–473.

Part V

Status and Path Forward

10

Summary and Outlook

Responsible resource usage and decarbonization are key thrust areas in all industrial sectors. Advances in the water, energy, chemicals, and materials sectors improve resource efficiency and facilitate decarbonization throughout the value chain and across the economy. The locally extreme conditions generated during hydrodynamic cavitation while operating under bulk ambient conditions can be harnessed for realizing beneficial physico-chemical transformations, intensifying a wide range of processes relevant to these key sectors, and developing innovative products. Hydrodynamic cavitation, therefore, holds immense potential and can be used as an effective process-intensification tool. Process intensification may be broadly defined as [1] the ability to obtain better results in terms of purity, conversion, and yield of the desired product by manipulating the rates of relevant transport processes and chemical reactions so as to enhance overall performance (more throughput, better quality, less energy consumption, less waste, safer, etc.). Hydrodynamic cavitation can be used to intensify a variety of processes: gas–liquid mass transfer and reactions, liquid–liquid emulsions, crystallization, nanoparticles, photocatalytic processes, cell disruption, and so on. In addition, it may also be used to develop products and processes which otherwise are not possible to realize without cavitation (see reviews by Asaithambi et al. [2] and Ashokkumar et al. [3]). It can also be used in conjunction (hybrid) with other complementary technologies [4]. Several such applications have been explored and evaluated in the past couple of decades. Despite the intense research and a handful of start-up companies, the full potential of hydrodynamic cavitation has not yet been achieved.

The lack of fundamental understanding of the complex physico-chemical processes occurring in hydrodynamic cavitation devices/reactors is one of the main reasons undermining its optimal exploitation. The quantitative understanding of the processes occurring around collapsing cavities—formation of radicals, their transport and reactions with surrounding liquid; thermal cracking and pyrolysis occurring in the vapor phase of cavities; the role of turbulence—inception of cavitation; influence of other factors like physical properties (density, viscosity, surface tension), the role of shear, dissolved gases, and operating parameters (temperature, pH, concentrations, presence of dispersed particles)—is still not adequate (see Chapters 2–5 and references cited therein). The lack of availability of devices with effective performance, the ability to handle multiphase flows, and a long working life with low

Hydrodynamic Cavitation: Devices, Design, and Applications, First Edition.
Vivek V. Ranade, Vinay M. Bhandari, Sanjay Nagarajan, Varaha P. Sarvothaman and Alister T. Simpson.

maintenance also holds back realization of the full potential of hydrodynamic cavitation. There are several applications which require a combination of hydrodynamic cavitation with other complementary technologies. Inadequate understanding and information available on the interaction of hydrodynamic cavitation and other complementary processes like photocatalysis [5–7] restrict the scope and number of realized applications. Several applications have been reported for ultrasonic cavitation. Most of these can be realized using hydrodynamic cavitation with much lower costs and higher productivity [8–10]. However, this requires systematic understanding of the design aspects (devices, experimental characterization, and computational modeling) and applications of hydrodynamic cavitation.

In this book, an attempt is made to provide a systematic basis for designing and evaluating different hydrodynamic cavitation devices/reactors and processes. Different types of cavitation devices, guidelines on designing the experimental setup, and different approaches of computational modeling are presented in the first part. Three major applications of hydrodynamic cavitation are then discussed in the second part. A brief overview of a few other applications of hydrodynamic cavitation is provided. An attempt is made to evolve generalized guidelines so that information in the book may be used for applications that are not included in this book. In this last chapter of the book, we summarize the current state-of-the-art on devices, experiments, and modeling (in Section 10.1), and the applications of hydrodynamic cavitation (in Section 10.2). An attempt is made to highlight some of the common pitfalls in developing effective applications of hydrodynamic cavitation. Some of our thoughts on outlook and future prospects are included.

10.1 Devices, Experimental Characterization, and Modeling of Hydrodynamic Cavitation

Appropriate design of the devices used for hydrodynamic cavitation is crucial for any successful application. In most of the published studies, however, known devices have been used on a trial and error basis for different applications rather than specifically designed devices to suit a particular application. The known hydrodynamic cavitation devices have been classified in Chapter 1. One way of classification is based on whether devices contain moving parts. The other is based on type of flow—axial flow or swirling flow used for generating cavitation. Hydrodynamic cavitation devices based on linear or axial flows are discussed in Chapter 2. The focus is on discussing key flow characteristics of the two most widely used devices, namely orifice and venturi. Though the discussion included in the Chapter 2 centers mainly on basic designs of orifice and venturi, several variations of basic designs have been proposed and used. A simple example based on orifice is schematically shown in Figure 10.1.

Over the years, several studies have been published, reporting experimental data on a variety of different orifice and venturi-based devices (for example, see recent review of Wang et al. [11]). These devices, based on linear or axial flows, however, exhibit disadvantages particularly with respect to their susceptibility to erosion and clogging. Cavitation devices based on swirling flows, particularly those

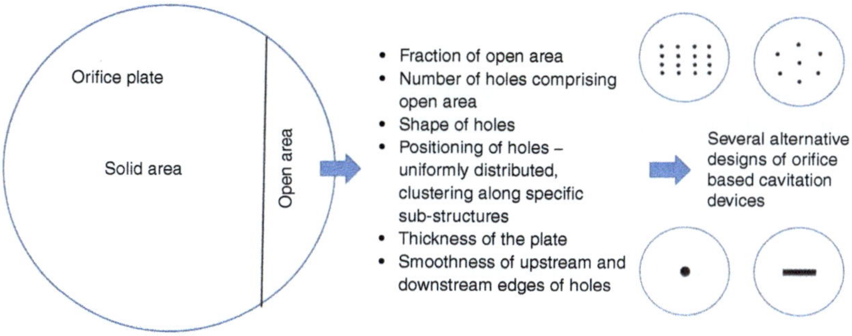

Figure 10.1 Generic designs of orifice-based cavitation devices.

without using any moving parts, look quite promising since they have significantly lower susceptibility to erosion (swirling flow moves collapsing cavities away from device walls and thereby shields the walls from erosion) and clogging (since the smallest diameter required to achieve hydrodynamic cavitation is significantly larger than that required for axial flows [12]). Both these consequences of using swirl flow devices would lead to virtually low/no maintenance, which is attractive from an industrial operation perspective. These devices have a maximum swirling velocity of about five to six times that of throat velocity. Chapter 3 discusses these attributes in detail with the help of computational fluid dynamics-based models. The key characteristics, design parameters, and advantages realized by using swirling flows are highlighted. Vortex diode-based cavitation devices appear to be quite promising in terms of their flow characteristics. Several methods are being developed to further improve the desirable flow characteristics of such vortex-based cavitation devices. For example, Ranade et al. [13] have disclosed the use of specific internals for tailoring the flow characteristics and thereby the inception and extent of hydrodynamic cavitation realized in such devices (see Figure 10.2).

One of the basic flow characteristics of hydrodynamic cavitation devices is the pressure drop versus the net flow. This is represented either based on discharge coefficients (typically used for orifice and venturi type of devices) or the Euler number

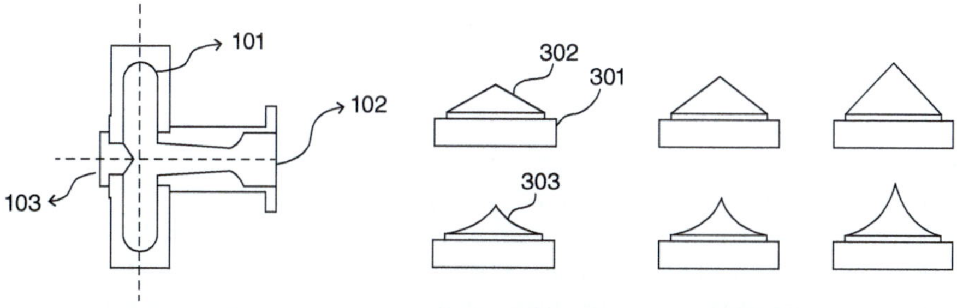

101: Vortex diode chamber, 102: Axial outlet pot, 103: Insert
301: Base of insert, 302: Conical insert, 303: Hyperbolic insert

Figure 10.2 Schematic of internals proposed for improving the performance of vortex-based cavitation devices (Source: From [13]).

(typically used for swirling flow-based devices). Despite numerous published studies, a generalized relationship for pressure drop and flow rate for different devices is not available. Even for the same type of devices, such a relationship over the entire range of Reynolds number of interest is rarely available. It should be noted that it is important to establish such a relationship over the range of Reynolds number for judicious choice of device sizing. The expected relationship between pressure drop (in terms of Euler number) and flow rate (in terms of Reynolds number) is shown schematically in Figure 10.3. It is expected that for a laminar regime, the Euler number is inversely proportional to the Reynolds number, and for a fully turbulent regime, the Euler number is more or less independent of the Reynolds number. For an intermediate regime, the relationship is rather complex and may be crudely approximated by linear proportionality between the Euler number and the Reynolds number, as shown in Figure 10.3.

It should be noted that conventional discharge coefficient used for orifice and venturi will exhibit an opposite trend to that shown in Figure 10.3; the discharge coefficient will increase with Reynolds number in the laminar region, will decrease with Reynolds number in the intermediate region, and will remain more or less constant in the fully turbulent region. Considering that many applications of hydrodynamic cavitation may involve viscous liquids (biomass slurries, liquid–liquid emulsions, etc.), it is essential that further research on cavitation devices reports full details of device configuration and characterizes it over the entire range of laminar, intermediate, and turbulent regimes.

Of the other important flow characteristics of cavitation devices is the flow rate or pressure drop beyond which cavitation occurs. Inception of cavitation

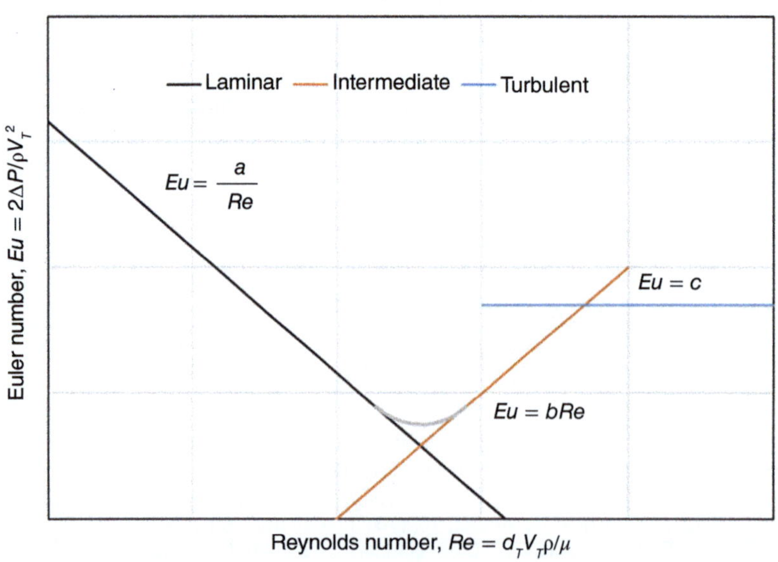

Figure 10.3 Relationship between pressure drop and flow rate represented in terms of Euler number and Reynolds number.

occurs when the lowest pressure generated in a cavitation device approaches vapor pressure of liquid. At low flow rates (in other words, at low pressure drops across the cavitation device), the lowest pressure generated may not be sufficiently low to generate cavities. As the flow rate increases, at a certain point, the lowest pressure generated within the system may approach vapor pressure, leading to an inception of cavitation. It is important to identify this condition to ensure the appropriate operation of hydrodynamic cavitation devices and processes. Conventionally, for linear or axial flow devices like orifice and venturi, cavitation inception is often characterized using the cavitation number, as discussed in Chapter 2. The cavitation number, Ca or σ, is defined as:

$$Ca = \sigma = \frac{P_d - P_v}{\frac{1}{2}\rho V_{max}^2} \tag{10.1}$$

where ρ is density of liquid, P is pressure, and V is velocity. The subscripts d, v, and max denote downstream, vapor pressure, and maximum velocity, respectively. The expression of cavitation number developed for linear/axial flow devices may be extended to devices based on swirling flows by using appropriate value of maximum velocity (which is usually a swirling velocity). The simplistic assumption suggesting that cavitation inception occurs when the cavitation number approaches the value of unity is, unfortunately, not very useful. Published studies, as comprehensively discussed by Šarc et al. [14], indicate that a wide variation in cavitation number at the inception occurs across different types of devices. The specific geometric details of the restriction play an important role, as these details determine the resulting velocity and pressure profiles in the cavitating region. Recently, Ranade et al. [15] have used acoustic signals from the cavitating device for clearly identifying inception. They have shown that various features extracted from the acquired acoustic signals can be used to identify cavitation inception. A sample of their results is shown in Figure 10.4.

The computational fluid dynamics (CFD)-based models discussed in Chapters 2 and 3 may also provide an insight into how the lowest pressure value and therefore cavitation inception is influenced by various design and operating parameters. The CFD models presented in these two chapters will also be useful in identifying the key design parameters and quantitatively steering the design of devices toward optimal characteristics. We hope that the two chapters will lead to development of newer and better hydrodynamic cavitation devices.

Despite continuous advances in computational flow modeling, it is still essential to experimentally evaluate the flow characteristics and performance of hydrodynamic cavitation devices. Key components of the typical experimental setup required to characterize and quantify the performance of hydrodynamic cavitation devices are discussed in Chapter 4. Specific suggestions on designing the experimental setup to avoid common pitfalls are made. A brief discussion on planning and executing such experiments on inception and quantification of performance is also included. Despite the proven advantages of design of experiments, few studies of hydrodynamic cavitation devices and processes are currently available. The generic and well-known techniques of design of experiments and response surface

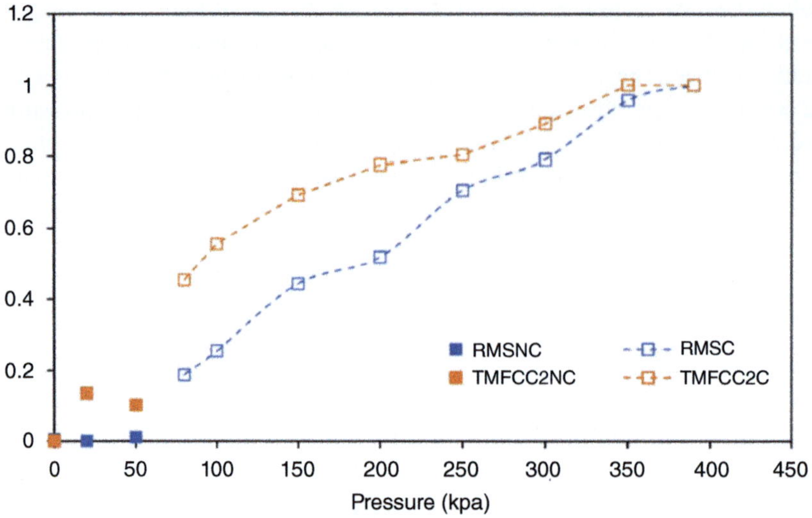

Figure 10.4 Use of normalized ($\emptyset_N = \frac{\emptyset - \emptyset_{min}}{\emptyset_{max} - \emptyset_{min}}$) RMS and transformed MFCC-2 for identifying cavitation inception (Source: From [15]/with permission of American Chemical Society).

methodology [16, 17] may be used for reducing the experimental efforts required to decipher the influence of key design and operating parameters on the overall performance of hydrodynamic cavitation. It is essential to report the details of the cavitation device and the experimental rig used. It is also useful to perform extensive device characterization (pressure drop versus flow characteristics, identification of inception and choking [if any], cavitation, etc.) prior to actual experiments. This will facilitate meaningful comparison of different devices, operating conditions, and processes. Though most of the laboratory studies will be based on the batch (with 100% recycle) mode of operation, it is important to process the experimental results based on the number of passes through the cavitation device rather than on time. The status of understanding on hydrodynamic cavitation can be greatly advanced by harnessing the recent advances in experimental techniques, associated electronics, instrumentation, and data processing. The ultra-fast dynamics of inception, growth, and collapse of cavities in the cavitation devices of interest can be investigated using the state-of-the-art ultrafast imaging systems which now offer an ability to capture 1 million frames per second [18, 19]. Recent advances in X-ray tomography [20, 21] will allow capture of cavitation in opaque devices. Such new techniques may be used to quantify the high-velocity jets and shear generated during cavity collapse. Understanding the generated jets and shear will help in improving the knowledge on non-reactive shear-based HC processes like breakage of liquid droplets or solid particles. Recent advances in dosimetric techniques for quantifying generated hydroxyl radicals [22, 23] will be useful to interpret experimental data and refine computational models. It is important to use targeted chemical probes (coumarin, terephthalic acid) to quantify radicals rather than non-specific chemicals such as KI to quantify the oxidizing radicals. It will

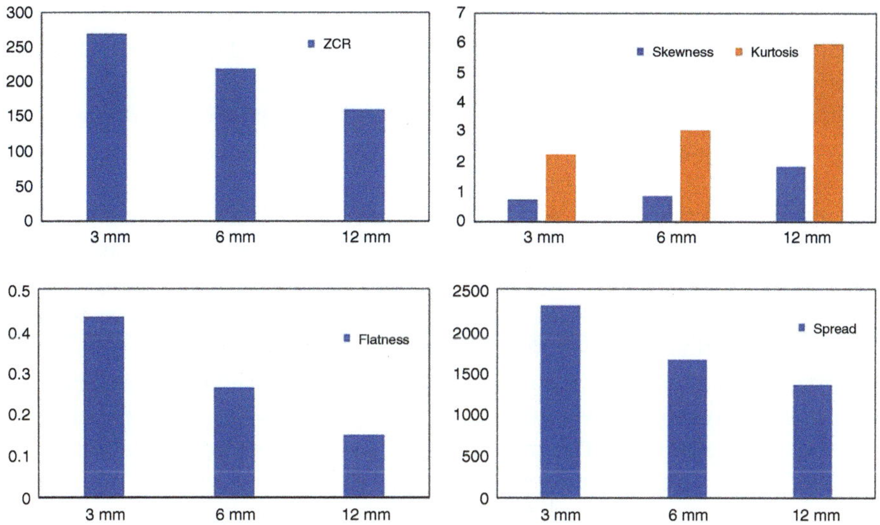

Figure 10.5 Influence of scale on the key characteristics of acoustic signals from vortex-based cavitation devices: size quoted below the bars indicates throat diameter (Source: From [15]/with permission of American Chemical Society).

be most useful if such new experimental data are collected and made accessible to modelers for developing, validating, and calibrating multiphase models of hydrodynamic cavitation. Recent emphasis on ensuring that the generated data is Findable, Accessible, Interoperable, and Reusable [24] will greatly facilitate further development and refinement of computational models of hydrodynamic cavitation.

The modeling frameworks needed to interpret and use the experimental data and evaluations are presented in Chapter 5. Various approaches—ranging from empirical models to rigorous multilayer computational models simulating performance of hydrodynamic cavitation devices and processes—are discussed. The semi-empirical framework-based per-pass degradation factor presented in Chapter 5 is useful for interpreting and using the experimental data for designing practical applications of hydrodynamic cavitation (see for example, Sarvothaman et al. [7], Nagarajan et al. [25]). The data-driven approach presented in Chapter 5 is also useful for reliable interpolations when relevant experimental/plant data are available. The data-driven approach was also shown recently to be useful for scale-up of vortex-based cavitation devices [15]. In this example, Ranade et al. used acoustic signals emanating from the vortex-based cavitation devices for extracting key features and related these extracted features with the performance at different scales, as shown in Figure 10.5 and Eq. (10.2).

The per-pass degradation factor ($\emptyset$) was related to the extracted feature, flatness (F):

$$\ln\left(\frac{\emptyset}{\emptyset_\infty}\right) \propto \left(\frac{F}{F_\infty} - 1\right) \tag{10.2}$$

where, subscript ∞ denotes infinite scale-up. Thus, the influence of device scale on performance is captured via the extracted feature F, from the acquired acoustic signals from different scales.

While such semi-empirical or fully data-driven models are useful, rigorous mathematical models based on first principles are necessary for making a priori predictions. The processes based on hydrodynamic cavitation essentially use generated shear, localized hot spots, and hydroxyl radicals for realizing various physico-chemical transformations. It is therefore essential to develop an ability to make a priori predictions of these as a function of device design and operating parameters. Though attempts are being made to develop a quantitative understanding of cavitation processes for more than a century (probably starting from Lord Rayleigh in 1917 [26]), quantitative simulations of overall performance of hydrodynamic cavitation devices/processes are still elusive without the use of adjustable parameters [27]. The main reason is the coexistence of relevant time scales and length scales spanning several orders of magnitude, ranging from radicals and molecules to micron-sized cavities and tens of centimeter-scale cavitation devices. For such a complex physico-chemical process, it is often beneficial and more efficient to develop mathematical and computational models in several stages or layers, which will exchange relevant information rather than a single monolithic comprehensive model trying to incorporate all the scales simultaneously. Sarvothaman et al. [27] have presented an early effort in this direction.

While it is practical to use separate models using a multilayer modeling framework, further research is essential to improve our ability to make quantitative predictions. Some areas which require specific attention with the goal of improving the state-of-the-art on modeling hydrodynamic cavitation are:

- Models for cavitation source terms: Better models for correctly representing micro-scale cavity dynamics into device-scale simulations are needed. The currently used models use overly simplified cavity dynamics for formulating cavitation source terms.
- Turbulence modeling: State-of-the-art understanding and models representing the influence of turbulence, vortices, and other coherent structures on inception and extent of cavitation are not adequate. This is especially relevant in the presence of dissolved gases or dispersed phase particles (solid, liquid, or gas).
- Models for representing the physical effects of cavity collapse—high-speed jets, shear, and hammer pressure generated by collapse: For quantitatively estimating influence of collapsing cavities on particle breakage or cleaning, quantitative estimation of high speed jets, shear and hammer pressure generated by cavity collapse is essential. Current models are validated for collapse of a single cavity. However, further work on extending these models to account for the influence of neighboring cavities, dispersed phase particles or wall on these quantities is important and essential.
- Models for representing source terms corresponding to the generation of hydroxyl and other species formed due to collapsed cavities and subsequent reactions caused by these species: The high temperature and high pressure generated by

cavity collapse lead to radical formation and other thermally driven reactions inside and near the vicinity of the collapsing cavity. The local temperature and pressure values are highly sensitive to local composition as well as individual species transport properties. The uncertainties associated with the estimation of these transport properties at extreme pressure and temperature generated during cavity collapse will also add to the overall uncertainties in the estimation of physico-chemical transformations resulting from collapsed cavities. Further work is needed on estimating and reducing these uncertainties.

The list is merely suggestive. The complexity of reactive multiphase flows occurring in hydrodynamic cavitation devices may greatly expand the list of issues on which further research is in progress or is needed. Apart from an appropriate formulation of model equations based on first principles, it is also essential to quantify the influence of numerical issues (grid spacing, time step, degree of convergence, and so on) on simulated results before one can use results obtained from a computational flow model for engineering applications. Ranade [28] has discussed various methods for verification and validation of computational models which may be used. For cases where resource constraints make it difficult to obtain grid-independent solutions, extrapolation of key quantities of interest (for example, inception point for a cavitation device) to zero grid spacing may be used. While further research on these aspects is expected to enhance the ability of current computational models to make a priori quantitative predictions, more often than not, current quantitative simulations rely on adjustable parameters for obtaining simulations closer to real- life performance. Significant research on the experimental side as well as on the modeling side is needed to identify relevant and minimal adjustable parameters. In such cases, the exercise of validation and calibration of the computational model becomes very important. In many applications of hydrodynamic cavitation, the data required for adequate validation are not available. In such cases, one has to rely on "indirect validation." Critical examination of known case studies and hands-on experience is needed to use computational models for practical applications in such cases.

It will also be worthwhile to explore new opportunities opened up by the recent advances in data science and machine learning for enhancing the capabilities of physics-based multi-scale models [29]. Availability of non-invasive sensors such as acoustic sensors may be harnessed for exploring the use of real-time data for estimating relevant parameters of physics-based computational models, to enhance the overall accuracy of first-principles-based models. This concept is shown schematically in Figure 10.6.

Such efforts may bridge the gaps in the current state-of-the-art and enable process innovations to accelerate realization of the full potential of hydrodynamic cavitation in the water, energy, chemicals, and materials sectors.

10.2 Applications of Hydrodynamic Cavitation

Despite the unavailability of predictive models and the inadequate understanding of underlying physico-chemical processes, hydrodynamic cavitation has been explored

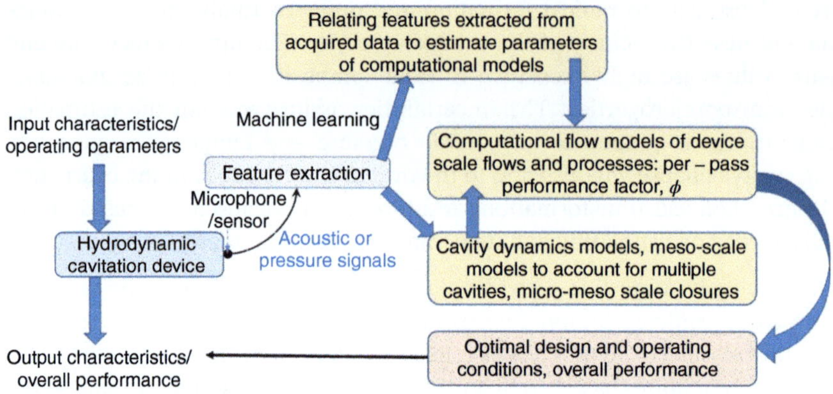

Figure 10.6 Conceptual framework for ensemble approach combining real-time data, machine learning and physics-based models for realizing high-fidelity computational models.

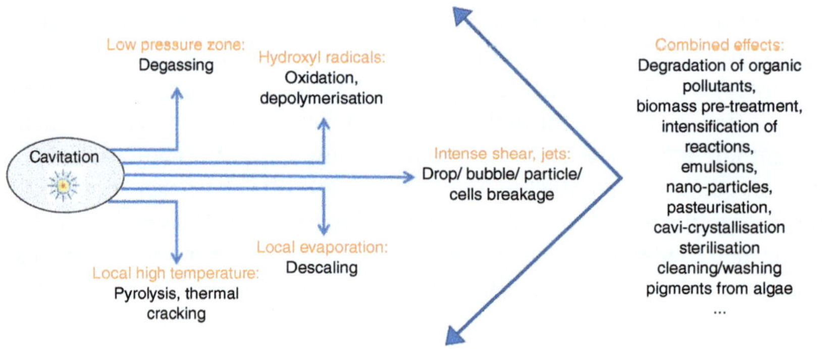

Figure 10.7 Physico-chemical effects of hydrodynamic cavitation and their applications.

for numerous applications in the water, energy, chemicals, and materials sectors. These applications harness low-pressure, localized high-pressure temperatures, local evaporation, intense shear, and hydroxyl radicals generated by hydrodynamic cavitation, as shown in Figure 10.7.

Hydrodynamic cavitation, alone or in conjunction with complementary techniques utilizing alternative energy sources or additional chemicals and catalysts, has shown significant promise in the disinfection of harmful microorganisms, decolorization of dye streams, COD removal, and in general, the treatment, recycle, and reuse of water. Key applications in the water sector—disinfection and wastewater treatment—were discussed in Chapters 6 and 7, respectively. Combining hydrodynamic cavitation with hydrogen peroxide or ozone has shown significant synergy, with a synergistic index of more than 2. This may be explored further to develop effective processes for handling difficult-to-treat organic dyes or pharmaceutical compounds. Photocatalysis may be synergistically combined for the removal of hydrophilic organic pollutants such as phenols/primary carboxylic acids, which are otherwise hard to tackle using hydrodynamic cavitation alone.

Understanding the nature of the pollutant, its chemistry around polarity, and its pKa (measure of acidity) are important in determining the complementary methods that assist hydrodynamic cavitation in hybrid processes.

While Chapters 6 and 7 present detailed discussion and the advantages of application of hydrodynamic cavitation for disinfection and removal of pollutants from water, large-scale applications in practice are still not common. The key issues discussed in Section 10.1—particularly the lack of availability of scale-up methodology—may be holding back applications. The application of the hybrid methodology for drinking water treatment could be particularly interesting for the commercial scales of operation, and for the advantages it offers compared to chemical methods of disinfection. It would be prudent to emphasize here that water disinfection using different forms of cavitation has been unsuccessfully explored since many decades, and now, there could possibly be an emerging practical solution in this regard using hydrodynamic cavitation. In the wastewater treatment area, especially for the removal of organic pollutants from industrial wastewaters, the technological lock-in of the existing effluent treatment plants often raises barriers in implementing hydrodynamic cavitation-based solutions. It is essential to develop appropriate "retrofit" or "drop-in" solutions using hydrodynamic cavitation for enhancing the overall effectiveness of an effluent treatment plant—either by reducing the cost of treatment or by enhancing the treatment capacity of the plant. Large beverage companies are looking for such solutions to complement their existing membrane-based treatment plants. Another area of potential application in existing effluent treatment plants is realizing distributed water treatment solutions instead of the conventional "end-of-pipe" solutions. In many manufacturing processes, small quantities of wastewater containing high pollutant load are generated—for example, washing of a reactor or process vessel. Instead of mixing such a stream to equalization tanks (as is done in a conventional end of-pipe treatment), it is more effective to treat such high pollutant-containing streams locally using a hydrodynamic cavitation process and then passing the treated stream to the equalization tank for further processing. Such an arrangement will greatly reduce practical and operational issues associated with the load fluctuations in a typical effluent treatment plant. Such distributed and localized treatment may also facilitate the recycle and reuse of water [30] and may also promote application of water pinch analysis to reduce water footprint [31–34]. A schematic of a potential intensified and distributed wastewater treatment plant replacing the conventional end-of pipe-approach is shown in Figure 10.8. The devices and processes based on hydrodynamic cavitation and the possibility of combining them with the other complementary techniques discussed in this book will be useful to realize similar effective water treatment solutions.

Applications of hydrodynamic cavitation to intensify processes for biofuels are discussed in Chapter 8 (biogas and bioethanol). Application to another biofuel—biodiesel production—is discussed briefly in Chapter 9. The key limitation in using biomass for biofuels is the complex structure of lignocellulose which restricts access to hemi cellulose and cellulose for biochemical and chemical transformations. Shear, hydroxyl radicals, and localized high-pressure and temperature

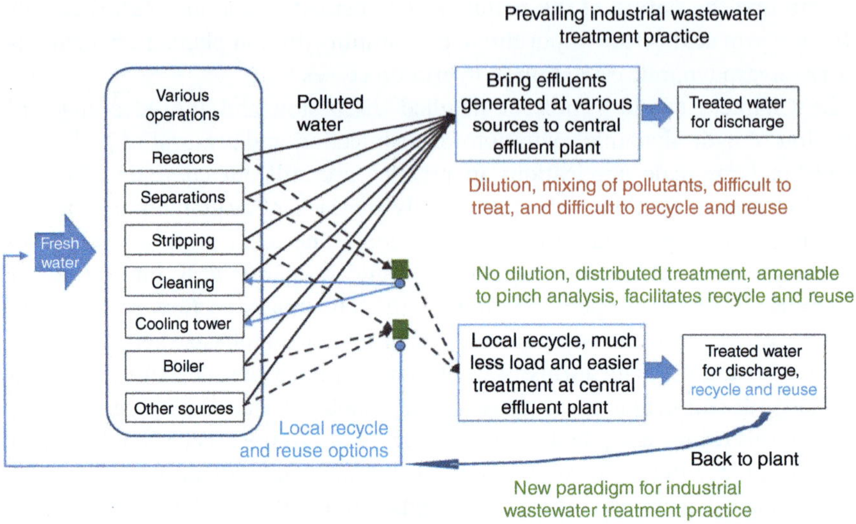

Figure 10.8 Distributed wastewater treatment by leveraging hydrodynamic cavitation.

zones created by hydrodynamic cavitation enhance the valorization potential of biomass by improving the rates of transformation processes and improving the utilizable fraction of the biomass. The key aspects of using hydrodynamic cavitation for pre-treating biomass feedstock of anaerobic digesters for enhancing the rate and quantity of biogas generation are discussed in Chapter 8. For bioethanol production, hydrodynamic cavitation needs to be combined with other chemical-based treatment (preferably with alkali) for improving the yield of sugars and subsequently of ethanol from biomass. It has been shown that in both these broad classes of applications, the energy required for implementing hydrodynamic cavitation is far less than the gain in energy (via enhancement in yield of biofuels). This is particularly true when the hydrodynamic cavitation process is implemented for slurries containing high-solids loading. In both the cases, dried and milled biomass has been preferred. However, there is immense scope for realizing hydrodynamic cavitation using fibrous biomass feedstock without milling. Cavitation devices which do not rely on small constrictions for generating cavitation are therefore key for such applications. Apart from the better cavitation devices, current state-of-the-art on the fundamental understanding of the interactions between the biomass and collapsing cavities needs to be improved for optimizing hydrodynamic cavitation-based pre-treatment. The requirements for further work in improving the state-of-the-art are shown schematically in Figure 10.9.

From a circular economy perspective, it is critical to push the limits of biomass conversion. For instance, as discussed by Nagarajan and Ranade [35], maximal utilization of hemicellulose toward biogas production can occur via hydrodynamically cavitated biomass. The whole digestate could either be used as a fertilizer or further fractionated to solids and liquids. In such a scenario, the liquid fraction could be either used as a fertilizer or recycled back to the digester. The solid residue,

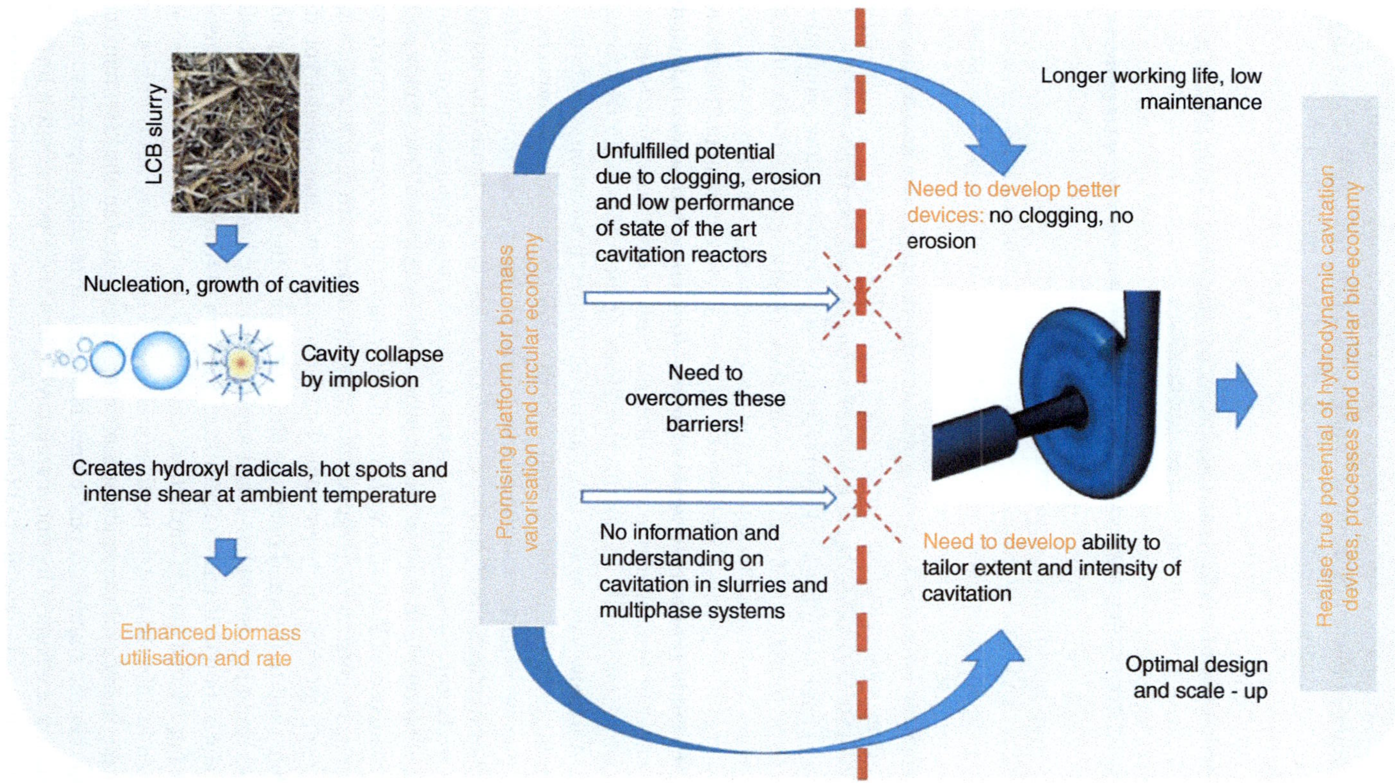

Figure 10.9 Approach needed for realizing step change in the state-of-the-art for applications of hydrodynamic cavitation for valorization of biomass.

rich in cellulose, could be used as a feedstock for the production of high-value biorenewables. The process intensification offered by hydrodynamic cavitation is helpful for achieving net neutrality via a circular economy approach [36].

Various other applications such as flotation, oxidative desulfurization, extracting bioactive compounds, biodiesel production, and beer brewing, are briefly discussed in Chapter 9. The applications were categorized based on fundamental processing as gas–liquid, liquid–liquid, and solid–liquid type applications. Either one or more physico-chemical effects arising due to hydrodynamic cavitation, shown in Figure 10.7, play a governing role in these applications. As an example of gas–liquid applications, flotation and aeration, were discussed in Chapter 9. Generating stable submicron-sized bubbles (dependent on shear for bubble breakage) is critical for these applications. Furthermore, hydrodynamic cavitation provides in situ aeration which can "wet" the mineral of interest and increase its hydrophobicity, to increase its "floatability" and thereby enhance its recovery performance from low-grade minerals. Liquid–liquid systems such as oxidative desulfurization and biodiesel production require intense mixing and very high liquid–liquid interfacial area to overcome the mass transfer limitations, so as to enhance the reaction rates and reduce the catalyst/oxidant/solvent loading that are often prohibitively expensive for industrial operations. In the case of oxidative desulfurization, hydroxyl radicals generated in the aqueous phase also play a role in reversing the polarity of sulfur compounds present in fuels. In the case of emulsification, shear plays a significant role in creating stable microemulsions. The solid–liquid systems discussed in Chapter 9 include applications like beer brewing, particle size reduction, and extraction of bioactives. Similar to the applications discussed in Chapter 8 (biomass pre-treatment), the predominant effect of hydrodynamic cavitation on these solid compounds is intense shear that can damage cells and enable the release of the intracellular compounds of interest. A narrow particle size distribution for catalyst production, synonymous with ball milling, has been reported with hydrodynamic cavitation and is expected to be a game changer in bulk catalyst synthesis, relevant to many processing industries.

Many other applications of hydrodynamic cavitation besides those discussed in this book are being investigated. For example, see various applications of hydrodynamic cavitation mentioned in recent reviews by Holkar et al. [4], Panda et al. [37], and Mancuso et al. [38]. These include new applications such as improving the heat transfer coefficient, reducing viscosity, liposome destruction, and so on. Beyond broad reviews on harnessing hydrodynamic cavitation, sector-specific reviews dedicated to food and beverage [2, 39, 40], bioactive compound extraction from plants and biomass [41], thermal applications [42], biodiesel synthesis [43], biomass valorization [36], and the petroleum industry [44] are available. Key issues in harnessing hydrodynamic cavitation for various applications and its role are summarized in Table 10.1.

Translating many of these applications being investigated in the research laboratories to practice requires a better understanding of relationships between key physico-chemical effects of hydrodynamic cavitation (as shown in Figure 10.7) and performance relevant to the application under consideration. Appropriately realizing these desired physico-chemical effects by manipulating the design of cavitation

Table 10.1 Harnessing hydrodynamic cavitation for a variety of applications.

Goal	Some of the key issues	Role of hydrodynamic cavitation
Enhance product quality	Impurities, side reactions, size and shapes of solid particles or droplets, functional performance ...	Intensify mixing, enhance heat and mass transport rates, manipulate particle size distribution, enhance reaction rates, enhance selectivity
Enhance process throughput	Safe and stable operation, lower energy consumption, lower emissions	Intensify transport rates, retrofit ("drop-in") solutions for debottlenecking
Reduce water footprint	Load fluctuations, small quantity of highly concentrated streams to large equalization tanks, large foot print	Distributed water treatment solutions, synergistically complement other water treatment technologies, small foot print
Reduce chemical consumption	Cost of additives, disposal and waste treatment, positive carbon footprint	Intensify mixing, maximize mass transfer, improve reaction rates
Valorize liquid and solid wastes	Accessing useful materials in the waste, slow processes, large footprint	Pre-treatment for enhancing accessibility, enhancing transport and reaction rates, small footprint

devices and the operating parameters, and possibly combining the hydrodynamic cavitation with other complementary techniques will be the key for successful realization of the true potential of hydrodynamic cavitation. The information presented in this book will be useful to identify and address key issues relevant to applications under consideration and to develop appropriate solutions based on hydrodynamic cavitation. It will also be useful for developing new applications and newer designs of cavitation devices. Recently, Doltade and Pandit [45] developed a novel hand pump with in situ hydrodynamic cavitation for improving the quality of water being pumped from a borewell. Several new ways of synergistically combining hydrodynamic cavitation with other complementary techniques are being developed. For example, synergy between hydrodynamic cavitation and microwave has been used for upgrading crude oil fractions [46]. Hydrodynamic cavitation has been combined with ultrasound for intensifying dye decolorization [47]. The use of plasma for water disinfection and water treatment is quite recent and has shown promise in decolorizing dye-based wastewater [48] within three passes. Martinez et al. [49] have recently discussed the state-of-the-art, new synthetic trends, the industrial potential, and a general perspective of organic sonochemistry. Many of the reactions mentioned therein can be intensified using hydrodynamic cavitation, with better scalability than ultrasound-based systems. Many other applications like catalyst synthesis [50], nutrient extraction [51], degassing [52], descaling [53], and nanoparticles [54] are being developed. Initiatives like the use of hydrodynamic cavitation for on-demand production of personalized emulsions [55] or for cavi-crystallization [56] augur well for expanding the application horizons of hydrodynamic cavitation. We believe that hydrodynamic cavitation has a great potential to beneficially impact resource efficiency, the development of new products, and decarbonization across many sectors. We hope that this book provides a framework and useful guidelines for harnessing hydrodynamic cavitation for a wide range of applications.

Notations

Ca	Cavitation number (–)
F	Flatness (–)
F_{∞}	Flatness at infinite scale-up
P_d	Downstream pressure of cavitation device (kPa)
P_v	Vapor pressure of liquid (kPa)
$\emptyset$	Per-pass degradation factor (–)
$\emptyset_{\infty}$	Per-pass degradation factor at infinite scale-up (–)
ρ	Density of liquid (kg/m^3)
σ	Cavitation number (–)
v_{max}	Maximum velocity in cavitation device (m/s)

References

1 Utikar, R.P. and Ranade, V.V. (2017). Intensifying multiphase reactions and reactors: strategies and examples. *ACS Sustainable Chemistry & Engineering* 5 (5): 3607–3622.

2 Asaithambi, N., Singha, P., Dwivedi, M., and Singh, S.K. (2019). Hydrodynamic cavitation and its application in food and beverage industry: a review. *Journal of Food Process Engineering* 42 (5): e13144.

3 Ashokkumar, M., Rink, R., and Shestakov, S. (2011). Hydrodynamic cavitation-an alternative to ultrasonic food processing. *Technical Acoustics/Tekhnicheskaya Akustika* (9): 1–10.

4 Holkar, C.R., Jadhav, A.J., Pinjari, D.V., and Pandit, A.B. (2019). Cavitationally driven transformations: a technique of process intensification. *Industrial & Engineering Chemistry Research* 58 (15): 5797–5819.

5 Suryawanshi, N.B., Bhandari, V.M., Sorokhaibam, L.G., and Ranade, V.V. (2017). Developing techno-economically sustainable methodologies for deep desulfurization using hydrodynamic cavitation. *Fuel* 210: 482–490.

6 Patil, P.B., Bhandari, V.M., and Ranade, V.V. (2021). Wastewater treatment and process intensification for degradation of solvents using hydrodynamic cavitation. *Chemical Engineering and Processing: Process Intensification* 108485.

7 Sarvothaman, V.P., Nagarajan, S., and Ranade, V.V. (2018). Treatment of solvent-contaminated water using vortex-based cavitation: influence of operating pressure drop, temperature, aeration, and reactor scale. *Industrial & Engineering Chemistry Research* 57 (28): 9292–9304.

8 Sawant, S.S., Anil, A.C., Krishnamurthy, V. et al. (2008). Effect of hydrodynamic cavitation on zooplankton: a tool for disinfection. *Biochemical Engineering Journal* 42 (3): 320–328.

9 Maddikeri, G.L., Gogate, P.R., and Pandit, A.B. (2014). Intensified synthesis of biodiesel using hydrodynamic cavitation reactors based on the interesterification of waste cooking oil. *Fuel* 137: 285–292.

10 Innocenzi, V. and Prisciandaro, M. (2021). Technical feasibility of biodiesel production from virgin oil and waste cooking oil: comparison between traditional and innovative process based on hydrodynamic cavitation. *Waste Management* 122: 15–25.

11 Wang, B., Su, H., and Zhang, B. (2021). Hydrodynamic cavitation as a promising route for wastewater treatment – a review. *Chemical Engineering Journal* 128685.

12 Simpson, A. and Ranade, V.V. (2019). 110th Anniversary: comparison of cavitation devices based on linear and swirling flows: hydrodynamic characteristics. *Industrial & Engineering Chemistry Research* 58 (31): 14488–14509.

13 Ranade, V.V., Kulkarni, A.A., and Bhandari, V.M. (2016). Vortex diodes as effluent treatment devices. US Patent 9,422,952.

14 Šarc, A., Stepišnik-Perdih, T., Petkovšek, M., and Dular, M. (2017). The issue of cavitation number value in studies of water treatment by hydrodynamic cavitation. *Ultrasonics Sonochemistry* 34: 51–59.

15 Ranade, N.V., Sarvothaman, V., and Ranade, V.V. (2021). Acoustic analysis of vortex-based cavitation devices: inception and extent of cavitation. *Industrial & Engineering Chemistry Research* 60 (34): 12752–12766.

16 Chitsaz, H., Omidkhah, M., Ghobadian, B., and Ardjmand, M. (2018). Optimization of hydrodynamic cavitation process of biodiesel production by response surface methodology. *Journal of Environmental Chemical Engineering* 6 (2): 2262–2268.

17 Dana, M., Sobati, M.A., Shahhosseini, S., and Ansari, A. (2020). Optimization of a continuous ultrasound assisted oxidative desulfurization (UAOD) process of diesel using response surface methodology (RSM) considering operating cost. *Chinese Journal of Chemical Engineering* 28 (5): 1384–1396.

18 Thoroddsen, S.T., Etoh, T.G., and Takehara, K. (2008). High-speed imaging of drops and bubbles. *Annual Review of Fluid Mechanics* 40: 257–285.

19 Versluis, M. (2013). High-speed imaging in fluids. *Experiments in Fluids* 54 (2): 1–35.

20 Fischer, F., Hoppe, D., Schleicher, E. et al. (2008). An ultra fast electron beam x-ray tomography scanner. *Measurement Science and Technology* 19 (9): 094002.

21 Fischer, F. and Hampel, U. (2010). Ultra fast electron beam X-ray computed tomography for two-phase flow measurement. *Nuclear Engineering and Design* 240 (9): 2254–2259.

22 Arrojo, S., Nerin, C., and Benito, Y. (2007). Application of salicylic acid dosimetry to evaluate hydrodynamic cavitation as an advanced oxidation process. *Ultrasonics Sonochemistry* 14 (3): 343–349.

23 De-Nasri, S.J., Nagarajan, S., Robertson, P.K.J., and Ranade, V.V. (2020). Quantification of hydroxyl radicals in photocatalysis and acoustic cavitation: utility of coumarin as a chemical probe. *Chemical Engineering Journal* 127560.

24 Stall, S., Yarmey, L., Cutcher-Gershenfeld, J. et al. (2019). *Make Scientific Data FAIR*. Nature Publishing Group.

25 Nagarajan, S. and Ranade, V.V. (2020). Pre-treatment of distillery spent wash (vinasse) with vortex based cavitation and its influence on biogas generation. *Bioresource Technology Reports* 11: 100480.

26 Rayleigh, L.V.I.I.I. (1917). On the pressure developed in a liquid during the collapse of a spherical cavity. *The London, Edinburgh, and Dublin Philosophical Magazine and Journal of Science* 34 (200): 94–98.

27 Sarvothaman, V.P., Simpson, A.T., and Ranade, V.V. (2019). Modelling of vortex based hydrodynamic cavitation reactors. *Chemical Engineering Journal* 377: 119639.

28 Ranade, V.V. (2002). *Computational Flow Modeling for Chemical Reactor Engineering*. Academic Press.

29 Radhakrishnan, R. (2021). A survey of multiscale modeling: foundations, historical milestones, current status, and future prospects. *AIChE Journal* 67 (3): e17026.

30 Ranade, V.V. and Bhandari, V.M. (2014). *Industrial Wastewater Treatment, Recycling and Reuse*. Butterworth-Heinemann.

31 Foo, D.C.Y. (2009). State-of-the-art review of pinch analysis techniques for water network synthesis. *Industrial & Engineering Chemistry Research* 48 (11): 5125–5159.

32 Mohammadnejad, S., Bidhendi, G.R.N., and Mehrdadi, N. (2011). Water pinch analysis in oil refinery using regeneration reuse and recycling consideration. *Desalination* 265 (1–3): 255–265.

33 Lima, A., Abreu, T., and Figueiredo, S. (2021). Water and wastewater optimization in a food processing industry using water pinch technology. *Sustainable Water Resources Management* 7 (5): 1–13.

34 Chin, H.H., Liew, P.Y., Varbanov, P.S., and Klemeš, J.J. (2022). Extension of pinch analysis to targeting and synthesis of water recycling networks with multiple contaminants. *Chemical Engineering Science* 248: 117223.

35 Nagarajan, S. and Ranade, V.V. (2019). Pretreatment of lignocellulosic biomass using vortex-based devices for cavitation: influence on biomethane potential. *Industrial & Engineering Chemistry Research* 58 (35): 15975–15988.

36 Nagarajan, S. and Ranade, V.V. (2021). Valorizing waste biomass via hydrodynamic cavitation and anaerobic digestion. *Industrial & Engineering Chemistry Research* 60 (46): 16577–16598.

37 Panda, D., Saharan, V.K., and Manickam, S. (2020). Controlled hydrodynamic cavitation: a review of recent advances and perspectives for greener processing. *Processes* 8 (2): 220.

38 Mancuso, G., Langone, M., and Andreottola, G. (2020). A critical review of the current technologies in wastewater treatment plants by using hydrodynamic cavitation process: principles and applications. *Journal of Environmental Health Science and Engineering* 18 (1): 311–333.

39 Arya, S.S., Sawant, O., Sonawane, S.K. et al. (2020). Novel, nonthermal, energy efficient, industrially scalable hydrodynamic cavitation – applications in food processing. *Food Reviews International* 36 (7): 668–691.

40 Pathania, S., Ho, Q.T., Hogan, S.A. et al. (2018). Applications of hydrodynamic cavitation for instant rehydration of high protein milk powders. *Journal of Food Engineering* 225: 18–25.

41 Wu, Z., Ferreira, D.F., Crudo, D. et al. (2019). Plant and biomass extraction and valorisation under hydrodynamic cavitation. *Processes* 7 (12): 965.

42 Gevari, M.T., Abbasiasl, T., Niazi, S. et al. (2020). Direct and indirect thermal applications of hydrodynamic and acoustic cavitation: a review. *Applied Thermal Engineering* 171: 115065.

43 Patil, A., Baral, S., and Dhanke, P. (2021). Hydrodynamic cavitation for process intensification of biodiesel synthesis – a review. *Current Research in Green and Sustainable Chemistry* 4: 100144.

44 Avvaru, B., Venkateswaran, N., Uppara, P. et al. (2018). Current knowledge and potential applications of cavitation technologies for the petroleum industry. *Ultrasonics Sonochemistry* 42: 493–507.

45 Doltade, S.B. and Pandit, A.B. (2021). Novel hydrodynamic cavitation based hand pump for disinfection of groundwater. *Environmental Quality Management* 1–10.

46 Ansari, K.B., Loke, N.H., Pandit, A.B. et al. (2015). Process intensification of upgradation of crude oil and vacuum residue by hydrodynamic cavitation and microwave irradiation. *Indian Chemical Engineer* 57 (3–4): 256–281.

47 Verma, P. and Samanta, S.K. (2018). Microwave-enhanced advanced oxidation processes for the degradation of dyes in water. *Environmental Chemistry Letters* 16 (3): 969–1007.

48 Abramov, V.O., Abramova, A.V., Cravotto, G. et al. (2021). Flow-mode water treatment under simultaneous hydrodynamic cavitation and plasma. *Ultrasonics Sonochemistry* 70: 105323.

49 Martínez, R.F., Cravotto, G., and Cintas, P. (2021). Organic sonochemistry: a chemist's timely perspective on mechanisms and reactivity. *The Journal of Organic Chemistry* 86 (20): 13833–13856.

50 Ciocanea, A., Vasile, E., Ionescu, V. et al. (2020). Second life application of automotive catalysts: hydrodynamic cavitation recovery and photo water splitting. *Metals* 10 (10): 1307.

51 Lee, A.K., Lewis, D.M., and Ashman, P.J. (2015). Microalgal cell disruption by hydrodynamic cavitation for the production of biofuels. *Journal of Applied Phycology* 27 (5): 1881–1889.

52 Li, H., Afacan, A., Liu, Q., and Xu, Z. (2015). Study interactions between fine particles and micron size bubbles generated by hydrodynamic cavitation. *Minerals Engineering* 84: 106–115.

53 Aiping, Z. and Yang, Y. (2010). Application of ultrasonic antiscaling and descaling and its cavitation effect mechanism. *Heilongjiang Electric Power* 5: 321–324.

54 Carpenter, J., Badve, M., Rajoriya, S. et al. (2017). Hydrodynamic cavitation: an emerging technology for the intensification of various chemical and physical processes in a chemical process industry. *Reviews in Chemical Engineering* 33 (5): 433–468.

55 Thaker, A.H. and Ranade, V.V. (2020). Towards 'factory in a box' for personalised emulsions on demand: harnessing vortex based hydrodynamic cavitation. *Formulation* 41.

56 Madane, K. and Ranade, V.V. (2022). Using vortex based cavitation device for tailoring particle size distribution in anti-solvent crystallisation. SSPC Report. University of Limerick.

Index

a

acoustic cavitation 6, 154, 156, 158, 159, 162, 174, 177, 184, 215, 228, 246, 256, 258, 270, 300
acoustic sensors 15, 80, 82, 323
advanced oxidation processes (AOPs) 153, 174, 195, 196, 217, 218, 220, 225, 229
anaerobic digestion (AD) 50, 195, 240, 241
artificial neural network (ANN) 13, 104–107, 113, 126, 271

b

Bachelor vortex model 54
ballast water treatment 181–183
baseline geometry 26, 37
beer brewing 301, 328
bioactive compound extraction 302–304, 328
biochemical methane potential (BMP) 253, 256, 258, 262, 263
biodiesel 3, 14, 156, 218, 239, 296–300, 325, 328
bioethanol 14, 239, 240, 243–250, 260, 263, 267, 269, 270, 301, 325, 326
biofuels 239, 240, 244, 269, 271, 325, 326
biogas 240
  lignocellulosic biomass 260–264
  production 250–264
  wastewater and sludge 253–259
biomass, pre-treatment method 239–271
bubble internal energy 111

c

Cavitating venturi (CV) 34, 36, 37, 61, 116
Cavitation Detection System 84
cavitation model 65, 120, 123, 127, 133
cavitation number 23, 25, 26, 34, 36, 154, 156–159, 174, 319, 330
cavitation yield 165, 179, 203, 207, 208, 293
cavity dynamics models 34, 108–114, 116, 117, 119, 121, 126, 322
cellulose 14, 241, 243, 244, 248, 251, 260, 262, 269, 271, 325, 328
chemical pre-treatment 241, 243, 244
computational fluid dynamics (CFD) models 10, 12, 13, 29, 33, 34, 35, 37, 50, 51, 55–59, 61, 66, 108, 116, 119–121, 127, 131–136, 148, 159, 317, 319
confined vortex flow devices 56
contraction coefficient 25
COVID-19 149, 304
critical cavitation number 26
crystallization 80, 301, 315

d

data driven models 104–106
data processing techniques 83
decarbonisation 251, 315, 330
discharge coefficient 24–28, 32, 33, 317, 318

Hydrodynamic Cavitation: Devices, Design, and Applications, First Edition.
Vivek V. Ranade, Vinay M. Bhandari, Sanjay Nagarajan, Varaha P. Sarvothaman and Alister T. Simpson.

disinfection 3, 194, 300
 bacterial 50, 53
 water 11, 13, 14, 50, 84, 86, 147–184, 195, 220, 324, 325, 329
distributed wastewater treatment 325, 326
dosimetric methods 81
downstream pressure 23, 25, 209–210
drinking water treatment 179–180, 325
drops, droplet size distribution 124, 125, 294, 295
DynaJet® system 53

e
Ecowirl® reactor 53
effluents 147, 195, 196, 198, 199, 212, 213, 224, 239
empirical models
 data driven models 104–106
 per-pass performance model 101–104
 pseudo-reaction kinetics model 98–100
emulsification 14, 289, 294–296, 328
Euler number 57, 81, 317, 318
Eulerian models 120
experimental set-up
 centrifugal pump 77
 cutter pump 77
 diaphragm pump 77–78
 gear pump 77
 holding tank 74–76
 in-line sensors 79–80
 lobe pump 77
 peristaltic pump 77
 piping arrangements/fittings 79
 progressive cavity pump 77
 submersible pump 77
 torque pump 78

f
fabrication tolerances 78
Fenton's processes 225
filtration technology 182
flatness 321
food sterilization 300–301
fossil fuels 239, 289

g
gas–liquid applications 283–289, 305, 328
gas phase equation of state (EoS) 109

h
harnessing hydrodynamic cavitation 14, 15, 328–330
hemicellulose 14, 240, 243, 244, 251, 260, 262, 271, 325, 326
heterogeneous systems 122–125
"home-made" devices 78
hybrid methodology 179, 325
hydrodynamic cavitation 3, 315
 application 3, 11
 beneficial physico–chemical transformations 3–4
 boiling 5
 devices 7–11, 316–323
 flow characteristics 317
 history 6
 liquid 5
 mathematical modelling 4
 types 6
hydrogen peroxide (H_2O_2) treatment 221–222
hydroxyl radicals 6, 13, 79, 81, 95, 96, 102, 110–111, 113, 117–119, 153, 155, 168, 175, 196, 197, 206, 208, 215, 293, 305, 320, 322, 324, 325, 328

i
inlet concentration of pollutant 200, 206
in-line sensors 73, 79–80
inception
 identification 80–84
industrial applications 84
 control experiment 87
 executing experiments/troubleshooting 88–89
 hydraulic test 86–87
 mixing/eliminating dead zones 88
 thermal test 87
inexpensive microphones 83
intensify processes 325

j

jet hammer pressure 113, 114

k

Keller–Miksis (KM) equation 109

l

lab scale batch hydrodynamic cavitation experiments 247–248
lignocellulosic biomass (LCB) 14, 95, 239, 242, 260–265, 270
  pre-treatment methods 241–242
linear/axial flows 23, 49
liquid–liquid applications 289
  emulsification 294–296
  food sterilization 300–301
  microalgal oil extraction 296–297
  oxidative desulphurization 289–294
  transesterification of oils 297–300
liquid–liquid systems 122, 328
liquid vapour pressure 23, 176

m

methanogenesis 250, 251
micro-scale processes 95–97, 107, 108, 126, 127
microalgal oil extraction 296–297
milling energy 241, 269
mineralisation 102, 195, 201, 217, 227
minimization of Gibbs free energy (MGFE) method 110, 111
modelling hydrodynamic cavitation 13, 322
multi-scale performance models 115
multiphase system 88, 283

n

net energy gains 244, 262, 264–269
non thermal plasma 219
non-invasive sensors 15, 323

o

operating pH 210–213, 222
orifice based devices 24–34, 41–44, 246
oxidative desulphurization 14, 289–294
ozone (O_3) treatment 222–224

p

particle size distribution (PSD) 88, 123, 262, 304, 328
particle size reduction 304–305, 328
passive cavitation devices 81
per-pass degradation coefficient 118, 200, 206
per-pass degradation factor 102, 118, 121, 199–203, 205, 206, 208, 209, 217, 218, 321
per-pass performance model 101–104, 117
peroxonation 220, 224–225
pH 14, 80, 97, 154, 174, 206, 210–213, 215, 222, 225, 226, 251
photocatalytic processes 226, 227, 315
physical pre-treatments 241
physico-chemical transformations 3, 4, 6, 7, 13, 95, 315, 322, 323
physics based models 108
  cavity dynamics models 108–114
  multi-scale/multi-layer models 114–122
plasma irradiation 219
pollutant concentration 102, 118, 200, 203, 206–208, 226
predictive models 126, 323
pressure drop (ΔP)
  across cavitation device 14, 208–209
  hydrodynamic cavitation 158
pressure fields 32
prevailing modelling approaches 97
pseudo-reaction kinetics model 98–100
pumped seawater 183

r

radial pressure gradients 58
Rankine vortex model 54
Rayleigh Plesset equation 108, 133, 134
Reynolds Averaged Navier Stokes (RANS) 33, 35, 132
Reynolds number 24, 25, 81, 318
rotor-stator hydrodynamic cavitation devices 49–52

s
saturated vapour pressure 32–33, 135
sewage wastewater treatment 180–181
shockwave pressure 111
singing vortex 57, 82
single bubble dynamics model 109
solid-liquid applications 301
- beer brewing 301
- bioactive compound extraction 302–304
- particle size reduction 304–305

soluble chemical oxygen demand (SCOD) 50
sparged gases 214, 284
steam explosion 241, 244
swirling flows 7–9, 12–13, 44, 49–67, 78, 161, 316–319
synergistic index 172, 218, 221, 222, 224, 225, 228, 229, 324

t
temperature, operating 213–214
3D Large Eddy Simulation (LES) models 34, 36
3D swirling flows 55
turbulence models 123, 133
turbulent pressure fluctuations 7, 81, 112, 135

u
ultrasonic cavitation 6, 7, 112, 116, 117, 213, 316
ultraviolet 153, 218

v
value-added chemicals 239
venturi based devices 8, 34–41, 43, 44, 316
visualisation method 81
vortex based cavitation devices 52–59, 119, 122, 124, 158, 205, 270, 317, 321
vortex throttle type devices 60

w
wall angle 39, 41
waste activated sludge (WAS) 50
wastewater treatment
- application 198–199
- augmentation 217–228
  - by catalyst based AOPs 225–228
  - chemical based AOPs 220–225
  - energy sources 218–220
- cavitation devices 197
- hot spot theory 201
- influence of device design 203–206
- operating parameters 206–217
- techniques 194

water disinfection 147
- conventional methods 149–154
  - drawbacks 152–153
  - emerging newer methods 153–154
- hybrid methods 165–170
  - aeration 169–170
  - conventional 168
  - hydrogen peroxide 168–169
  - ozone 169
- hydrodynamic cavitation 148, 154
  - devices/reactors 159–162
  - kinetics 162–165
  - principle 154–158
  - state of the art 158–159
- natural oils 170–174
  - conventional *vs.* hybrid processes 174–176
  - cost calculation 178–179
  - cost comparison 177–178
  - process economics 177–179
  - temperature effects 176–177

water per-capita 193, 194
water-borne diseases 147
water-borne viruses 147

x
x-ray tomography 320
xylose 243

z
zero crossing rate (ZCR) 84, 86